Conducting Research in Human Geography: Theory, Methodology and Practice

ROB KITCHIN AND NICHOLAS J. TATE

PEARSON

Prentice
Hall

Harlow, England • London • New York • Boston • San Francisco • Toronto
Sydney • Tokyo • Singapore • Hong Kong • Seoul • Taipei • New Delhi
Cape Town • Madrid • Mexico City • Amsterdam • Munich • Paris • Milan

Pearson Education Limited
Edinburgh Gate
Harlow
Essex CM20 2JE
England

and Associated Companies throughout the world

Visit us on the World Wide Web at:
http://www.pearsoned.co.uk

First published 2000

ISBN 0 582 29797 4

British Library Cataloguing in Publication Data
A catalogue record for this book is available from the British Library

Library of Congress Cataloging-in-Publication Data
A catalog record for this book is available from the Library of Congress

10 9 8 7 6
09 08 07 06 05

Typeset by 35 in 9.5/11pt Times
Printed in China
EPC/06

CONDUCTING RESEARCH IN HUMAN GEOGRAPHY

Contents

Preface   ix
Acknowledgements   xii

Chapter 1   Thinking about research   **1**
1.1   What is research?   1
1.2   Why do research?   2
1.3   What's unique about geographic research?   3
1.4   What choice of approaches have you got?   4
1.5   Which approach is best?   19
1.6   Summary   26
1.7   Questions for reflection   26
Further reading   27

Chapter 2   Planning a research project   **28**
2.1   Introduction   28
2.2   Choosing a research topic   28
2.3   Narrowing the focus   30
2.4   Linking theory and practice   32
2.5   Research design   34
2.6   Choosing a method to generate data   39
2.7   Choosing a method to analyse your data   41
2.8   Conducting group research   42
2.9   Managing and piloting a research project   42
2.10 Summary   43
2.11 Questions for reflection   44
Further reading   44

Chapter 3   Data generation for quantitative analysis   **45**
3.1   Introduction   45
3.2   Classifying data types and measurement scales   45
3.3   Generating primary quantitative data   47
3.4   Sampling, estimation and distribution   53
3.5   Obtaining and using secondary data   60
3.6   Summary   67
3.7   Questions for reflection   69
Further reading   69

Chapter 4   Preparing, exploring and describing quantitative data   **70**
4.1   Introduction   70
4.2   Data pre-processing and checking   70
4.3   Computer programming   76
4.4   Using the MINITAB statistical package   77
4.5   Initial data analysis   82

	4.6 Probability	95
	4.7 Transforming data	105
	4.8 Summary	107
	4.9 Questions for reflection	107
	Further reading	107

Chapter 5 Analysing and interpreting quantitative data   108

	5.1 Introduction	108
	5.2 Classifying tests	109
	5.3 Tests of significance	109
	5.4 Choosing the right test	112
	5.5 The tests	114
	5.6 Parametric tests	115
	5.7 Non-parametric tests	139
	5.8 What do the test results tell you?	152
	5.9 Summary	155
	5.10 Questions for reflection	155
	Further reading	155

Chapter 6 Spatial analysis   156

	6.1 Introduction	156
	6.2 Maps	156
	6.3 Geographical Information Systems	164
	6.4 Current issues in the use of GIS for socio-economic applications	175
	6.5 Sources of digital spatial data	182
	6.6 Planning and implementing analysis using GIS	184
	6.7 Spatial statistics	187
	6.8 Summary	209
	6.9 Questions for reflection	209
	Further reading	210

Chapter 7 Producing data for qualitative analysis   211

	7.1 Introduction	211
	7.2 Qualitative approaches	211
	7.3 Primary data production	212
	7.4 Specific approaches to producing qualitative data	224
	7.5 Secondary sources of qualitative data	225
	7.6 Summary	228
	7.7 Questions for reflection	228
	Further reading	228

Chapter 8 Analysing and interpreting qualitative data   229

	8.1 Introduction	229
	8.2 Description, classification, connection	231
	8.3 Transcribing and annotation	236
	8.4 Categorising qualitative data	238
	8.5 Splitting and splicing	244
	8.6 Linking and connecting	246
	8.7 Corroborating evidence	251
	8.8 Analysing qualitative data quantitatively	253
	8.9 Summary	256

8.10 Questions for reflection   256
Further reading   256

Chapter 9   Analysing qualitative data using a computer   257
9.1   Introduction   257
9.2   Getting started and description   258
9.3   Classifying qualitative data using a computer   260
9.4   Connecting qualitative data using a computer   266
9.5   Summary   269
9.6   Questions for reflection   269
Further reading   269

Chapter 10   Writing-up and dissemination   270
10.1 Introduction   270
10.2 Writing as part of the research process   270
10.3 Writing a final report   271
10.4 Presenting your research as a talk   286
10.5 Before submitting your work   287
10.6 After submitting your work   287
10.7 Summary   289
10.8 Questions for reflection   289
Further reading   289

Chapter 11   Final words   290
11.1 An overview of conducting research in human geography   290
11.2 Coping with problems   291
11.3 Independent research   291
11.4 Epilogue   292
11.5 Summary   292

Appendix A   Tables   293

**Appendix B   Annotation for the interviews with respondents
              B, C and D   308**

Appendix C   Addresses   315

References   317
Index   326

For our parents

Preface

This book is concerned with the process of undertaking a research project. It is designed to guide you, the student, through all aspects of the research process, from the nurturing of your ideas and the development of a proposal, to the design of an enquiry, the generation and analysis of data, the drawing of conclusions from your results, and the presentation of your findings. In essence, the aim is to take you from a novice researcher towards being a competent investigator capable of independent enquiry.

The book has been written because, in the course of teaching and research, we have noted the absence of a single book which covered all the ingredients needed to produce a good human geography researcher. Instead we have referred students to a mixture of reference books, often with a quantitative or qualitative bias, none of which we felt to be fully adequate.

We appreciate that as novice researchers you may be apprehensive about a number of different matters. For example, you might be concerned about the generation of data, especially if it involves interviewing one or more individuals, or about the analysis or interpretation of data, or possibly the use of statistics or a computer. This book is specifically written with these concerns in mind. As any seasoned academic will tell you, research is never simple or easy; there are always problems that need to be solved or hurdles to clear. Even experienced, professional researchers can become unstuck, or encounter totally new situations that hinder their progress. However, with foresight, appropriate planning and the right skills, the majority of problems encountered can be overcome in a sensible and uncomplicated manner. As we demonstrate throughout this book, the task need not be as daunting as it might seem at first. Research can be challenging, but it can also be rewarding and exciting.

We have taken some care to try to make all the description and explanation as simple as possible, and to use detailed step-by-step examples derived from real research in human geography to provide essential background knowledge and structure to your learning experience. Many types of research and research techniques are covered, but as with any book that tries to cover the whole breadth and depth of research in a discipline, certain compromises have had to be made. This book is no exception. We provide a broad overview of the theoretical underpinnings and detail some of the main research methodologies currently being used in contemporary human geography. We refer you to other texts where appropriate. As with any textbook, it is only advice that can be given – the best source of understanding and learning is through actually doing research. This book will guide you around some of the worst pitfalls and give you sufficient practical advice to get you started. Ultimately, we claim responsibility for the content of the book and any remaining errors or oversights.

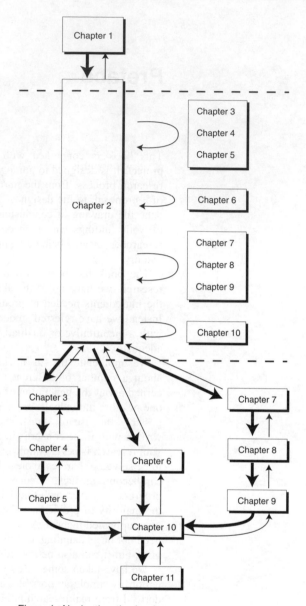

Figure i   Navigating the book.

Navigating the book

This book is designed to be both comprehensive and easy to follow. It can be either read as a linear text (i.e., starting at the beginning and reading to the end) or dipped into selectively. What is relevant will depend on the nature of your research project, and doubtless portions of the text will be inappropriate in this context. Although the exact route will be up to you, our suggested method of use is outlined schematically in Figure i. We advise that you read the whole of Chapters 1, 2 and 10. These chapters cover material that is central to all projects. The remaining chapters should be visited twice, first when reading Chapter 2 and then again whilst actually generating and analysing your data. The first visit is to help ensure that the project is planned rigorously. All stages of a project need to be considered before data is generated and analysed so that problems that might be encountered at a later date can be identified and strategies to combat them implemented. The second visit will help guide you through the details of implementing specific techniques.

The data sets that we use as examples can all be downloaded from Pearson Education's web site: http://www.awl-he.com (you will need to search for this book and the companion web site is linked from the marketing page there), as can GASP (a computer package that has been written to perform spatial statistics). All the data sets are named as their respective table number in the text (e.g., the data in Table 4.1 is saved in the file Table41.dat).

Acknowledgements

A number of people have provided constructive advice concerning the contents of this book and/or supplied the example data sets and unpublished material we have used throughout. We would like to take this opportunity to thank Mark Blades, Paul Boyle, Mike Bratt, Chris Chatfield, John Coshall, Nick Cox, Robin Flowerdew, Reg Golledge, Alastair Grant, Ian Gregory, Keith Halfacree, Robin Haynes, Phil Hubbard, Dan Jacobson, Nuala Johnson, Iain Lake, David Livingstone, Paul Longley, Andrew Lovett, David Mark, Dave Martin, Victor Mesev, Stan Openshaw, Julian Parfitt, Charlie Richards, Steve Royle, Ian Shuttleworth, John Spain, Bob Taft, Chris Taylor, Nick Williams, Peter Williams, Heather Winlow, Neil Wrigley, and the Cartographic Unit in the School of Geosciences, The Queen's University of Belfast, for the preparation of the figures; Mike Wood for the idea behind Figures 6.5/6.6.

The computer program GASP, used in Chapter 6, was developed and programmed by Mike Bratt and Rob Kitchin. We are grateful to Mike for his time and effort. We would also like to acknowledge the work and patience of Matthew Smith and Shuet-Kei Cheung at AWL, and Sally Wilkinson for commissioning the book.

This book was written while both authors were lecturers at the School of Geosciences, Queen's University of Belfast.

Rob Kitchin and Nick Tate
Belfast, September 1998

Copyright permissions

We are grateful to the following for permission to reproduce copyright material:

Sage Publications Inc for figure 2.1 from p. 17 of C. Marshall and G.B. Rossman (1995) *Designing Qualitative Research*; Arnold for figure 2.2 from p. 85 of R.M. Downs (1970) 'Geographic space perception: past approaches and future prospects' in *Progress in Geography* **2**: 65–108; Simon Jones and the data archive at the University of Essex for figure 3.6; Routledge for figure 3.7 from D. Martin (1996) *Geographic Information Systems*; the American Statistical Association for figure 5.6 from F.J. Anscombe (1973) 'Graphs in statistical analysis' in *American Statistician* **27**: 17–21; John Wiley & Sons Inc for figure 6.1 from A.H. Robinson et al. (1995) *Elements of Cartography*; Humphrey Southall for figure 6.2; John Wiley & Sons Inc for figure 6.3 from chapter 8 by Mike Batty in P. Longley et al. (eds) (1998) *Geocomputation: A Primer*; Addison Wesley Longman Ltd. for figure 6.4 from D. Dorling & D. Fairbairn (1997) *Mapping: Ways of Representing the World*; Addison Wesley Longman Ltd for figure 6.7 from C. Jones (1997) *GIS and Computer Cartography*; L.H.

Systems Ltd for figure 6.9; Iain Lake and the Ordnance Survey for figure 6.17; the Ordnance Survey for figure 6.20; the Ordnance Survey and Taylor & Francis for figure 6.22 from S. Openshaw and S. Alvanides' chapter in A. Frank et al. (eds) (in press) *Life and Motion of Socio-economic Units*; Blackwell Publishers for figure 6.31 from D. Ebdon (1985) *Statistics in Geography*; M. Goodchild for figure 6.33 from M. Goodchild (1986) 'Spatial Autocorrelation' in *Concepts and Techniques in Modern Geography* **47**; the Ordnance Survey for figures 6.34 and 6.35; Allen & Unwin for table 3.3 from J. Silk (1979) *Statistical Concepts in Geography*; John Wiley & Sons Limited for table 3.4 from A.S.C. Ehrenberg (1975) *Data Reduction: Analysing and Interpreting Statistical Data*; Chris Dixon for table 3.5; the Economic and Social History Society of Ireland for table 4.3 from M. Turner (1984) 'Livestock in the agrarian economy of counties Down and Antrim from 1903 to the famine' in *Irish Economic and Social History* **11**: 19–43; Pion Ltd for table 4.4 from A.D. Clif and J.K. Ord (1973) *Spatial Autocorrelation*; the Provincial Treasury and Government of Prince Edward Island, Canada for table 4.5; Blackwell Publishers for table 5.4 from D.Z. Sul and J.D. Wheeler (1993) 'The location of office space in metropolita' in *Professional Geographer* **45**; the Royal Geographical Society for table 5.7 from N.J. Williams, J. Sewel and F. Twine (1986) 'Council house allocation and tenant incomes' in *Area* **18**; the Royal Geographical Society for table 5.9 from P.R. Waylen and A.M. Snook (1987) 'Success in the football league, 1921–1987' in *Area* **22**; Blackwell Publishers for table 5.7 from J. Coshall (1988) 'The nonparametric analysis of variance and multiple comparison procedures in geography' in *Professional Geographer* **40**; John Wiley & Sons Inc for table 6.1 and box 6.3 from P. Longley et al. (eds) (1999) *Geographical Information Systems: Principles, Techniques, Management, Applications*; the American Society for Photogrammetry and Remote Sensing for table 6.3 from V. Mesev (1998) 'The use of census data in urban image classification' in *Photogrametric Engineering and Remote Sensing (PE & RS)* **64**(5): 431–38; Taylor & Francis for table 7.1 from R. Tesch (1990) *Qualitative Research: Analysis Types and Research Tools*; Sage Publications Inc for tables 8.2 and 8.3 from R. Kitchin (1997) 'Exploring spatial thought' in *Environment & Behavior* **29**(1): 123–56; David Martin for box 3.12 from D. Martin (1993) *The UK Census of Population 1991*; NCGIA for box 3.14 from H. Calkins (1990) 'Unit 8 socioeconomic data' in M.F. Goodchild and K. Kemp (eds) *NCGIA Core Curriculum*; Kenneth E. Foote for box 6.5 from K.E. Foote and D.J. Hubner (1995) 'Error, accuracy and precision' in *The Geographer's Craft Project* at the University of Texas at Austin.

The census data used in the preparation of Figures 6.22 are © Crown copyright. The digital boundaries used in Figures 6.22 are Crown and ED-LINE Copyright and were provided as part of the ESRC/JISC 1991 Census of population initiative.

Portions of MINITAB Statistical Software input and output contained in this book are printed with permission of Minitab Inc. MINITAB is a trademark of Minitab Inc. in the United States and other countries and is used herein with the owner's permission.

Whilst every effort has been made to trace owners of copyright material, in a few cases this has proved impossible, and we would like to take this opportunity to apologise to any copyright holders whose rights we may have unwittingly infringed.

CHAPTER 1	# Thinking about research

This chapter covers	**1.1** What is research?
	1.2 Why do research?
	1.3 What's unique about geographic research?
	1.4 What choice of approaches have you got?
	1.5 Which approach is best?
	1.6 Summary
	1.7 Questions for reflection

1.1   What is research?

Research is the process of **enquiry** and **discovery**. Every time you seek the answer to a question you are undertaking a small piece of research. For the human geographer, research is the process of trying to gain a better understanding of the relationships between humans, space, place and the environment. The human geography researcher, by carefully generating and analysing evidence, and reflecting upon and evaluating the significance of the findings, aims to put forward an interpretation that advances our understanding of our interactions with the world. This book is about how to undertake successful research and aims to provide sound, practical advice and ideas that will help you become a confident, capable human geographer.

Although a research project might at first seem daunting, it should be remembered that we are all capable of conducting research. Many research skills are commonplace such as the ability to ask questions, to listen and to record the answers. The secret of successful research is to develop and harness those skills in a productive manner using careful planning and design. As long as you plan your research carefully, almost any problem can be approached and answered in a sensible way. This is not to say that there is a 'magic formula' that makes research easy. Undertaking a research project, although challenging

and stimulating, can be intensely frustrating, confusing and messy. Many first-time researchers run into all sorts of problems: they do not know where to begin; they do not know how to design an effective research strategy or the options available to them; they do not know which method of data generation or analysis is best or the full range of options available; they are unsure as to how to interpret or write up their findings. To make things more complex, research is rarely just a process of generating data, analysing and interpreting the results. By putting forward answers to research questions you are engaging in the process of debate about what can be known and how things are known. As such, you are engaging with **philosophy**.

As we will see, there are many ways of approaching each particular question and the research process is not divorced from **theory**. Theory, methodology and practice are intimately and tightly bound. Your beliefs as a person are going to affect the approach you take to study and also the conclusions you might draw – if every question only had one definitive answer, then there would be no debates, no different political parties, and libraries would contain far fewer books! However, we do all have different beliefs concerning how research should be undertaken and the exact nature of a problem and, as a result, our understanding of the world, the people, creatures and plants which inhabit it, is constantly changing and evolving as more and more studies are undertaken.

The aim of this book is to make the process of conducting research easier and more rewarding by guiding you through the research process from the choice of a research topic to the presentation of your results.

1.2   Why do research?

Given that research is not always easy, why should you want to undertake it? What is your motivation? Research provides us with a picture of specific aspects of the world. By undertaking a piece of research you are helping to contribute to world knowledge. You might feel that your research project will do little for the world other than help you pass your course. However, student projects have contributed to policy issues and at the very least make clear to the groups being researched or associated agencies that there might be a need for greater understanding of an issue. Perhaps more importantly, the undertaking of a study as part of a course will help later in the workplace where you might be expected to collate, analyse and interpret data, often at short notice. Such research skills are increasingly important in the workplace. For professional researchers the reasons for undertaking research usually centre upon five main motivations (see Box 1.1).

Human geographers undertake research for all the reasons presented in Box 1.1, often in combination with each other. It is possible, for example, to link four together. In a large study you might start with some **exploratory** investigations to determine which variables or factors are important. Next, you might try to **describe** the phenomena and how they are related. You might follow this by seeking to **explain** what caused the phenomena, using this information to make a **prediction** about future outcomes. For example, if we were interested in why people migrate to new, relatively unknown areas, the four could be linked in the following way:

- *Explore* possible reasons why people might want to move (perhaps these might include improvement of economic status, quality of life, better schooling and other services, family and relatives).
- *Describe* the patterns of migration in an area based upon the factors found during exploration.
- *Explain* the patterns of migration identified when describing the exploratory factors.
- *Predict* possible future migrations based upon the explanation of current patterns of migration.

Box 1.1   Reasons for undertaking a study

1   Exploration
- To investigate little-understood phenomena
- To identify/discover important variables
- To generate questions for further research
2   Explanation
- To explain why forces created the phenomenon in question
- To identify why the phenomenon is shaped as it is
3   Description
- To document and characterise the phenomenon of interest
4   Understanding
- To comprehend and understand process, interaction, phenomenon and people
5   Prediction
- To predict future outcomes for the phenomenon
- To forecast the events and behaviours resulting from the phenomenon

Source: Adapted after Marshall and Rossman 1995: 78.

Given these reasons for conducting research, you may still be lacking in motivation. It may be the case that you are uninterested in conducting research, a task that you have to fulfil only as part of a course. If this is the case then your motivation should be driven by a desire to do as well as possible, and to think about how the skills developed might help you gain employment on completion of the course. One way to try and generate some enthusiasm for conducting a project is to research a topic which you find interesting (see Section 2.2). This does not mean that the topic should be of personal relevance, only that you are interested in understanding a phenomenon or situation better. If you still find yourself unmotivated then Blaxter *et al.* (1996: 13) suggest that you might find some inspiration by:

- changing your research project to a more interesting topic;
- focusing on the skills you will develop through undertaking the project;
- incorporating within the research some knowledge acquisition of relevance to you;
- seeing the research project as part of a larger activity, which will have knock-on benefits for your course and future career.

1.3 What's unique about geographic research?

As this book is designed to help you conduct research in **human geography**, what is peculiar to research from a *geographical* perspective? What separates human geography from the other social sciences? Defining geography is a task fraught with difficulty. For decades, geographers have been struggling with their identity, with no clear consensus as to what geographers *are*, what geographers *do*, and *how* they should study the world. We have tried, over a number of years, to get our own students to think about what geography, and in particular human geography, is and what it concerns. When asked, most students will either stare back blankly or have a stab at something which usually includes the words 'people', 'environment', 'world' and 'interaction'. Defining what geographers *do* can be even more difficult. To help our students, we ask them to consider the following party scenario outlined by Peter Gould (1985):

Party-goer:	What do you do for a living?
Geography reveller:	I teach geography at the University.
Party-goer:	Oh. What do geographers do exactly?

Next, we ask the students to take the role of the *geography reveller* and to give an answer. After, we ask them to summarise what they think the *party-goer* previously suspected a geographer might do. Judging from the responses we have received, the latter task is often the easier to complete. The terms 'geography' and 'geographers' seem to defy easy definition. Indeed, Holt-Jensen (1988) reports that the general public hold three common misconceptions regarding what geography *is* and what geographers *do*. First, to many people, geography is the encyclopaedic collection of knowledge relating to places and geographic facts (e.g., longest river, biggest town). Second, many people consider geography to be anything relating to maps, with geographers as the cartographers and collectors of information for these maps. Third, many people consider geography to be about writing travel descriptions at both the local and global scale. So, if these conceptions are wrong, what is geography and what do geographers do?

As stated, there is no clear consensus amongst professional geographers as to what constitutes their discipline. A number of different definitions of geography can be found depending upon where you look.

These definitions do, however, all generally revolve around the same themes: place, space, people, environment (see Box 1.2). Haggett (1990) suggests that geography is difficult to define because of its historical development as an area of study. Indeed, he contends that geography's identity crisis is a result of its puzzling position within the organisation of knowledge, straddling the social and natural sciences. This is a result of the history of geographic thought, which can be traced back to classical Greek scholars who viewed humanity as an integral part of nature. Geography thus consisted of a description of both animate and inanimate objects. By the time geography became a university subject in the late nineteenth century, academic studies had already been divided into the natural and physical sciences on the one hand, and the humanities and social sciences on the other. Geography, with its natural and social constituents, had to be slotted into this existing inappropriate structure. The fitting of geography into the traditional academic organisation has proved uncomfortable and has caused a search for an identity that fits more snugly. Johnston (1985) thus suggests that geographers have sought to constantly refine and redefine their discipline in order to demonstrate its intellectual worth. Indeed, Livingstone (1992) contends that geography is elusive to define because it changes as society changes – geography as a practice has changed throughout history, with different people still attaching salience to different interpretations:

Geography . . . has meant different things to different people at different times and in different places.

In other words, there are many different geographies, some new, some old. All have slightly different emphases and some are more popular than others. As such, the variability of definition in Box 1.2 is due to the way that the definers cast geography. For example, Hartshorne (1959) saw geography as an **idiographic science** (that is, its main emphasis is description) whereas Yeates (1968) saw geography as a **nomothetic science** (that is, its main emphasis is explanation and law-giving). Unwin (1992) suggests that definitions also vary depending on whether we try to define geography as simply 'what geographers do' (academic), as 'what geographers study' (vernacular), or in terms of its methodology or techniques. Whilst it is difficult to pin down a clear definition, it is clear that at present, the totality of geographical research and expertise is diverse, covering both the natural and social sciences, and the interaction of the two (Table 1.1).

Box 1.2   Defining geography

Mackinder (1887: 143):
'I propose therefore to define geography as the science whose main function is to trace the interaction of man [*sic* – see Box 1.15] in society and so much of his environment that varies locally.'

Hartshorne (1959: 21):
'Geography is concerned to provide accurate, orderly, and rational description and interpretation of the variable characters of the Earth's surface.'

The Concise Oxford Dictionary (1964: 511):
'Geography, *n*. Science of the earth's surface, form, physical features, natural and political divisions, climate, productions, population, etc. (*mathematical*, *physical* and *political*, ~, the science in these aspects); subject matter of ~; features, arrangement, *of* place; treatise or manual of ~.'

Yeates (1968: 1):
'Geography can be regarded as a science concerned with the rational development, and testing, of theories that explain and predict the spatial distribution and location of various characteristics on the surface of the earth.'

Dunford (1981: 85):
'Geography is the study of spatial forms and structures produced historically and specified by modes of production.'

Haggett (1981: 133):
'[Geography is] the study of the Earth's surface as the space within which the human population lives.'

Johnston (1985: 6):
'Literally defined as "earth description", geography is widely accepted as a discipline that provides "knowledge about the earth as the home of humankind".'

Haggett (1990):
'Geographers are concerned with three kinds of analysis:
• Spatial (location): numbers, characteristics, activities and distributions.
• Ecological: the relationship between humans and environment.
• Regional: the combination of the first two themes in areal differentiation.'

Geography Working Group's Interim Report (1990):
• 'Geography explores the relationship between the earth and its peoples through the study of place, space, and environment. Geographers ask questions where and what; also how and why.
• The study of place seeks to describe and understand not only the location of the physical and human features of the Earth, but also the processes, systems, and interrelationships that create or influence those features.
• The study of space seeks to explore the relationships between places and patterns of activity arising from the use people make of the physical settings where they live and work.
• The study of the environment embraces both its physical and human dimensions. Thus it addresses the resources, sometimes scarce and fragile, that the Earth provides and on which all life depends; the impact on those resources of human activities; and wider social, economic, political and cultural consequences of the interrelationship between the two.'

Gale (1992: 21):
'Geography, for me, is about how we view the world, how we see people in places.'

In this book, we are concerned with conducting research in human geography. However, defining human geography is as fraught with difficulties as is defining geography in general. Definitions vary for all those reasons stated above. For our purposes, we have taken human geography to refer to the *study of society in relation to space and place*. As such, it includes all elements, bar physical geography, that are listed in Table 1.1. We have made the distinction between physical and human geography for two reasons: first, to keep the book manageable; and second, because in general the study of people and human-made objects requires different research techniques from the study of natural phenomena. Indeed, some would argue that there are clear philosophical and methodological differences between human and physical geography research.

1.4   What choice of approaches have you got?

Philosophy aims at the logical clarification of thoughts. . . . Without philosophy thoughts are, as it were, cloudy and indistinct: its task is to make them clear and to give them sharp boundaries.

(Wittgenstein, 1921, quoted in Ragurman, 1994)

Many geographers doubt that philosophical issues are actually relevant to geographic research. [However] no research (geographic or otherwise) takes place in a philosophical vacuum. Even if it is not explicitly articulated all research is guided by a set of philosophical beliefs. These beliefs influence or motivate the selection of topics for research, the selection of methods for research, and the manner in which

Table 1.1 Types of geographic study.

Human geography	Physical geography	Mixed human and physical geography	Other
Cultural geography Economic geography – Employment – Location theory – Manufacturing – Marketing – Retailing – Services – Trade Gender studies Rural geography – Rural economy – Rural planning – Rural population and change Industrial geography – Location – Organisation – Regional development – Technological change Medical geography Urban geography – Urban economy – Urban housing – Urban morphology – Urban politics – Urban population – Urban renewal – Urban retailing – Urban sociology – Urban theory, models, systems Political geography – Electoral geography – Geopolitics Population geography – Demography – Population change – Population migration Recreational geography – Leisure – Sport – Tourism Historical geography – Countryside – Industry – Population – Towns Social geography – Ethnicity – Social theory – Socio-economic status Transport geography – Rural/urban	Biogeography – Vegetation studies – Zoogeography Climatology – Applied – Climatic change – Microclimatology – Synoptic climatology Ecology Geomorphology – Applied – Arid – Coastal – Fluvial – Glacial – Karst – Slopes – Weathering Hydrology – Applied – Runoff – Water quality Meteorology Quaternary environments – Archaeology – Landform evolution – Paleoecology – Sediments Soils	Agricultural geography – Agricultural policy – Agricultural systems Development studies – Agrarian – Urban planning – Policy studies Regional geography Environmental studies – Conservation – Environmental change – Mineral resources – Environmental impact assessment – Environmental management – Environmental perception – Environmental quality – Environmental systems Hazards Planning – Economic – Environmental – Regional – Urban planning Resource geography – Energy – Fishing and forestry – Mineral resources – Water resources	Applied geography Education and geography Geographical information systems (GIS) – Cartography – Image analysis – Photogrammetry – Remote sensing Geographical thought – History – Methodology – Philosophy Quantitative methods – Computers – Mathematical techniques – Statistical techniques Theoretical geography

Source: 1993 listings of the Institute of British Geographers and Association of American Geographers study groups.

completed projects are subjected to evaluation. In short, *philosophical issues permeate every decision in geography.*
(Hill, 1981: 38, our emphases)

Since human beings started to record and observe the world there have been differences in opinion on how research should be conducted. Over the centuries, philosophers have argued about:

- **Ideology**: the underlying social or political reasons or purpose for seeking knowledge.
- **Epistemology**: how knowledge is derived or arrived at; the assumptions about how we can know the world (What can we know? How can we know it?).
- **Ontology**: the set of specific assumptions underlying a theory or system of ideas (what can be known).
- **Methodology**: a coherent set of rules and procedures which can be used to investigate a phenomenon or situation (within the framework dictated by epistemological and ontological ideas).

Human geographers have been involved in such debates and, as a result, there are a number of schools of thought on the best way to approach the relationship between society, space, place and environment. Indeed, Cloke *et al.* (1992) argue that contemporary human geography is extremely diverse, both in the topics investigated (as we have seen in Table 1.1) and in the diversity of approaches and methods of enquiry. The arguments for and against each approach are often quite complex, using carefully selected and what often seems like ambiguous and over-complicated language. Indeed, Ragurman (1994: 244) argues that the net effect of complex philosophical debates upon the student is 'often a lot of apprehension, disenchantment and an uneasy feeling of being lost in a philosophical wilderness'.

Whilst it is tempting to dismiss **philosophy**, or to try and avoid it because it seems difficult, the reality of conducting research is that *you cannot avoid it.* As Hill (1981) discusses, your research aims to provide answers to questions. In doing so, you will be claiming to know something about a particular situation or phenomenon, or even the world in general. All such claims raise ideological, epistemological, ontological and methodological questions about why the study was conducted and whether such claims are warranted. Understanding philosophical approaches is important for two reasons. First, it will help you understand what other researchers have done and why. Second, it will help you find an approach on which to base your own research and will provide the theoretical context in which to justify your findings.

For the purpose of this book we have tried to draw out and simplify the dominant issues and debates concerning how to approach research in human geography. The aim, however, is not to provide a comprehensive account of underlying philosophies. Rather we aim to provide a basic flavour of geographic thought and to stimulate you to explore the theoretical nature of how to conduct research. To do this, we detail 12 different approaches that have gained some currency in geographic thought over the past 30 years. Some have received more support than others but we leave it to you to decide which approach has the most personal appeal. It must be appreciated that *these approaches are a great deal more complex than can be detailed in one chapter.* To gain a deeper understanding of all the arguments, nuances and relationships between different positions, and to understand the history and development of each approach, you ought to refer to some of the texts recommended in the Further reading section at the end of this chapter. It must be noted that, within this discussion, whilst we try to give a respectful and objective assessment of each school of thought, we are not completely impartial.

Unwin (1992) uses Habermas's taxonomy of the different types of science to structure his discussion of approaches within geography, and we follow his lead. Habermas (1978) divided science into three different varieties: **empirical–analytical**, **historical–hermeneutic** and **critical**. These differ fundamentally from each other in a number of respects in relation to how knowledge and human action is mediated. He suggests that knowledge within each type is mediated through a series of *interests* (technical, practical and emancipatory), developed within differing *social media* (work, language and power), and expressed through different *forms* (material production, communication, and relations of domination and constraint) (Unwin, 1992). Essentially, the approaches to science differ because of varying opinions on what purpose knowledge should serve and how it should be constructed and represented (e.g., epistemology, ontology and methodology). We appreciate that this material is difficult and some confusion may arise because others have used alternative taxonomies to discuss approaches to research (for example, phenomenology, existentialism and idealism are often discussed under the heading 'humanistic approaches'). However, time invested at this stage is time well spent as it allows your study to be better grounded in theory. Boxes 1.3–1.15 are designed to allow you to quickly contrast the different approaches, but they should be used in conjunction with the text and other recommended reading, as they provide oversimplified, caricature accounts.

1.4.1 Empirical–analytical science

Empiricism

> **Box 1.3   Empiricism**
>
> *In a nutshell*
> - Empiricism refers to the school of thought where facts are believed to speak for themselves and require little theoretical explanation.
> - Empiricists hold that science should only be concerned with objects in the world and seek factual content about them.
> - Normative questions concerning the values and intentions of people are excluded from study as it is claimed we cannot scientifically measure them.
> - A source of primary data is closed-question questionnaires (see Chapter 3).
>
> *Example study of poverty*
> Facts about poverty would be collected and presented for interpretation by the reader (e.g., indices of poverty – social welfare recipient, housing tenure, etc.).

Empiricism is based around the notion that science can only be concerned with **empirical questions**. Empirical questions concern how things are in reality, where reality is defined as the world which can be sensed (see Ayer, 1969). Empiricists hold that science should only be concerned with objects in the world and should only seek factual content. As such, all knowledge is derived from the evidence provided by the senses and processed in an inductive fashion. **Normative questions** concerning the values and intentions of a subject(s) are excluded as we cannot scientifically measure them. Holt-Jensen (1988: 87) provides the following example to illustrate the difference between empirical and normative questions. 'How *are* the available food resources distributed between the inhabitants of the world?' is an empirical question. 'How *should* the available food resources be distributed between the inhabitants of the world?' is a normative question. As such, empiricist research is merely a presentation of the facts as gathered and determined by the objective researcher. It is important not to confuse the terms empirical and empiricism. The term 'empirical' refers to the collection of data

for testing, whereas the term 'empiricism' refers to the school of thought just described, where facts are believed to speak for themselves and require little theoretical explanation (May, 1993).

Positivism

> **Box 1.4   Positivism**
>
> *In a nutshell*
> - Positivists argue that by carefully and objectively collecting data regarding social phenomena, we can determine laws to predict and explain human behaviour in terms of cause and effect.
> - Like empiricists, positivists reject normative and metaphysical (relating to being) questions that cannot be measured scientifically.
> - Positivism differs from empiricism because it requires propositions to be verified (logical positivism) or hypotheses falsified (critical rationalism) rather than just simply presenting findings.
> - Sources of primary data are closed-question questionnaires and surveys (see Chapter 3).
>
> *Example study of poverty*
> Poverty is explained through testing a hypothesis by collecting and scientifically testing data related to poverty (e.g., statistically testing whether poverty is a function of educational attainment).

Comte (1798–1857) established the concept of **positivism** as a reaction against the 'negative philosophy' of pre-revolution France. Comte argued that the latter tradition was speculative in nature and was based upon emotion and romantic notions of considering alternative utopias. As such, it was neither practical nor constructive because it did not concern itself with material objects and given circumstances (Holt-Jensen, 1988). Just as the empiricists argued that we should not be engaging with normative questions, positivists argued that we should avoid metaphysical questions as they are unscientific, **metaphysics** being defined as that which lies outside, or is independent of, our senses and relates to questions of being. Unwin (1992) notes that Comte used the term 'positive' to refer to the actual, the certain, the exact, the useful and the relative rather than the imaginary, the undecided, the imprecise, the vain and the absolute. As such, Comte demanded the formulation of theories which could be tested and verified using a union of methods. Positivism thus differs from empiricism because it requires experience to be verified rather than just simply presented as fact (Johnston, 1986a). Comte's hope was that positivism would provide society with knowledge so that **speculation** could be avoided. In addition to providing **laws** for nature, Comte believed that there were laws of society and social relationships which, although more complex, could be discovered using the same principles (i.e., sociology). Positivists thus argued that by the careful and **objective** collection of data regarding to social phenomena, we could determine laws to predict and explain human behaviour in terms of cause and effect.

Although there are various versions of positivism, contemporary positivism can, in the main, be divided into two streams of thought: **logical positivism** based upon verification, and **critical rationalism** based upon falsification. Logical positivism was developed by the Vienna school in the 1920s and was intended to combine British empiricism with traditional positivism (Holt-Jensen, 1988). The Vienna school defined precise scientific principles and used formal logic to verify theories and make statements of knowledge based upon the axioms produced. The formal laws constructed, in turn, led to the formation of new questions to be *verified* against reality. In contrast to Comte, the Vienna school accepted that some statements could be verified without recourse to experience. As such, logical positivism is based upon a distinction between analytical statements and synthetic statements. Gregory (1986a) describes **analytical statements** as *a priori* propositions whose truth was guaranteed by their internal definitions, e.g., tautologies. These constituted the domain of the formal sciences, logics and mathematics. In contrast, **synthetic statements** are propositions whose truth still had to be established empirically through testing the verification of hypotheses (see Section 2.4).

Critical rationalism was put forward by Karl Popper as an alternative to logical positivism. He argued that the truth of a law does not depend upon the number of times it is experimentally observed or verified, but rather on whether it can be *falsified* (Chalmers, 1982). Scientific validation should not proceed along the lines of providing confirmatory evidence but rather by identifying circumstances which may lead to the rejection of the theory. If no situation can be found where the law does not hold, then the law can be said to be corroborated, although its validity has not been confirmed. Popper's approach has been criticised as being virtually impossible to implement (Sayer, 1992) and has not been adopted by many human geographers (Gregory, 1986b).

In geography, positivism has been most closely associated with the use of quantitative methodologies. Until the early 1950s geographical studies had been largely descriptive and regional in nature, and it was at this point that geographers such as Schaefer (1953) started to argue that research needed to become more scientific in nature. Schaefer, for example, advocated the adoption of a logical positivist approach. Throughout the late 1950s and 1960s geography underwent a quantitative revolution as geographers sought to replace description with explanation, individual understanding with general laws, and interpretation with prediction (Unwin, 1992). As such, central concerns were **space**, **quantification** and **theory building**, and throughout the 1960s there was increasing adoption and usage of quantitative methodologies. Whilst a number of geographers who advocated quantitative methods were positivists, for many positivism was implicit in their work rather than explicitly recognised, and many studies were merely empirical and inductive (Gregory, 1986b). It was not until Harvey's (1969) seminal work *Explanation in Geography* that geographers really started to examine questions of how and why knowledge was produced. Subsequently, in the 1970s the implicit adoption of positivism came under attack and new modes of enquiry were developed as a reaction to its increasing use in geography. However, it must be noted that quantitative methods are not solely used by positivists and the use of such methods does not make a piece of research positivistic in nature. Rather it is the adoption of the underlying principles of **objectivity** and **formal logic**.

Positivism has been criticised for a variety of reasons, and from a number of quarters. Gregory (1986a) divides these attacks into three main fronts. First, positivism has been criticised for its **empiricism**. It is argued that positivists underestimate the complex relationship between theory and observations and in particular the difficulty in separating the effects of phenomena that are interrelated. Second, positivism is criticised for its **exclusivity** and the assumption that methods of the natural sciences can be effectively used to explain social phenomena. As such, positivism fails to recognise that spatial patterns and processes are bound up in economic, social and political structures (Cloke *et al.*, 1992). In addition, it is argued that mathematical language filters out social and ethical questions and fails to recognise that spatial patterns and processes are reflected in, and are reflections of, the perceptions, intentions and actions of human beings (Cloke *et al.*, 1992). Third, positivism is criticised for its **autonomy**. Positivism's arguments that science should be neutral, value-free and objective have been widely rejected and it is argued that positivism creates a false sense of objectivity by artificially separating the observer from the observed (Cloke *et al.*, 1992).

1.4.2 Historical–hermeneutic science

Behaviouralism

Box 1.5 Behaviouralism

In a nutshell
- Behaviouralism acknowledges, explicitly or otherwise, that human action is mediated through the cognitive processing of information.
- Behaviouralist seeks to model spatial behaviour by explaining spatial choice and decision making through the measurement of people's ability to remember, process and evaluate geographic information.
- Sources of primary data are closed-question questionnaires (see Chapter 3) and specialised, psychologically based tests of knowledge (see Golledge and Stimson, 1997).

Example study of poverty
Poverty is explained through the scientific testing of a hypothesis which examines the behavioural decision making of poor people and/or people in positions of power (e.g., statistically testing whether poor people have low levels of self-esteem, and if so, how this relates to job-seeking behaviour).

Behavioural approaches (often misguidedly referred to as *perceptual* approaches) are the variety of approaches which acknowledge, explicitly or otherwise, that human action is mediated through the **cognitive processing** of environmental information. Behaviouralism was initially conceived in psychology as a reaction against the positivistic school of thought of behaviourism, which views behaviour in terms of stimulus–response, in which the cognitive processes or consciousness has little part. Behaviouralism, alternatively, assumes that actions are mediated by cognitive processes. By the end of the 1960s many geographers were starting to become dissatisfied with the stereotyped, mechanistic and deterministic nature of many of the quantitative models being developed, as they realised that not everyone behaved in the spatially rational manner advocated by positivists. In response, some geographers turned to behaviouralism as an alternative.

Behavioural geography is based upon the belief that the explanatory powers and understanding of geographers can be increased by incorporating behavioural variables, along with others, within a decision-making framework that seeks to comprehend and find reasons for overt **spatial behaviour**, rather than describing the spatial manifestations of behaviour itself (Golledge, 1981). It is argued that superficial descriptions of the natural, human or built environments are not enough, and for both an understanding and an explanation of geographic phenomena an insight into 'why' questions is needed so that investigations become process-driven (Golledge and Rushton, 1984). By the early 1970s behaviouralism was increasingly being adopted by researchers to study a number of different themes, and in recent years behavioural studies can be found in work relating to retailing, migration, housing, industrial location, travel, leisure and tourism, spatial behaviour, disability, planning, geographic education and cartography. Throughout the 1970s, however, it became clear that behavioural geography was actually manifesting itself in two very different forms. On the one hand, there were those who were concerned with incorporating behavioural variables in spatial models, **analytical behaviouralism**, and on the other, those who rejected spatial analysis outright and were concerned with 'sense of place', values, morals and **phenomenological inquiry** (see next section). The net result was that by the late 1970s, the field was beset by internal division and conflict, and by the beginning of the 1980s, behaviouralism, initially conceived as a reaction to the excesses of conventional spatial science, was being depicted as merely a logical 'outgrowth' of the commitment to positivism enshrined in the quantitative revolution.

Like positivism before it, analytical behaviouralism was roundly criticised as mechanistic, dehumanising and ignorant of the broader social and cultural context in which decision makers operate. Analytical behaviouralism thus overemphasises empiricism and methodology at the expense of worthwhile issues and philosophical content. It also acknowledges a dichotomy between subject/object and fact/value, thus failing to 'conceive life in its wholeness' (Eyles, 1989). As a result, behaviouralism was criticised for dehumanising and depersonalising the people and places studied, ignoring the contours of experience and systematically detaching individuals from the social contexts of their actions. Other critics are concerned with the adoption of psychological theory and practices, arguing that the close links to cognition lead to problems of measurement, analysis and generalisation. A related issue concerns the danger of psychologism, the fallacy of explaining social phenomena purely in terms of the mental characteristics of individuals. Some contend that by concentrating upon the individual, behaviouralism is susceptible to building models inductively so that outcomes can only be treated as a sum of parts. In addition, some geographers argue that societal and institutional constraints are the dominant factors affecting spatial behaviour and therefore behaviouralism has little utility. As a result, behaviouralism reinforces the status quo by failing to study the dominant issues and concentrating upon idealism rather than materialism. Many of these attacks on behaviouralism were doubly destructive because they came from disillusioned behavioural researchers themselves (e.g., Bunting and Guelke, 1979).

Geographers such as Golledge (1981) have vehemently defended behaviouralism, suggesting that criticisms are based upon misunderstandings. With Helen Couclelis (Couclelis and Golledge, 1983) he has tried to refute the arguments that behaviouralism was just an extension, or outgrowth, of positivism. They argue that although born in the positivist tradition and reflecting some underlying principles, behaviouralism has significantly progressed beyond positivism. For example, they suggest that the critical tenet of the non-existence of the unobservable was dropped, weakening reductionism and physicalist interpretations of human behaviour. Spatial behaviour thus became differentiated from behaviour in space. Transactional and constructivist positions were also adopted from developmental psychology and destroyed the tenet of the scientist as a passive observer of objective reality. This, they argue, has hastened the demise of the positivist positions of objectivism, the non-scientific importance of values and beliefs, and the separation of value and fact. It is thus assumed that the mind and world are in constant dynamic interaction and therefore the *a priori* world of the positivist position is rejected.

Golledge, whilst acknowledging the weakening position of behaviouralism within contemporary debate, argued with Kevin Cox (who had by this time moved to a more critical position – see later) that behavioural geography could avert becoming an irrelevance through an evolution created by the expansion of relevant variables and/or a switch from the emphasis upon the individual as an isolated decision maker to an individual as caught up in a 'web' of social constraint (Cox and Golledge, 1981). Walmsley and Lewis (1993) argue that such evolutions have now

occurred. As such, many of the well-rehearsed arguments used to condemn the behavioural approach are viewed as being anachronistic, irrelevant and outdated. We will have to wait and see to determine whether behavioural ideas are reintegrated into contemporary debates. However, it is probably fair to say that, just like positivism, many studies are still behavioural in nature, particularly in retail, consumer and migration studies.

Phenomenology

Box 1.6   Phenomenology

In a nutshell

- Phenomenology rejects the scientific, quantitative approaches of positivism and behaviouralism.
- Instead phenomenology suggests that we concentrate upon understanding rather than explaining the world.
- The goal of phenomenology is to reconstruct the worlds of individuals, their actions, and the meaning of the phenomena in those worlds to understand individual behaviour, without drawing upon supposed theories.
- Sources of primary data are in-depth interviews with people who have experienced the phenomena in question (see Chapter 7).

Example study of poverty

To understand poverty it is suggested that we need to reconstruct the world of people who are poor (e.g., we need to try and see the world through the eyes of poor people). This might be attempted by talking to them about their life experiences.

The development of phenomenology is usually attributed to Husserl (1859–1938). Husserl argued that because positivists ignored the question of their own involvement in the research process, they could not fully know the world. To Husserl the distinction between **object** (world) and **subject** (humans) was problematic. He sought to overcome this dualism and to provide a powerful, rigorous and alternative philosophy to positivism so far lacking in humanist thought (Cloke *et al.*, 1992). Phenomenology was designed 'to disclose the world as it shows itself *before* scientific inquiry, as that which is pre-given and presupposed by the sciences. It seeks to disclose the original way of being prior to its objectification by the empirical sciences' (Pickles, 1985: 3). Here, Pickles is suggesting that the way we conduct science influences our conclusions and that what we should be seeking is not the rose-tinted view positivism offers but the view before the scientific glasses are put on. Pile (1993: 24) thus describes phenomenology as a 'people-centred form of knowledge based in human awareness, experience and understanding . . . the study of, and conscious reflection on, the meaning of being human, of being located in time and space'. As such phenomenology is based upon three assumptions: '[1] that people should be studied free of any preconceived theories of suppositions about how they act. [2] The search for **understanding** or appreciation of the nature of an act is the goal of social science, rather than explanation. [3] that for people the world exists only as a **mental construction**, created in acts of intentionality. An element is brought into an individual's world only when he or she gives it meaning because of some intention towards it' (Johnston, 1986a: 62−63). Johnston (1986a) further explains that the goal of phenomenology is to reconstruct the worlds of individuals, their actions, and the meaning of the phenomena in those worlds to understand behaviour. In contrast to scientific approaches, which treat phenomena as external objects which can be studied objectively, phenomenology recognises subjectivity and demands that we reflect on our own consciousness of things in our experience to come to a deeper understanding of the world (Relph, 1981). As such, phenomenology seeks to disclose and elucidate what we experience and how we experience it.

For phenomenologists, we actively constitute the knowledge to be had from objects in the world, rather than just passively accessing and using them. Husserl explains that objects must be understood as objects for human subjects; as objects that human subjects experience/are conscious of; and as objects towards which humans always intend to use or interact with (Cloke *et al.*, 1992). Phenomenologists then must strip away the 'rose-tinted glasses' to really reveal the true **essences** of objects to human subjects. As such, phenomenology openly acknowledges metaphysics and the need for **reflection**. Although a reaction against positivism, it should be realised that Husserl was also concerned with producing laws. In contrast to positivism, however, Husserl's concerns were with **metaphysical laws** that governed the workings of human spirit (the inner world of human being). Thus, phenomenology is about not only understanding behaviour, but enriching life by increasing human awareness (Johnston, 1986a).

Husserl's phenomenology has not been successfully adopted by geographers because its emphasis on pure reflection of essences leads to methodological difficulties (Unwin, 1992). As such, Husserl's phenomenology is a personal transcendental exercise that is reflective and leads to individual understanding (Unwin, 1992). Instead Husserl's arguments have been used most effectively to attack positivism as a philosophy and it has been left to those succeeding Husserl to soften his strict approach by forwarding alternative phenomenologies that talk less of trying to transcend the everyday and more about studying the everyday meanings etched into the **lifeworlds** of particular peoples, societies and cultures (Cloke *et al.*, 1992). Spiegelberg (in Johnston, 1986a) suggests two methodologies to gain access to absolute knowledge that resides in consciousness. The first is imaginative self-transposal, which requires imagining the world from the perspective of the other person using clues of how to achieve this from first-hand perception and other documentary evidence. The second is cooperative encounter and exploration in which the investigator and subject embark on a joint exploration of the latter's lifeworld. Here, meaning and knowledge are sought through communication based upon trust and respect. Not unsurprisingly, phenomenology is criticised for its reliance on the subject to be able to communicate their interpretations and meanings, and the ability of the investigator to interpret such communications.

A number of geographers, such as Buttimer (1976), Ley (1977), Relph (1976), Seamon (1979) and Tuan (1974), have adopted variations of softer approaches which are characterised by their search for **meaning** (Ley, 1977). These geographers saw phenomenology as a viable alternative to the peopleless and dehumanising positivistic and behavioural approaches being adopted. Tuan (1971), for example, argued that a phenomenological approach combined with elements of existentialism can 'tease out the "essences" of certain "geographical concerns" residing in the deepest psychological, emotional and existential attachments that all human beings hold for the spaces, places and environments encircling them' (Cloke *et al.*, 1992: 75). In essence, the approach emphasises 'the social construction of places, taking into account such aspects as their emotional, aesthetic and symbolic appeal' and seeks to reflect the ties between individuals and the environment (Unwin, 1992: 148). Smith (1981) suggests a phenomenological methodology in which material is gathered through participant observation and through reflecting and selecting observations and experiences gathered transactionally (through contact with subjects). This produces an empathetic understanding of behaviour (Johnston, 1986a). In recent years there has been an increased interest in this transactional approach (Aitken and Bjorklund, 1988; Aitken, 1991). **Transactionalism** represents the view that an understanding of person-in-environment contexts must take on board an appreciation of on-going transactions between the person and the environment, based upon both past events and future expectations (Aitken and Bjorklund, 1988).

Existentialism

Box 1.7   Existentialism

In a nutshell
- Existentialism is based on the notion that reality is created by the free acts of human agents, for and by themselves.
- Whereas phenomenology is primarily concerned with meaning, existentialism also concerns values.
- Existentialism focuses upon how individuals come to create and place meaning to their world and how they subscribe values to objects and to others.
- Sources of primary data are in-depth interviews; ethnography; and participant observation (see Chapter 7).

Example study of poverty
Poverty is understood by trying to gain insight into how people who are poor come to know, ascribe meaning, and interact with the world (e.g., interviewing poor people about how they decide how much money to spend on different things).

Existentialism is based on the notion that reality is **created** by the free acts of human agents, for and by themselves (Johnston, 1986a). It differs from phenomenology by positing that there are no general essences, pure consciousness or ultimate knowledge, and in its fundamental concern with what Buttimer (1974) describes as 'the quality of life in the everyday world'. As such, there is no *single* essence of humanity, but rather each **individual** creates and forges their own essence from existence. Whereas phenomenology is primarily concerned with meaning, existentialism also concerns **values**. It focuses upon how individuals come to create and place meaning to their world and how they subscribe values to objects and others. A basic premise is that humans are alienated and detached from non-human things, and thus constantly seek to ' "make things meaningful" so as to fill the "existential void" (the complete

lack of meaning) at the heart of the human condition' (Cloke *et al.*, 1992: 76). The researcher's job is to seek to understand the process of making the world meaningful as it is these processes by which we come to know, and behave in, places. Samuels (1978, 1981) has been most vocal in a call for an existentialist approach within human geography. This approach effectively takes a historical perspective and 'endeavours to reconstruct a landscape in the eyes of its occupants, users, explorers and students in the light of historical situations that condition, modify or change relationships' (Samuels, 1981: 129). Samuels suggests that in creating their own essence, or identity, individuals define themselves spatially through building a relationship with the environment. Thus, the landscape is a biography of that creation (Johnston, 1986a).

Idealism

Box 1.8   Idealism

In a nutshell
- Idealism posits that, ontologically, the real world does not exist outside its observation and representation by the individual.
- Whereas existentialism focuses upon reality *as being*, idealism views reality as a construction of the mind.
- Idealism seeks to explain patterns of behaviour through an understanding of the thoughts behind them.
- Sources of primary data are in-depth interviews and ethnography (see Chapter 7).

Example study of poverty
Poverty is understood by trying to gain insight into how poor people think about poverty and the world they live in (e.g., interview poor people on what it feels like to be poor, why they think they are poor, and how they see themselves in relation to the rest of society).

Idealism posits that the real world does not exist outside its observation and representation by the individual (Johnston, 1986a). Whereas existentialism focuses upon reality *as being*, idealism views reality as a construction of the **mind** (Unwin, 1992). Thus, idealism holds that the world can only be known indirectly through ideas with knowledge based on **subjective experiences**. As such, idealism seeks to explain rational actions through an understanding of the thoughts behind them (Guelke, 1974). In this context all knowledge is entirely subjective and used to develop personal theories which serve to guide actions and allow the world to be negotiated and understood. Actions and the social world, therefore, are mediated and created through knowledge rather than being simply conditioned (May, 1993). If we wish to understand human decision making, idealists argue that we should be studying the decision makers and the personal theories that guide them (Johnston, 1986a). In order to study such personal theories idealists aim to recover the guiding mental constructs and ideas through empathetic understanding (Jackson and Smith, 1984). This is in contrast to behaviouralists, who try to determine knowledge, perceptions and attitudes to explain human decision making. Idealism differs from phenomenology in that it does not entirely reject explanation but aims to secure an **objective reconstruction** of thought rather than a subjective description of 'lifeworlds' (Jackson and Smith, 1984). Guelke (1974, 1981) has been the main proponent of idealistic studies within geography. He suggests that an idealist geography would study the mental activity which underlies human activity within the geographic world and could provide the basis of a rejuvenated regional geography by grouping together people who share similar world views.

Pragmatism

Box 1.9   Pragmatism

In a nutshell
- Pragmatism suggests that, rather than focus on individuals, attention should be paid to society and to the interaction of individuals within society.
- Pragmatists argue that understanding must be inferred from behaviour and rooted in experiences, not knowledge.
- By exploring the lives of people within a community it is hoped that the nature of the beliefs and attitudes which shape society will be uncovered.
- Sources of primary data are ethnography and participant observation (see Chapter 7).

Example study of poverty
Poverty is understood by observing how individuals in society interact to produce conditions which sustain destitution (e.g., examining whether poor people remain poor because they live in a cycle of crime, undereducation, low self-esteem, etc.).

Pragmatism, whose origins are predominately identified with the North American writers Peirce (1839–1914),

Dewey (1859–1952) and James (1842–1910), is concerned with the construction of meaning through **practical activity** (Gregory, 1986c) attempting to ground philosophical activity in the practicalities of everyday life (Smith, 1984). Pragmatism thus tries to understand the world through the examination of practical problems, believing that studying a particular real-world situation is important for providing both theoretical understanding and practical solutions (Frazier, 1981). Pragmatism rejects '**value-free**' research, instead arguing that research should address, and be used to solve, problems. Pragmatists envisage knowledge as an 'essentially fluid and intrinsically fallible process of "self-correcting enquiry"' (Gregory, 1986c: 366). Pragmatism is thus a theory of knowledge, experience and reality that maintains 'that thought and knowledge are biologically and socially evolved modes of adaptation to and control over experience and reality' (Thayer, in Johnston, 1986a: 60). Here, it is recognised that all knowledge is evaluative of future experiences and that thinking functions experimentally in anticipation of future experiences and consequences of actions. Johnston (1986a) thus explains that society changes as beliefs are actualised and that individuals make choices based upon meanings attached to various possibilities which are evaluated in terms of their utility and usefulness. Knowledge is thus achieved only through **experience** and a trial-and-error process of activity, based upon attitudes and beliefs, as we search for the 'truth'. Thus pragmatism is based within experience and is concerned 'with understanding and resolving the conflicts to which a fluid and uncertain world gives rise' (Jackson and Smith, 1984: 72). As such, understanding must be inferred from behaviour and rooted in experience, not knowledge.

Pragmatism, in contrast to other humanistic approaches, suggests a focus on **society** and the interaction of individuals within society – Rorty (1980) suggests that knowledge is merely an on-going **conversation** between us all. Pragmatism avoids many irresolvable philosophical questions 'by examining instead both constitution and application of knowledge in everyday life' (Cloke *et al.*, 1992). Geographers' interest in pragmatism stems from the work of the Chicago sociologists Park and Burgess earlier this century. They suggest that 'social interaction through which attitudes and beliefs are learned and developed occurs in places . . . [and] the construction and reconstruction of society – the search for truth – is a spatially situated activity' (Johnston, 1986a: 61). These sociologists in the main used **ethnographic techniques** (see Chapter 7) such as participant observation to explore the lives of local residents. Frazier (1981) reports that pragmatists use a scientific approach, using a **deductive–predictive approach** where a theory is formulated and then tested. However, pragmatists also refute replicability, acknowledging that retesting may lead to different answers which they attribute to the problems of observing a situation and changing social context. As such, because the world is constantly changing, positions must be constantly retested and re-evaluated. Jackson and Smith (1984) report that pragmatic research is also often participatory in nature, to allow the nature of the beliefs and attitudes which shape society to be uncovered, and to determine how they are being reconstructed through application. As such, pragmatism is **action-orientated** and **user-orientated**.

1.4.3 Critical science

Marxist approaches

Box 1.10   Marxism

In a nutshell
- Marxists suggest that society is structured so as to perpetuate the production of capital.
- Marxists are concerned with investigation of the political and economic structures that underlie and reproduce capitalist modes of production and consumption.
- To do this Marxists suggest that we need to consider how conditions might be under different social conditions to highlight how society operates.
- A source of primary data is observation (see Chapter 7), but Marxism also re-examines secondary data sources with analysis consisting of determining the dialectical (how one affects the other) relationship between societal structures and individuals.

Example study of poverty
Poverty is explained through the examination of how poor people are exploited for capital gain (e.g., examining whether poor people are poor because it is in the interests of capital to retain unskilled, low-wage jobs rather than distribute fully corporate profit).

Simply stated, Marxism is a 'system of thought . . . which claims that the state, through history, has been a device for the exploitation of the masses by a dominant class and that class struggle has been the main agent of historical change' (Peet and Lyons, 1981:

207). As with the approaches so far discussed, there are many different versions of Marxism. They all share, however, a *critical* approach to modern society which aims not only to study, but also to **change**, social processes. Marxist approaches seek to do this by exposing the inherent **injustices** within present social relations which they argue are the result of the economic bases of **capitalism**. They argue that social relations are constrained within regulating capitalist structures. These structures exist as a means of enforcing and reproducing wealth for a minority of the population through the **exploitation of labour**. Contemporary Western society is thus characterised by a capitalist '**mode of production**' as the means people employ to sustain themselves. Within this mode there are inherent contradictions that need to be exposed, so that unfair social relations enshrined in the class system can be overthrown.

Within a **structuralist** framework there are three levels of analysis: '(1) the level of appearances, or the *superstructure*; (2) the level of processes, or the *infrastructure*; and (3) the level of imperatives, or the *deep structure*' (Johnston, 1985: 220). Methods of analysis are **dialectical** in nature, examining from an economic and political perspective the processes of change and dynamics within society, and exposing the three (hidden) levels of structures which regulate the uneven nature of society. Dialectics is a method of seeking knowledge through a process of continuous questioning and answering, with one answer providing the basis for a subsequent question (Unwin, 1992). By dialectically examining the transformations in the mode of production, it is hoped that an understanding of social change can be developed through the identification of historically determined laws.

Marxist approaches within geography emerged at approximately the same time as humanistic approaches, and similarly were a reaction against the growth of spatial science (positivism) within the discipline. Whereas humanistic approaches criticised spatial science because of its disregard of human agency, Marxists argued that it failed to recognise the economic and political constraints imposed upon spatial patterns by the way in which society worked. Further, they suggested that positivistic methods restricted analysis to how things actually *seemed* to be, rather than considering how they *might* be under different social conditions (Cloke *et al.*, 1992). A Marxist geography seeks to identify how social relations vary over space and time in order to reproduce and sustain the modes of production and consumption, to suggest alternative futures, and to offer **political resistance** (Peet and Lyons, 1981). Whilst a 'pure' form of Marxism, **historical materialism**, gained favour in the late 1970s, by the early 1980s attention had moved to other structuralist approaches such as Gidden's **structuration theory** and **political economy**. Both these approaches relaxed the emphasis upon structure to incorporate ideas of **agency**. Unwin (1992) contends that structuralist approaches have most commonly been applied to four main areas of geographical study: an historical geography of transition form feudalism to capitalism; urban geography; regional inequalities and industrial restructuring; and the Third World.

Realism

Box 1.11   Realism

In a nutshell

- Realists are concerned with the investigation of the underlying mechanisms and structures of social relations, and identifying the 'building blocks' of reality.
- Rather than studying the communication and interaction between people, realism seeks the underlying mechanisms of policy and practice that made these possible in the first place.
- Realism concerns the identification of how something happens (causal mechanisms) and how extensive a phenomenon is (empirical regularity).
- Realists want to find out what produces changes, what makes things happen, what allows or forces changes.
- Sources of primary data include a mix of quantitative (see Chapter 3) and qualitative (see Chapter 7) techniques.

Example study of poverty

Poverty is understood by trying to determine its root causes through an examination of the mechanisms underlying how society operates (e.g., examine whether poverty exists because of the uneven development of modernisation).

Realism shares with positivism the aim of explanation rather than understanding. However, here the similarities end. Whilst realists believe that there is a 'real' world that exists independently of our senses, perceptions and cognitions, in contrast to analytical–empirical approaches, realists argue that the social world does not exist independently of knowledge and that this knowledge, which is partial or incomplete, affects our behaviour (May, 1993). They argue that

the task of research 'is not simply to collect observations but to explain these within theoretical frameworks which structure people's actions' (May, 1993: 7). Realists are, therefore, concerned with the investigation of the underlying **mechanisms** and **structures** of social relations, and with identifying the 'building blocks' of reality. Rather than studying the communication and interaction between people, realism seeks the underlying mechanisms of *policy* and *practice* that made these possible in the first place (May, 1993). As such, realism is concerned with the identification of how something happens (**causal mechanisms**) and how extensive a phenomenon is (**empirical regularity**) (Unwin, 1992). Realists want to find out 'what *produces* changes, what *makes* things happen, what *allows* or *forces* changes' (Sayer, 1985: 163). Unlike positivism which posits a closed system of discrete events that can be tested with specific hypotheses, realism presents an alternative by 'assuming a stratified and differentiated world made up of events, mechanisms and structures in an open system where there are complex, reproducing and sometimes transforming interactions between structure and agency whose recovery will provide answers to questions posed about processes' (Cloke *et al.*, 1992: 146). Realism, then, does not deny *human agency*, although it does emphasise that behaviour is constrained by *economic* processes (Johnston, 1991). Rather, individuals make decisions within an infrastructure that they are unaware of. As a result, the infrastructure is both constraining and enabling; it restricts yet stimulates choice (Johnston, 1991).

Sayer (1985, 1992) in particular has championed the cause of realism within geography. He suggests that geographers can engage with four different types of realist study. The first is *abstract*, theoretical research concerned with structures and mechanisms; it concentrates on developing a theory that might explain circumstances or lead to possible scenarios. The second is *concrete*, practical research focusing upon events and objects produced by structures and mechanisms and thus seeks to explain a circumstance or scenario. The third is *empirical generalisations* concerned with the establishment of the regularity of events. The fourth is *synthesis research* which combines all of these types of research in order to explain entire subsystems. Sayer also describes how realist research can be undertaken at the local scale using **intensive research** aimed at producing causal explanations, and at the regional scale using **extensive research** aimed at producing descriptive generalisations. Intensive research consists of trying to determine the processes and conditions both *necessary* (object/subject needed for a process or situation to arise, e.g., gunpowder has the necessary casual power to explode) and *contingent* (object/subject needed to activate or release causal powers, e.g., gunpowder needs a spark) that underlie the production of certain events or objects by studying individuals in their contexts, using qualitative methodologies such as interactive interviews and ethnography (Cloke *et al.*, 1992). Extensive research consists of trying to determine the generality or commonality of characteristics and processes in relation to a wider population using quantitative methodologies such as questionnaires. Although asking different types of questions, using different methodologies, both types of research are still seeking to explain the phenomena in terms of the underlying mechanisms and structures which dictate their pattern and form.

Postmodernism

Box 1.12 Postmodernism

In a nutshell
- Postmodernists argue that, so far, modernist meta-narratives (all the other approaches so far described) that seek universal truths by examining the associations and relationships between people and places, have failed to adequately account for differences within society.
- Postmodernism is based upon the notion that there is no one answer, that no one discourse is superior or dominant to another, and that no-one's voice should be excluded from dialogue.
- Postmodernists argue that there is no one absolute truth and that there is no truth outside interpretation.
- Postmodernism, rather than seeking 'truth', offers 'readings' rather than 'observations', and 'interpretations' rather than 'findings', and seeks intertextual relations rather than causality.
- Primary analysis consists of deconstructing (teasing apart) culture and societal practices.

Example study of poverty
Poverty is understood by trying to deconstruct and read the various ways in which poverty is constructed and reproduced in society, and how poor people are excluded from society (e.g., examine the ways in which poor people are excluded from society through unequal power relations).

Brown (1995) suggests that postmodernism is something everyone has heard of, but no-one can quite explain. This is because, as Peet and Thrift (1989) suggest, postmodernism is a confusing term which represents a combination of different ideas. At one level, postmodernity refers to a new way of **understanding** the world. It is a revolt against the rationality of modernism, and represents a fundamental attack on contemporary philosophy (Dear, 1988). At another level, postmodernity refers to *an object of study* – postmodernity is the study of the temporal and spatial organisation, and the complex interaction, of economic, social, political and cultural processes in the late twentieth century. In this framework, postmodern culture is often presented as an *alternative* to modernist visions of society, which are presented as fundamentally flawed and structurally weak (Poster, 1995). In this discussion, we only examine postmodernism as an approach.

So far, we have discussed approaches which are based within modernity (e.g., positivism, Marxism). Modernism concerns the search for a unified, **grand theory** of society and social knowledge and seeks to reveal **universal truths** and meaning. Dear (1988) argues that this has led to a variety of internally consistent but mutually exclusive approaches. Postmodernism challenges modernist thinking by examining epistemological independence and challenging 'truth claims'. In essence, postmodernism is based upon the notion that there is no one answer, no one discourse that is superior or dominant to another, and that no-one's voice should be excluded from dialogue (Dear, 1988). Postmodernists therefore argue that there is no one **absolute truth** and that there is no truth outside interpretation. As such, postmodern approaches represent a shift from ways of knowing and issues of truth, to ways of being and issues of reality.

Postmodern thinking is thus concerned with developing an attitude towards knowledge, methods, theories and communication, and posits that we move away from questions relating to the 'things actually going on . . . to questions about how we can find out about, interpret and then report upon these things' (Cloke *et al.*, 1992). Here, 'the very possibility of acquiring knowledge or giving an account of the world is called into question' (Lyon, 1994: 11). Within postmodern approaches, organised, objective science is replaced by a postscience which acknowledges the position of scientist as **agent** and **participant**. Essentially, there is a broad-gauged reconceptualisation of how we experience and explain the world around us which includes focusing attention upon alternative discourses and meanings rather than goals, choices, behaviour, attitudes and personality; the dissolution of disciplinary boundaries; and a re-emphasis on that which has largely been ignored by modernist scholars – namely the excluded, marginal and repressed (Rosenau,

1992). Postmodernism thus offers 'readings' rather than 'observations', 'interpretations' rather than 'findings', seeking intertextual relations rather than causality (Rosenau, 1992).

Postmodernism has been criticised as being little more than a form of critique – an intellectual, speculative, 'self-seeking cynicism' (Lyon, 1994: 77) with little substance. Critics argue that the dominant bases of the modernist agenda – enquiry, discovery, innovation, progress, internationalisation, self and economic development – are, however, still the principles underlying Western society and modernist approaches are still most appropriate to study them (Berman, 1992). As such, postmodern critique should merely be used to *improve* modernist methods, to make them more robust and to widen their scope, not to replace them. Such a radicalised, modernist approach has been put forward by both Habermas and Giddens. They both treat modernity as an incomplete project, and like postmodernists challenge the foundationalist approaches to science but insist that a hermeneutic approach to social science can still yield realistic results within a more critical framework (Lyon, 1994). Other approaches, such as feminism, also try to reframe modernist thought within a more emancipatory or reflexive framework rather than move to a new postmodern position.

Poststructuralism

Box 1.13 Poststructuralism

In a nutshell
- Poststructuralists argue that the relationship between society and culture is mediated through language: humans are configured and given cultural significance through language.
- The way we live our lives within society, the constraints and empowerment that operate, take effect in language.
- Poststructuralists propose that the way to gain an understanding of the factors that shape our lives is to deconstruct the multiple messages being conveyed to us by the objects we encounter.
- The primary mode of analysis is the deconstruction of language.

Example study of poverty
Poverty is understood through examination and deconstruction of exclusionary practices of the society as expressed through cultural practices and articulated in language (e.g., examine the extent to which cultural norms and myths feed into exclusionary processes which seek to marginalise poor people from material wealth).

Poststructuralists argue that the relationship between society and space is mediated culturally through **language**. In contrast to postmodernism, much of the focus is upon the individual, and methodological and epistemological issues, rather than society and cultural critique (Rosenau, 1992). For poststructuralists, a human being is configured and given cultural significance through language (Poster, 1995). As such, the way we live our lives within society, the constraints and empowerment that operate, take effect in language. Therefore, if we are to understand the relationship between space and society we need to explore the positioning of an individual in relation to language and how the individual is configured by language. Such an approach examines society by interpreting and **deconstructing** cultural dissemination to gain understanding. Peet and Thrift (1989: 23) explain that poststructural work assumes

that meaning is produced in language, and not reflected by it; that meaning is not fixed but is constantly on the move . . . and that subjectivity does not imply a unified, and rational human subject but instead a kaleidoscope of different discursive practices. . . . the kind of method needed to get at these conceptions will need to be very supple, able to capture the multiplicity of different meanings without reducing them to the simplicity of a simple structure.

Researchers then should focus on textuality, narrative, discourse and language as these do not just reflect reality but actively construct and constitute reality. As language precedes us and exceeds us so that it is something into which we are initiated and which governs our actions and thoughts, when each of us reads a text, views a landscape or a building we see and interpret them in different ways (Brown, 1995). Poststructuralists thus propose that the way to gain an understanding of the social, cultural, political and economic factors that shape our lives is to deconstruct the **multiple messages** being conveyed to us by the objects we encounter. Deconstruction is a technique for 'teasing out the incoherencies, limits and unintentional effects of a text' (Cloke *et al.*, 1992: 192).

Feminism

Box 1.14 Feminism

In a nutshell
- Feminism suggests that science is dominated by, and reflects the position of, men. Some feminist geographers extend this specifically to white, wealthy, Western men.

Box 1.14 (cont'd)

- Feminists suggest that there needs to be renegotiation of the role and structure of institutions and the production of knowledge.
- There needs to be a renegotiation of power relations within society so that how we come to know the world is more reflective of the people living in it.
- Sources of primary data include a mix of quantitative (see Chapter 3) and qualitative (see Chapter 7) techniques.

Example study of poverty
Poverty is understood by trying to adopt more emancipatory and empowering approaches that allow poor people to express experience and knowledge (e.g., ask poor people how they think that society should be reconfigured into a more just system).

Box 1.15 Sources of sexism within research

- *Androcentricity*: Viewing the world from a male perspective: e.g., when a test or other research instrument is developed and tested on males, and then assumed to be suitable for use with females.
- *Overgeneralisation*: When a study deals with only one sex but presents itself as generally applicable, e.g., a study dealing solely with mothers which makes statements about parents.
- *Overspecificity*: When single-sex terms are used when both sexes are involved: e.g., many uses of 'man' either by itself or as in 'chairman'.
- *Gender insensitivity*: Ignoring sex as a possible variable.
- *Double standards*: Evaluating, treating or measuring identical behaviours, traits or situations by different means for males and females: e.g., using female-derived categories of social status for males (or vice versa). This may not be inappropriate in a particular study but nevertheless could lead to bias which should be acknowledged.
- *Sex appropriateness*: A commonly used and accepted form of 'double standards': e.g., that child rearing is necessarily a female activity.
- *Familism*: A particular instance of 'gender insensitivity'. Consists of treating the family as the smallest unit of analysis when it would be possible and appropriate to treat an individual as the unit.
- *Sexual dichotomism*: Another instance of 'double standards': treating the sexes as two entirely different distinct social groups rather than as groups with overlapping characteristics.

Sources: Eichler 1988, Robson 1993: 64.

Since the early 1980s there has been a slow growth in feminist approaches within human geography, accompanied by feminist critique of geographical enquiry. These critiques have attacked all forms of geographical enquiry in two main ways. First, feminists have argued that geographical research largely ignores the lives of women and the role of **patriarchy** in society. They seek to redress this balance through specific studies of the everyday lives of women. Second, they have criticised the ways in which research is conducted, arguing that knowledge is predominantly produced by men and as a result represents men's views of the world. They thus argue that how we come to know the world is structured through the lenses of a *'male gaze'*. Some would suggest that this 'male gaze' is an implicit expression of the dominant ideology, a form of concealment, aimed at reproducing current **power relations**. As such, academic research has implicitly (and sometimes explicitly) worked to exclude and silence those in subordinate positions. Feminists suggest that the predominance of patriarchy within society can be observed by considering the range of **sexist practices** within current research (see Box 1.15). Feminist geographers have thus adopted an epistemology which challenges conventional ways of knowing by questioning the concept of 'truth', validating 'alternative' sources of knowledge such as autobiographical accounts and subjective experience, and acknowledging the **non-neutrality** and power relations within research (Women in Geography Study Group, 1997).

Unhappy with the primacy given to scientific methodology, feminist researchers have been instrumental in the reassessment of how to investigate research questions. In particular, they have advocated trying to understand the world through personal experience, forwarded the renegotiation of power relations between researcher and researched seeking a more **emancipatory** and **empowering** approach, and challenged such conventions as individual writing and writing style. Further, they have critiqued the practices and structures of academia as a whole, seeking to destabilise current patriarchal institutions and institutional practices. Like Marxism, feminism is a **political project** but one which seeks to address patriarchy rather than class. As such, feminists seeks emancipatory goals and social change for all those in the research process (Women in Geography Study Group, 1997).

The Women in Geography Study Group (1997) detail that feminist geography is involved in challenging traditional research approaches in three main ways. First, it challenges the formulation of theories by suggesting that traditional categories, definitions and concepts need to be rethought. Second, it examines the validity of methods (and associated theories) used for examining geographical issues. Third, it questions the basis by which issues are selected as worthy of geographical enquiry. This reassessment of conducting research has led to the formulation of a feminist methodology which is characterised by a search for a mutual understanding between researcher and researched (Katz, 1994). This methodology focuses thought upon four issues: ways of *knowing*; ways of *asking*; ways of *interpreting*; ways of *writing*. Within each of these issues the researcher is encouraged to reflect upon their own position, as well as that of the researched, and to acknowledge and use these reflections to guide the various aspects of the research process. In general, feminist researchers, rather than developing new techniques, use traditional methods but within a new frame of reference which aims to be more **reflexive** and **representative** of research subjects and consistent with feminist goals. At present, feminist research is largely restricted to female scholars, although men are increasingly recognising and adopting feminist approaches, and the ideas and practices are being used to underpin research concerning other oppressed groups such as disabled people and ethnic minorities.

1.5   Which approach is best?

Only you can provide the philosophical answers that will have meaning and lasting value in your future work as a geographer.
(Hill, 1981: 38)

It would be a grave mistake for us to try to prescribe any one of the approaches we have discussed as the best for you or your research. As discussed earlier, we all have our own beliefs about the world and the right and wrong ways to do things, including undertaking research. The aim of the discussion was to demonstrate the diversity of views in contemporary human geography. It is for you to decide which school of thought best reflects your beliefs in how a question should be approached and answered. This is not an easy task and may require a great deal of reflection. Whilst we have provided a list of different approaches, you should not treat the list as a shopping exercise. Do not, after reading each label, be tempted just to select an approach. Rather, you should think through a set of attendant questions to

help you determine which approach is most suitable to your views. Graham (1997) suggests that three good questions to ask relate to naturalism, realism and structure/agency. Other questions relate to the research strategy, the purpose of research, the production of knowledge and the nature of theories. In general, these questions are highly interrelated. For the purposes of this book we will discuss them separately and in brief. You should refer to the books recommended for further reading to gain a wider understanding of these issues. For reference, we have summarised Boxes 1.3–1.15 into Table 1.2.

1.5.1   Attendant questions

Naturalist or anti-naturalist?

Graham (1997) explains that naturalism concerns the nature of research. A naturalist approach would suggest that we can research the social world in much the same way that we can research the natural world. For example, we could adapt methods used in biology, chemistry or physics and apply them to human geography. Anti-naturalism, as the name implies, suggests that such an adaptation is invalid. Essentially, the difference concerns the use of the scientific method designed to measure empirical evidence and discover laws. Anti-naturalists argue that the scientific method is flawed because it seeks to explain the social world through the testing of laws. They suggest that this fails to acknowledge non-empirical evidence relating to thoughts, desires and values. As such, they contend that the social world is best approach through the seeking of understanding and interpretation, as humans do not follow set rules and patterns. In our discussion, in general, naturalists would be empiricists, positivists and analytical behaviouralists, and anti-naturalists everything else.

Inductive or deductive?

Associated with the naturalist/anti-naturalist debate are questions concerning the research strategy and methodology used to generate data. Inductive and deductive reasoning concern the logical processing of information and data into knowledge and how theory and practice are connected. At a basic level, using inductive reasoning means that the research comes before the theory. Here, theoretical propositions are generated from the data by identifying regularities. Alternatively, using deductive reasoning means that the theory comes before the research. Here, research

Table 1.2 Approaches in human geography.

Type of science	School of thought	Description	Example study concerning poverty	Main methodology
Empirical–analytical (technical, work, material production)	Empiricism	Empiricism refers to the school of thought where facts are believed to speak for themselves and require little theoretical explanation. Empiricists hold that science should only be concerned with objects in the world and seek factual content about them. As such, normative questions concerning the values and intentions of people are excluded as we cannot scientifically measure them.	Facts about poverty would be collected and presented for interpretation by the reader, e.g., indices of poverty – social welfare recipient, housing tenure, etc.	Presentation of experienced facts.
	Positivism	Positivists argue that by carefully and objectively collecting data regarding social phenomena, we can determine laws to predict and explain human behaviour in terms of cause and effect. Like empiricists, they reject normative and metaphysical (relating to being) questions that cannot be measured scientifically. Positivism differs from empiricism because it requires propositions to be verified (logical positivism) or hypotheses falsified (critical rationalism) rather than just simply presenting findings.	Poverty is explained through testing a hypothesis by collecting and scientifically testing data related to poverty, e.g., statistically testing whether poverty is a function of educational attainment.	Verifying factual statements: surveys, questionnaires, secondary analysis of other quantified data sets.
Historical–hermeneutic (practical, language, communication)	Behaviouralism	Behaviouralism acknowledges, explicitly or otherwise, that human action is mediated through the cognitive processing of information. It seeks to model spatial behaviour by explaining spatial choice and decision making through the measurement of people's ability to remember, process and evaluate geographic information.	Poverty is explained through the scientific testing of a hypothesis which examines the behavioural decision making of poor people and/or people in positions of power, e.g., statistically testing whether poor people have low levels of self-esteem, and if so, how this relates to job-seeking behaviour.	Verifying statements: surveys, questionnaires, specialised testing.
	Phenomenology	Phenomenology rejects the scientific quantification of positivism and behaviouralism. Instead it suggests that we concentrate upon understanding rather than explaining the world. The goal of phenomenology is to reconstruct the worlds of individuals, their actions, and the meaning of the phenomena in those worlds to understand individual behaviour, without drawing upon supposed theories.	To understand poverty it is suggested that we need to reconstruct the world of people who are poor, e.g., we need to try and see the world through the eyes of poor people. This might be attempted by talking to them about their life experiences.	In-depth interviews; ethnography.

Approach		Poverty	Methods
Existentialism	Existentialism is based on the notion that reality is created by the free acts of human agents, for and by themselves. Whereas phenomenology is primarily concerned with meaning, existentialism also concerns values. It focuses upon how individuals come to create and place meaning to their world and how they subscribe values to objects and to others.	Poverty is understood by trying to gain insight into how people who are poor come to know, ascribe meaning, and interact with the world, e.g., interviewing poor people about how they decide how much money they spend on different things.	In-depth interviews; ethnography; participant observation.
Idealism	Idealism posits that, ontologically, the real world does not exist outside its observation and representation by the individual. Whereas existentialism focuses upon reality *as being*, idealism views reality as a construction of the mind. As such, idealism seeks to explain patterns of behaviour through an understanding of the thoughts behind them.	Poverty is understood by trying to gain insight into how poor people think about poverty and the world they live in, e.g., interview poor people on what it feels like to be poor, why they think they are poor, and how they see themselves in relation to the rest of society.	In-depth interviews; ethnography.
Pragmatism	Pragmatism suggests that, rather than focusing on individuals, attention should be paid to society and the interaction of individuals within society. Pragmatists argue that understanding must be inferred from behaviour and rooted in experiences, not knowledge. By exploring the lives of people within a community, it is hoped that the nature of the beliefs and attitudes which shape society will be uncovered.	Poverty is understood by observing how individuals in society interact to produce conditions which sustain destitution, e.g., examining whether poor people remain poor because they live in a cycle of crime, under-education, low self-esteem, etc.	Ethnography; participant observation.
Historical materialism	Marxists suggest that society is structured so as to perpetuate the production of capital. They are concerned with investigation of the political and economic structures that underlie and reproduce capitalist modes of production and consumption. To do this they suggest that we need to consider how society might operate under different social conditions.	Poverty is explained through the examination of how poor people are exploited for capital gain, e.g., examining whether poor people are poor because it is in the interests of capital to retain unskilled, low-wage jobs rather than distribute fully corporate profit.	Dialectics; observation; interpretation of secondary sources.
Critical (emancipatory, power, relations of domination and constraint)			
Realism	Realists are concerned with the investigation of the underlying mechanisms and structures of social relations, and identifying the 'building blocks' of reality. Rather than studying the communication and interaction between people, realism seeks the underlying mechanisms of policy and practice that made these possible in the first place. As such, realism concerns the identification of how something happens (causal mechanisms) and how extensive a phenomenon is (empirical regularity). Realists want to find out what produces changes, what makes things happen, and what allows or forces changes.	Poverty is understood by trying to determine its root causes through an examination of the mechanisms underlying how society operates, e.g., examine whether poverty exists because of the uneven development of modernisation.	Mixed qualitative and quantitative.

Table 1.2 (cont'd)

Type of science	School of thought	Description	Example study concerning poverty	Main methodology
	Postmodernism	Postmodernists argue that, so far, modernist metanarratives (all the approaches above) that seek universal truths by examining the associations and relationships between people and places, have failed to account adequately for differences within society. Postmodernism is based upon the notion that there is no one answer, that no one discourse is superior or dominant to another and that no-one's voice should be excluded from dialogue. They therefore argue that there is no one absolute truth and that there is no truth outside interpretation. Postmodernism, rather than seeking 'truth', offers 'readings' not 'observations', 'interpretations' not 'findings', seeking intertextual relations rather than causality.	Poverty is understood by trying to deconstruct and read the various ways in which poverty is constructed and reproduced in society, and how poor people are excluded from society, e.g., examine the ways in which poor people are excluded from society through unequal power relations.	Deconstruction of culture and societal practices.
	Poststructuralism	Poststructuralists argue that the relationship between society and culture is mediated through language: humans are configured and given cultural significance through language. As such, the way we live our lives within society, the constraints and empowerment that operate, take effect in language. Poststructuralists propose that the way to gain an understanding of the factors that shape our lives is to deconstruct the multiple messages being conveyed to us by the objects we encounter.	Poverty is understood through examination and deconstruction of exclusionary practices of the society as expressed through cultural practices and articulated in language, e.g., examine the extent to which cultural norms and myths feed into exclusionary processes which seek to marginalise poor people from material wealth.	Deconstruction of language.
	Feminist critiques	Feminism suggests that science is dominated by, and reflects the position of, men, specifically white, wealthy, Western men. They suggest that there needs to be renegotiation of the role and structure of institutions and the production of knowledge. Here, there is a renegotiation of power relations within society so that how we come to know the world is more reflective of the people living in it.	Poverty is understood by trying to adopt more emancipatory and empowering approaches that allow poor people to express experiences and knowledge, e.g., ask poor people how they think that society should be reconfigured into a more just system.	Mixed qualitative and quantitative.

N.B. This table provides over-simplified, caricature accounts of each philosophical approach. Each approach is far more complex, often containing several competing versions (e.g., there is more than one type of positivism). The table is designed to illustrate the diversity in ideas underlying how to research a phenomenon.

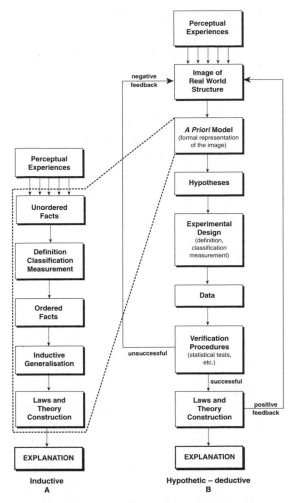

Figure 1.1 Inductive and deductive research strategies (source: adapted from Holt-Jensen 1988).

is undertaken to test the validity of a theory – whether a theory can be verified or falsified. For example:

- **Induction** Produce data on the spread of a disease and then construct a theory which explains disease diffusion.
- **Deduction** Construct a theory on disease diffusion and then generate data to test whether the theory is valid.

Inductive and deductive research strategies are elaborated in Figure 1.1. In general, although not always (e.g., empiricists), naturalists favour a deductive research strategy. Anti-naturalists use a mix of inductive and deductive research strategies dependent on approach.

Value-free or action-orientated?

Ideology concerns the purpose of research, of seeking knowledge. Each approach has an underlying ideology. As we noted earlier, Habermas suggests that approaches generally fall into three categories. He suggests that each category has a different purpose. Empirical–analytical research aims to *explain* the geographical world. Historical–hermeneutic research aims to *understand* the geographical world of its inhabitants. Critical research aims to be emancipatory, seeking to *change* the socio-political landscape for the better. Within the first two sets of approaches, research generally claims to be value-free. That is, research is neutral, with no specific political agenda. It just seeks to know the answer to a question or provide solutions to technical problems. Instead, it is left to the readers of the research to draw their own socio-political conclusions concerning the findings. Action-orientated research refutes the notion of value-free research. Instead, action-orientated research seeks explicitly to change the world either by addressing a specific social or practical problem (e.g., pragmatism) or through a political project aimed at changing social relations (e.g., Marxism and feminism). Within this framework, it is argued that research should have some wider purpose than just to add to our knowledge of the geographic world: it should aim to change it for the better.

Objective or situated?

The value-free/action-orientated debate concerns whether knowledge production should have a political/social purpose. A related issue concerns whether knowledge can be gained in an objective and neutral fashion. Both analytical (naturalist) and humanistic (anti-naturalist) approaches adopt methodologies where the researcher is the expert, an objective recorder and observer of the world who neutrally carries out the study. Some commentators are now starting to challenge this objectivity, suggesting that knowledge is in fact situated. That is, knowledge is not given, just there waiting to be discovered. Rather, knowledge is constructed through how we investigate and examine the world. Here, research is seen as a social activity that is affected both by the enthusiasm and motivation of the researcher and by the context in which the research takes place: no matter how impartial the researcher feels they are, they come to the research with a certain amount of 'baggage' – preset ideas, theoretical persuasions, personal interests, etc. Sample (1996) argues further that although, in theory, the research design is chosen to address the situation and questions under investigation, in reality the research design often suits the interests or speciality of the researcher. As such, research is researcher-orientated, based around the desires and agendas of the researcher rather than the subject(s) of the research. As Susan Hanson (1992: 573) suggests:

> your context – your location in the world – shapes your view of the world and therefore what you see as important, as worth knowing; context shapes the theories/stories you concoct of the world to describe and explain it.... Knowledge is contingent on beliefs and values.

Highly related to the positioning of the researcher is the positioning of the research. Just as it is argued that knowledge production is situated within the beliefs and values of the researcher, it is also suggested that knowledge is situated within cultural ideologies. **Cultural ideologies** concern the general, unwritten laws concerning what is permissible within society. As such, it is now argued that research does not take place in a social void. It takes place within a **social context** and is framed by societal expectations of what can be researched and how it should be researched. Society expects us to hold certain ethical and moral standards when conducting research. This means that research is framed within those standards to gain acceptability. Further, there are still '**taboo**' subjects such as sex and death (see Section 2.5.4). Research practice and focus are thus socially positioned. This does not mean that we have to accept the dominant mode of positioning – we can try to find positions that we think are more just (although we might have difficulty in getting 'mainstream' society to accept our research findings).

The adoption of a feminist approach by a number of women academics represents one particular example of a group of scholars seeking to challenge both what should be studied and also how research should be conducted. Whilst initially feminist work was received sceptically, the social positioning of feminist research is now slowly gaining social acceptance. These challenges to convention are, however, still finding resistance within geography, as the debates between Mona Domosh (1991a, 1991b) and David Stoddart (1991), and Linda Peake (1994) and Peter Gould (1994a, 1994b), demonstrate. We can illustrate the effects of the researcher's position and the positioning of the research with some hypothetical examples (see Box 1.16).

Geography or geographies?

Related to the production and situatedness of knowledge are arguments concerning the nature of theories. In the main, approaches, whether they be naturalist or anti-naturalist, seek to find an order within society. As such, they seek to provide a unified, grand theory of society and social knowledge and to reveal universal truths and meaning. In other words, they are trying to find a theory that has a universal commonality. Such theories (e.g., Marxism) are called **grand narratives**. These narratives suggest that there is one geography, one universal 'truth'. In contrast to this position, others would argue that the search for a totalising theory, applicable in all cases, is a misguided venture. Rather it should be recognised that, due to the complexity of society and the individuality of the people who constitute it, no one theory can fully account for all social events. Further, as knowledge is situated there are many possible geographies. In general, it is the postmodern and feminist approaches that reject the notion of grand narratives, although it should be noted that many feminists would still advocate a modernist agenda, but one that recognises patriarchy.

Realism or anti-realism?

Graham (1997) explains that the realist/anti-realist debate is quite complex and concerns the question of existence: the issue of validity in claiming that something exists. Realists argue that a 'real' world exists regardless of conceptions of it, that the world and the people living in it have a material existence beyond what we think about the world. This might seem like common sense. Anti-realists (or metaphysical idealists) contend that the world exists only in the mind – reality is constituted in thought – and there is no logical reason to suggest that it has material existence beyond our thoughts. This idea might at first be more difficult to accept. However, the more you think about this notion (and philosophers have spent a great deal of time debating this issue) the more the idea does have some appeal. This debate is important because it concerns how we can gain knowledge of the world. Can we only study how people come to construct their world or is there a real world that we come to know? Clearly, if you believe that people construct their own world then your research approach will be different from the approach you would adopt if you believed that a real world exists independently of people. In human geography, humanistic methods

Box 1.16   Case examples of the effects of the researcher's position and the positioning of the research

Example 1: Positivism and poverty

A positivist researcher would claim to be neutral or completely objective, having no vested interest in the research. Further, they would claim that their research takes up no social or political positioning. They are just trying to explain certain conditions, allowing the data to speak for themselves through statistical testing. In other words, their level of interpretation is limited in that it is constrained to just detailing what the statistical tests show. They place no value judgements on the findings of the tests and instead make new hypotheses as to why they achieved a particular result. These hypotheses then form the basis for further studies to determine their validity. In relation to trying to explain poverty, a positivist might statistically test the association between some measure of poverty (such as income) against some measure of educational achievement (such as GCSE results). If the test found a significant relationship between the variables (measures) used, they would conclude that poverty is in some way related to educational achievement. They then might suggest why this might be the case. For example, those with low educational attainment might be constrained to low-skilled, low-paid, employment. This could form the basis for a new study that would test this. The researcher would make no value judgements on this finding or suggest social policy changes.

Example 2: Humanistic approaches and race

In contrast to the positivist, a humanistic researcher tries to get much closer to the researched, seeking to build up relationship with those being studied, to gain their trust and confidence. Like the positivists, many humanists would claim to be neutral and objective, with their research having no social positioning. Whereas the positivist is trying to objectively study conditions using empirical measures, humanists place more emphasis on subjective experiences, values and opinions. For example, in relation to racism, rather than testing empirical measures which we might associate with racism, like the number of violent acts against black people and unemployment in the dominant group, humanists would be much more interested in interviewing the attackers about the personal motivations behind such attacks, trying to tease out the basis of racist violence. Again, the conclusions would largely consist of interpreting the findings and setting up future research questions rather than making social policy recommendations.

Example 3: Feminist research and gender

Feminist researchers generally adopt a socio-political stance that explicitly recognises that the knowledge from their research is constructed and produced. They therefore recognise that their research is not neutral or objective but is framed within the personal context of the researcher and the social context of the research environment. In contrast to positivists and humanists, feminists are much more likely to draw socio-political interpretations based upon the evidence they have generated. As such, their research has an implicit (and often explicit) agenda aimed at feeding in to and changing social relations within society. Here, they use a range of methods, particularly humanistic methods, to explore the ways in which women are, and have been, oppressed through patriarchal relations.

Example 4: Participatory action research and disability

One particular example that disabled researchers have been exploring, that recognises the social production and positioning of social research, is participatory action research (PAR) (see Barnes and Mercer, 1997; Kitchin, in press). This method tries to renegotiate the power relations in the research and to empower and give the researched much more say in research about them. Oliver (1992), in relation to disability research, is particularly critical of current 'social relations of research production' which he argues disenfranchises, disenchants and alienates the subjects of research, distancing them from the work of the researcher. He argues that current professional expert models of research perpetuate the status quo with the researcher taken as the expert, the harbinger of specialised knowledge and the controller of the research process, and the research subjects taken as subordinate. Such expert models disempower the research subjects, placing their knowledge into the hands of the researcher to interpret and make recommendations on their behalf. PAR is an attempt to address the problems of representiveness and unequal power arrangements between researcher and researched within social research by fully integrating research subjects into the research process (Whyte, 1991). Here, the role of the academic becomes enabler or facilitator: the academic takes a supportive position and seeks to inform and impart knowledge and skills to the research subjects.

are generally more anti-realist, although there can be considerable interplay with other approaches. Again, you are referred to other texts

Structure or agency?

Changing tack slightly, another way to evaluate which approach most suits your views is to consider your understanding of how society works (Graham, 1997). The structure–agency debate within geography has been in progress since the late 1970s. It essentially concerns the extent to which **social actions** are constrained by **social structures**. At one end of the spectrum, some approaches consider that social actions are highly structured and largely outside the control of individuals (e.g., Marxism). At the other end of the spectrum, other approaches recognise individuals as completely self-autonomous, free to act as they like (e.g., behaviouralism). Other approaches accept that there is a play-off between structure and agency, recognising that individuals make their own decisions but that these decisions are framed within broader structures (e.g., political economy – see Peet and Thrift, 1989).

1.5.2   Choosing an approach

Whilst choosing an approach is a difficult task, it is important that you come to some sort of a basic decision before you start any research, as your choice will have great bearing upon how you approach your research topic. Work through the questions posed in this section, and go back through those approaches detailed in the previous section. You might find that some of your answers point to one approach and others to different approaches. If this is the case do not worry. There are many different variations of each philosophy and to a certain degree you can mix-and-match. At this point in your careers, we suggest that you do not fret over the fine details but adopt an approach which *best* matches your views. All professional human geographers have wrestled with philosophical questions at some point in their careers and many change their mind, or make individual alterations to suit their own unique, personal viewpoints. As a novice researcher, it is anticipated that you will have undertaken little practical research, and it is only with experience that you will determine which approach suits your views best. You should not be afraid that once you have adopted a particular position, you have to stick with your decision for the rest of your researching days. Many world-renowned

human geographers have changed their minds about how research should be conducted and also what the focus of research should be. For example, in 1969 David Harvey published a book which adopted a positivistic stance, entitled *Explanation in Geography*, but by 1973 when he published *Social Justice in the City* he had adopted the completely opposing position of historical materialism! On a personal note, we are constantly changing and fine-tuning our ideas as we read more and hear the views of others at conferences. Remember that research, as well as being about discovery and understanding, is also a self-learning process where your ideas and thoughts develop and continue to take shape.

1.6   Summary

After reading this chapter you should:

- understand what conducting research means;
- have a basic understanding of different types of research;
- have a basic understanding of different philosophical approaches in geographical research;
- be able to decide which approach best suits your own beliefs.

In this chapter, we have explored the research process and the different ways to approach a problem or question. You should now have a basic understanding of different philosophical approaches available to you. You will need to think carefully about each philosophical approach before considering how you might research your chosen topic. This is because theory and practice are intimately entwined. Given the importance of starting from a position of strength, we suggest that you work through the final section again, evaluating each position in relation to your own ideas and beliefs before moving on. In the next chapter we explore the practicalities of planning your project and how to choose a topic, design a research strategy, link theory and methodology, and choose methods of data generation and analysis.

1.7   Questions for reflection

- *Why should you want to do research?*
- *How would you define geography and what geographers do?*

- *Why does philosophy matter when conducting a research project?*
- *On first impressions, which particular philosophical position towards research do you subscribe to and why?*

Further reading

Bird, J. (1993) *The Changing Worlds of Geography: A Critical Guide to Concepts and Methods*, 2nd edition. Clarendon Press, Oxford.

Cloke, P., Philo, C. and Sadler, D. (1992) *Approaching Human Geography: An Introduction to Contemporary Theoretical Debates*. Paul Chapman, London.

Gregory, D. (1981) *Ideology, Science and Human Geography*. Hutchinson, London.

Harvey, M.E. and Holly, B.P. (eds) (1981) *Themes in Geographic Thought*. Croom Helm, London.

Holt-Jensen, A. (1988) *Geography: History and Concepts*, 2nd edition. Paul Chapman, London.

Johnston, R.J. (1986) *Philosophy and Human Geography: An Introduction to Contemporary Approaches*, 2nd edition. Edward Arnold, London.

Johnston, R.J. (1991) *Geography and Geographers: Anglo-American Human Geography since 1945*, 4th edition. Edward Arnold, London.

Johnston, R.J., Gregory, D. and Smith, D.M. (1992) *The Dictionary of Human Geography*, 3rd edition. Blackwell, Oxford.

Ley, D. and Samuel, M. (1978) *Humanistic Geography*. Croom Helm, London.

Livingstone, N. (1993) *The Geographical Tradition*. Blackwell, Oxford.

Unwin, T. (1992) *The Place of Geography*. Longman, Harlow.

Women in Geography Study Group (1997) *Feminist Geographies: Explorations in Diversity and Difference*. Longman, Harlow.

CHAPTER 2 Planning a research project

This chapter covers

2.1 Introduction
2.2 Choosing a research topic
2.3 Narrowing the focus
2.4 Linking theory and practice
2.5 Research design
2.6 Choosing a method to generate data
2.7 Choosing a method to analyse your data
2.8 Conducting group research
2.9 Managing and piloting a research project
2.10 Summary
2.11 Questions for reflection

2.1 Introduction

In our experience, the most difficult part of any project is to get the research process started; to decide upon the exact focus of study and then to plan a strong research design. There always seem to be so many things to think about and what is required often seems daunting. This chapter has two aims. The first aim is to help you start thinking about conducting your own research project. As such, the chapter introduces you to all facets of the research process (see Figure 2.1) and outlines the main issues you need to consider when both planning and conducting a research project. The second aim is to provide a textual map that will guide you through the rest of the book and help you make decisions about your project in a constructive manner. As such, this chapter should not be read in isolation from the rest of the book. Instead, this chapter should be used like a 'boomerang launcher', forming a platform from which you explore other chapters before returning to base.

We suggest that you work through the whole of this chapter in sequence, branching out to other relevant sections when required. In this way, you will address essential considerations which relate to all the basic facets of the research process and you will gain a holistic view of your project. This is important because projects planned and executed in an *ad hoc*

manner, for example choosing a topic, then generating data, then seeking a way to analyse the data generated, often run into difficulties due to oversights and unforeseen problems. Planning all stages before generating any data reduces the risk of later problems by highlighting potential weak links that may cause the research process to break down. In other words, to ensure a good and valid piece of research, it is important that in planning your project all stages in the research process receive careful thought and attention before any data are generated. Once you have planned your research, and it is time to generate and analyse your data, you will need to return to the following chapters for guidance on how to successfully implement specific methods and techniques.

2.2 Choosing a research topic

The first consideration in any piece of research is the choice of a focus for your research. Given the broad scope of human geography and the fact that nearly everything has a geographic context, you can draw your focus from practically any facet of human life. Choosing which facet, however, can be tricky. One of the keystones to conducting successful research is to examine a topic in which you are **interested**. Choosing a topic for the sake of expediency or because it

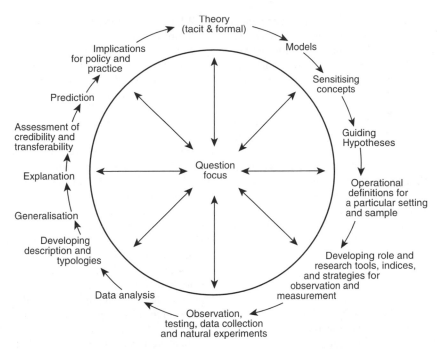

Figure 2.1 A model of the research cycle (source: Marshall and Rossman 1995: 23, reprinted by permission of Sage Publications Inc.).

seems easy or because you think the topic is 'trendy', or a safe bet, will almost certainly lead to a poorly executed and uninteresting study.

Therefore, before 'diving in', think about the sort of phenomena or situations you find of interest and which might make for a coherent and worthwhile study. Maybe one of your lecturers has inspired you with their analysis of a research problem. Maybe you see something every day on the way to university that annoys or frustrates you. Perhaps a certain situation is receiving much media coverage which you think will be interesting to probe further. Maybe you have just read a particular paper or book which has raised a number of questions that you feel need answers. However, your research does not have to be socially relevant or focus on a contemporary problem. For example, it could be an historical study.

If, at first, your subject area does not seem geographical in nature do not worry. Think about how geography might matter in the chosen context. Many lines of geographical enquiry can be suggested by applying key geographical concepts, such as space, time or pattern, to the issue, phenomenon or situation at hand. For example:

- To what extent is the issue, phenomenon or situation *differentiated* across space, place and time?

- Are *spatial processes* important to any distribution, pattern and activities arising?
- How does the issue, phenomenon or situation *relate to the surrounding area*?

It is worth mentioning that although there are considerable attractions in choosing a subject area or issue in which you are **personally involved** – either because you take part in the activity (e.g., football) or because you are a member of that group (e.g., a fan) – there are problems associated with personal involvement. The first is the difficulty in detaching yourself from the subject matter. Obviously in some approaches (see Section 1.4) this is not a problem. However, there is a danger that you may be biased in your viewpoint and in the selection of criteria or issues to study. You may fail to notice pertinent questions or issues because of the inability to step back from a situation and fully assess the circumstances. For example, you might have strong emotional feelings about specific topics such as the government's transport policy or issues relating to gender or nationality. In other words, there is a danger of your looking at a situation through rose-tinted or blinkered glasses or lending your value judgements to the research agenda. This is not to say that you should avoid doing research with strong personal relevance, but that you should be aware of the problems you might encounter whilst undertaking the study. Obviously, there are advantages to studying a subject matter with which you have personal involvement, such as genuine enthusiasm, motivation, insider contacts and personal knowledge.

Whatever the subject, the study must have a clear set of aims and be **focused** onto one or two particular aspects rather than being broad and general in nature. The presence of a clear objective cannot be overstressed and is the key to completing a successful research project. If your research topic is too broad, it will be difficult to conduct any in-depth analysis, and the project will also quickly become unmanageable. The potential to become side-tracked increases as does the likelihood of confusion. As such, broad issues should only provide the context for your study. Most studies will and should start with an idea that is too big. Fine-tuning your research within a broad context helps to ensure a good piece of research. You should aim to say a lot about a specific subject, rather than a little about a broad subject. This leads to the next stage of narrowing your research focus.

2.3   Narrowing the focus

As we have discussed, finding a focus to your project is of critical importance. It gives your study purpose and allows the formulation of **specific questions**. However, narrowing the focus further is not as easy as it might at first seem. You have to be careful to identify a topic within your broad area of interest not only that will keep you interested, but that you will be able to generate data on, and which is not too narrow in its focus. For example, if you have a general interest in retail geography, you need to decide which particular aspect of that topic you find most interesting. You may decide to concentrate your study upon out-of-town shopping centres. However, there are many questions you could ask in relation to out-of-town shopping, ranging from:

- How does the building of an out-of-town shopping centre affect local shopping patterns?
- How does the building of an out-of-town shopping centre affect local traffic rates?
- Why did the local council approve the construction/location of an out-of-town shopping centre when they had strong evidence from an impact assessment study that it would lead to shop closures in the town centre?

- What are the impacts of an out-of-town shopping centre on 'community spirit'?
- Are out-of-town shops discriminatory against those without a car or who live off bus routes?
- How are out-of-town shopping centres changing the urban infrastructure?
- How can we use GIS for the optimal location of an out-of-town shopping centre?

Clearly it is beyond the scope of one study to investigate all these questions. Instead, several projects would be needed, each with just one or two of these questions forming the central core of investigation.

To aid the identification of specific questions relating to your general interest, we suggest that at this stage you visit the **library** to see what work other researchers have undertaken. Whatever the topic, somebody, somewhere, will almost certainly have undertaken some research on it or a similarly related topic (see Box 2.1 for strategies to identify and track down sources of information). **Reading** through some of the literature will provide you with firmer ideas and help you narrow your focus. Box 2.2 provides details of the questions you should ask when reading about your topic area. Answering these questions will help you to draw out the main themes within each study. The process of going through books and papers might seem like a tedious and unnecessary task, especially if you already have a firm idea about what you want to research. However, you should become familiar with the writings of other researchers

---

Box 2.1   Identifying and tracking down sources of information

- **Reading lists** from supervisors.
- **Bibliographies** of papers and books already found.
- Searching the **library catalogue**.
- Using other library catalogues using the **Internet** (e.g., NISS Gateway).
- Searching *GeoAbstracts* (CD-ROM).
- Searching a **citation index** (e.g., *Social Science Citation Index* using BIDS (in UK)). In general, a username and a password will be needed.
- Searching the **Internet** (use a search engine such as InfoSeek to search for 'keyword' information).
- **Inter-library loans** (if the book or article is not available in your own library you can ask your local library to borrow it from a central reserve – a small fee is usually charged).

Source: Abridged from Flowerdew 1997: 46–56.

Box 2.2   Questions to ask of your information sources

- What motivated the study?
- What were the main research questions and why?
- Which main sources of information did they refer to? Are they of relevance to my study?
- What methods of data generation and analysis did they use and why?
- Where was the study carried out and who were the sample population?
- Were particular problems encountered or ethical issues raised? Will these potentially affect my study?
- What were the main findings? How do these agree with other material?
- What were the main conclusions?
- Did they suggest possible future studies? If so, are any of them suitable as the basis of my study?

Box 2.3   Why it is important to read the work of other researchers

- You gain a deeper understanding of the relevant issues.
- You place your research into the broader context of other studies.
- You gain an insight into how to go about undertaking the study in terms of both data generation and data analysis.
- You can fully justify the need for your study.
- You do not end up needlessly replicating another study (of course some studies might need replicating if you have reason to suspect their credibility!).
- Conflicting findings in the literature can be a fertile source for research projects.

for a number of reasons (see Box 2.3). Remember, it is no coincidence that it is called 'reading' for a degree!

Making this leap from general to specific can be tricky, but as you become more familiar with the literature you should be able to start to identify gaps in what is known and be able to formulate questions to which you would like to know the answers. For example, Box 2.4 provides a framework which you might apply to your research ideas to narrow the focus of your study. Essentially, you are trying to determine the **main elements** of your research project: the one or two things that will distinguish your study from others. By framing your study within these questions you ensure that your research project will have a *degree* of originality often required by university courses. Completely original research is difficult and rare. You are probably setting your sights far too high if you aim to undertake a ground-breaking study. If you are still having difficulty finding a specific focus, ask yourself the following questions:

- What is the specific objective of my research project?
- Why am I doing a project on this subject?
- What are the important issues?
- Who will potentially be affected by the research project?
- Who will potentially benefit?
- What things might change as a result?

If, at this point, you cannot clearly and unambiguously state the research objective using just a single sentence, you should probably be asking yourself whether the project is attempting to achieve too much or whether it is too complicated. If this is the case then you need to go through the process of narrowing your focus again. In general, you are seeking to concentrate on one or two specific questions which your research will address. Asking **relevant questions** is what 'good' research is all about. Although there are literally hundreds of thousands of interesting and important questions to be researched and answered, specific questions can be divided into just a few types. Table 2.1 provides some examples of these questions by focusing upon disabled people's use of the urban environment.

An important point to remember is to keep your questions **clear** and **simple**. Students often feel that they have to dress up their questions to make them sound more scientific or academic. For example, we might dress up a question to read:

'In studying the socio-spatial processes that underlie access provision for disabled people are we always entrenched in multiple interpretations, the politics of representation and what Giddens calls the "double hermeneutic"?'

Without prior knowledge concerning the politics of representation or what the 'double hermeneutic' is, a lay person will have no idea what the question means. There is a good chance that the writer also doesn't fully understand the question, especially when they have only used these terms because they have seen them in a journal or because their lecturer uses them a lot. In this example, the question is probably best rewritten as

---

Box 2.4   Research focus

Madeup (1993) investigated this topic and raised the question regarding the role of *x*. I'll investigate the role of *x*.

e.g., Madeup (1993) investigated traffic flows in an urban area and questioned the role of pedestrianisation. I'll investigate the effects of pedestrianisation on traffic flows.

Madeup (1993) investigated this topic and found out that ... but they ignored the possible effects of *x*. I'll investigate the effects of *x*.

e.g., Madeup (1993) investigated traffic flows in an urban area and found that it varied as a function of time of day ... but (s)he ignored the possible effects of traffic management. I'll investigate the effects of traffic management on traffic flows.

Madeup (1993) investigated this topic at location *x* and found that ... I'll see if the same is true for *y*.

e.g., Madeup (1993) investigated traffic flows in London and found that it varied as a function of urban density. I'll see if the same is true for Bristol.

Madeup (1993) investigated this topic and stated that the dominant controls on *x* were *y* and *z*. This may be wrong. I'll test this.

e.g., Madeup (1993) investigated traffic flows in an urban area and stated that the dominant controls were urban density, traffic management and time of day. I'll test this.

Madeup (1993) investigated this topic and found that ... I wonder if things have changed since then?

e.g., Madeup (1993) investigated traffic flows in an urban area and found that it varied as a function of traffic management. The city has been redeveloped. I wonder if things have changed since then?

Madeup (1993) investigated this topic by method A. I wonder whether you get different results using method B. I'll use method B and compare my results with theirs.

e.g., Madeup (1993) investigated traffic flows by counting the number of vehicles passing a location in a minute period once every hour. I wonder whether I will get different results by counting the numbers of vehicles at more regular intervals. I'll count the numbers of vehicles over a minute every ten minutes and compare my results with theirs.

Since Madeup (1993) did their study a new data set has become available. I wonder if the new data set supports Madeup's conclusions.

e.g., Since Madeup (1993) did their study the Department of Transport has conducted its own study. Both data sets are available for further analysis. I will investigate whether the new data set supported Madeup's conclusions.

Madeup's (1993) results conflicted with those of Invented (1997); why is this?

e.g., Madeup (1993) found that traffic flows varied as a function of traffic management; Invented (1997) concluded that traffic flows varied as a function of urban form. Which is correct?

Source: Adapted from Parsons and Knight 1995: 34.

'When studying access provision for disabled people is the researcher fairly and correctly representing the views of disabled people?'

Writing in a complex style or dressing up the questions does not make the research more valid. Writing in a clear, eloquent style and demonstrating a well thought-out and executed study is more important. To achieve this, it is best to keep the question clear and easily understandable so that a lay person will know intuitively exactly what you are trying to understand or explain. We are not denying that it is easy to be drawn into language that is complex, and which may have multiple meanings, but where possible you should try to avoid this.

Once you have decided upon a topic and your specific question(s), the difficulty is then to turn your research idea into a coherent, consistent and valid piece of research. The first task is to place your idea into a theoretical context.

2.4   Linking theory and practice

As we have already noted in Chapter 1, *research is not atheoretical*. Whereas we discussed the philosophical bases of research in Chapter 1, in this section we are concerned with specific theory relating to a particular issue. Here, theory is taken to be a set of

Table 2.1 Question types.

General form	Sample questions
What is A like?	What is interacting in the urban environment like for disabled people?
What does A mean?	What do we mean by 'access for all' and 'barrier-free environments'?
Is A like B?	Are the planning needs of disabled people the same as those of the able-bodied community?
Is A different from B?	Do planners and disabled people agree on how the urban environment should be designed?
Is A better than B?	Is the urban planning for disabled people better in Labour council areas than in Conservative run areas?
Are A and B related?	Is there a relationship between the size of a town and the quality of urban planning for disabled people?
Does A affect B?	Do poor design and low accessibility of an environment decrease usage by disabled people?
Does A cause B?	Do disabled people use this area more as a direct result of its redevelopment to make it more accessible?
Is A located where B is min/max?	Are special day care centres located in the areas of primary need?
How are A and B minimised simultaneously?	Can we maximise the accessibility of the environment whilst minimising the expense?
Why does A support B?	Is a disabled access project receiving funding from the state for political reasons rather than genuine commitment to the access needs of disabled people?

Source: Adapted from Smith 1987: 311 and Parsons and Knight 1995: 38–39.

explanatory concepts that are useful for explaining a particular phenomenon, situation or activity. These concepts offer certain ways of looking at the world and are *essential* in defining a research problem. As Silverman (1994: 1) notes, '*without theory, there is nothing to research*'. Whenever research takes place we are either assessing the validity of a theory (using a deductive approach) or trying to construct a theory (using an inductive approach). Your reading of the literature should have provided you with a firm set of ideas about your specific topic. These ideas now need to be formulated into a theory that your research will then test *or* provide the context for the research that will underpin the construction of your theory. A theory generally extends beyond the answer(s) to your specific question(s) and places them into a wider context. We suggest that a useful way to construct your theory, either prior to the research (deduction) or after the research (induction) (see Section 1.5), is to build a **conceptual model**.

A conceptual model is a diagrammatic version of a theory which demonstrates processes, concepts and relationships. Whilst your theory does not have to be expressed diagrammatically, we find that sketching out ideas helps to place ideas in context and give expression to abstract and complex thoughts. Figure 2.2 is the conceptual framework developed by Downs (1970) to underlie a deductive study of cognitive mapping research. In this schema the boxes represent concepts and the directional arrows the links between these concepts. In his explanation, Downs details that the starting point for the schema is taken as the real world, which is the source of information. This information is filtered through a system of perceptual receptors which are essentially the five main senses. Meaning is given to the information through an interaction between the individual's value system and their stored 'image' or cognitive map knowledge of the real world. The remaining filtered information is then used to update the cognitive map knowledge and to formulate a behaviour decision. This decision leads either to a reiteration of the whole process, creating another search for information from the real world until sufficient information has been acquired or some time/cost limitation acts to constrain the search, or to overt behaviour. As a result of the latter, the real world undergoes a change, fresh information becomes available and the whole process begins again. Using this schema to underlie deductive research, we could now develop some hypotheses to test its validity.

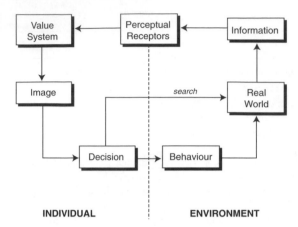

Figure 2.2 Downs' (1970) conceptual schema of cognitive mapping research.

Hypotheses link theory and practice in **deductive research**. They are used to assess the validity of a theory. A hypothesis is a statement of what should be the case if a certain prediction is true. If, after testing, the hypothesis is disproved, then the theory can be dismissed, altered for retesting or declared less useful. In essence, hypotheses are formal statements of your research questions. A full discussion concerning the development and use of hypotheses is developed in Section 5.3.

Turning practice into theory using an inductive approach utilises **inductive reasoning**. This consists of making theoretical inferences from the data generated based upon identifying regularities. If, for example, we were studying the diffusion of AIDS using an inductive approach we would generate data and then look for patterns within the data. Induction therefore works from the level of the particular (e.g., individual AIDS cases) to make inferences at the general level (e.g., AIDS diffusion).

2.5 Research design

Now that we have a focus for the research, and *if appropriate*, designed a theory and hypotheses to be tested, it is time to plan the practicalities of generating data. There is much to think about. If we use an analogy of the process of constructing a building, we are moving from the choice of what type of building we want to construct to decisions regarding the process of construction. To take the analogy further, if we miss out this stage and progress straight to constructing the building without adequate planning,

then there is a good chance we will run into problems at a later date or construct a building that is structurally weak or flawed. Similar to the role of architect and builder, we need to think carefully through our project. Rushing in to generate data without first thinking about a number of related issues, such as whether the methodology chosen will generate data that will allow you to assess, or construct, your theory, or how long the study will take or cost, will lead to a poor piece of work.

The research design phase holds all the parts and phases of a research project together. It is difficult to provide a foolproof guide as to how to design your research strategy, but we have endeavoured to give as much constructive advice as possible. Always remember that there is no set way to conduct research – no magic or all-pervasive formula. As such, your **imagination** and **creativity** are two of the most important assets to bring to any project. In the next few subsections we discuss various critical issues you need to think about whilst deciding upon methods of data generation and analysis.

2.5.1 Issues of validity and reliability

All good studies aim to be **valid** and **reliable**. Your project should be no different. Validity concerns the soundness, legitimacy and relevance of a research theory and its investigation. For an idea or theory to be tested *or* become an accepted proposition, its theoretical and practical aspects must fulfil basic validity requirements (Russ and Schenkman, 1980). As Silverman (1994) argues, issues of validity and reliability apply just as much to qualitative-based studies as they do to quantitative-based studies. In other words, qualitative research has to be more than 'telling convincing stories'. It has to be rigorous in nature so that its conclusions can be accepted more definitively. There are a number of different validity issues that you need to consider when conducting your study. In general, these can be classified into validity that relates to theoretical or practical issues. Types of validity relating to **theory** concern the integrity of the theoretical constructs and ideas that support and provide the foundations for empirical research. Types of validity relating to **practice** concern the soundness of the research strategies used in the empirical investigation and the integrity of the conclusions that can be drawn from a study (see Box 2.5).

An issue related to validity is **reliability**. Reliability refers to repeatability or consistency of a finding and is of particular importance in approaches that

---

Box 2.5   Types of validity

- **Validity relating to theory**

Content validity refers to the content and definitional strength of terms in a field. For example, a field has low content validity when there is little agreement on the focus of research or the terms used by its practitioners.

Face validity concerns the justification for study. Studies are often described as having weak face validity when they have little practical or theoretical relevance to real-world situations or scenarios.

Conceptual validity relates to the correct marriage of theory and methodology, so that the research becomes philosophically sound and adopts appropriate methodologies of data generation and analysis. For example, if an empirical study is not based explicitly within a theoretical framework it can be described as having weak conceptual validity.

- **Validity relating to practice**

Construct and analytical validity both relate to the methodological integrity of a study.

Construct validity concerns whether data generation techniques are sound, measuring the phenomenon they are supposed to without introducing error or bias; they are telling you about what you want to know.

Analytical validity concerns whether the correct method of data analysis has been chosen, leading to results that represent the data truly. Clearly if you have chosen the

wrong method of data analysis then you might end up drawing inappropriate conclusions.

Ecological and internal validity both relate to the integrity of the conclusions drawn from a study.

Ecological validity is concerned with the inferences that can be made from the results of a study. In psychology, this generally refers to the extent to which environments experienced by the respondents in a scientific investigation have the properties they are supposed to have, or are assumed to have by the investigator (e.g., laboratory vs natural environment). In geography, however, this generally concerns the problems of inferring characteristics of individuals from aggregate data referring to a population. There are a number of other fallacies associated with ecological validity concerning the problems of generalisation across circumstances, times and areas. For example, can we extend our conclusions from one study to other people living in different places?

Internal validity concerns whether the results from a study can be interpreted in different ways; can different conclusions be drawn from the same results?

Ecological and internal validity become weak when practically related validities, such as construct validity, are weak.

Sources: Bronfenbrenner 1979; Coolican 1990; Howard *et al*. 1973; Johnston 1986b; Turk 1990.

---

utilise a deductive strategy of inquiry. Golledge and Stimson (1997) describe three kinds of reliability: (1) **quixotic reliability**, where a single method of observation continually yields an unvarying measurement; (2) **diachronic reliability**, which refers to the stability of an observation through time; and (3) **synchronic reliability**, which refers to the similarity of observations within the same time period. Within a deductive approach, it is generally accepted that the more consistent a finding the more weight can be attributed to the finding. For example, if a result is consistently found it is often referred to as a law. While consistent findings are quite common in the physical sciences, they are much rarer in the social sphere.

2.5.2   Research ethics

Every project undertaken has associated **ethical issues**. This is particularly the case when people are involved directly but also when secondary sources such as the census are used. Research ethics are concerned

with the extent to which the researcher is **ethically and morally responsible** to her/his participants, the research sponsors, the general public, and her/his own beliefs. Clearly there may be a conflict of interests between these four groups, leading to an ethical dilemma. For example, consider the scenarios in Box 2.6 and try and decide on which course of action you would take. Do you think there are any right answers to the questions posed in Box 2.6? These sorts of questions are generally difficult to answer, and the only person who can decide their answers is yourself, depending upon your own **moral code**.

Researchers can and do answer these sorts of ethical dilemma questions differently. For example, some medical researchers are willing to carry out experiments upon animals whereas others are vehemently opposed. Similarly, some policy analysts point to the benefits of accurate prediction of the outcome of a policy initiative, whereas others warn against possible infringements of privacy and civil liberty. Despite the individual basis of ethical dilemma questions,

---

Box 2.6   Ethical questions in the research process

- **Is it ethical to research a situation or phenomenon covertly?**   You are undertaking research on a sensitive topic which may put an informant at risk of losing their job. You gain access to a possible informant through a friend. Do you tell the informant about the nature of your research before you interview them?

- **Is it ethical to get information for your own purposes from, or conduct research for, people or institutions you dislike or distrust?**   You are approached by a tobacco company to conduct research on the most cost-effective site for a new distribution centre. You are offered a large research contract; do you accept the offer?

- **Is it ethical to see a need for help and not respond to it directly?**   You are conducting a study on the geography of old age and are interviewing the residents of a retirement home. You witness what you believe to be systematic abuse of the residents. Do you intervene immediately, report the incidents to the relevant authorities at a later date, or turn a blind eye to the situation?

- **Is it ethical to 'pay' people trade-offs for access to their lives and beliefs?**   You are a white researcher studying racial discrimination in the labour market. You are having problems getting people to speak on the record and take part in your study. To encourage people from ethnic minorities to take part in your study, do you offer to provide their lobbying groups access to a restricted government database which you have clearance to access?

- **Is it ethical to 'use' people as allies or informants in order to gain entry to other people or exclusive surrounds?**   You can gain access to a restricted data source by pretending to be a friend or ally of somebody else. Do you 'use' this source to gain access?

- **Is it ethical to accept help from a source with an alternative motive?**   You are examining the planning process relating to disabled access. The planning officer is called out to a meeting. The secretary dislikes her/his boss and offers you access to her/his personal files. Do you accept the offer?

Box 2.7   Questionable practices

- Involving people without their knowledge or consent.
- Coercing them to participate.
- Withholding information about the true nature of the research.
- Otherwise deceiving the participant.
- Inducing them to commit acts diminishing their self-esteem.
- Violating rights of self-determination (e.g., in studies seeking to promote individual change).
- Exposing participants to physical or mental stress.
- Invading their privacy.
- Withholding benefits from some participants (e.g., in comparison groups).
- Not treating participants fairly, or with consideration, or with respect.
- Failing to protect a participant's confidentiality or anonymity.

Source: Adapted from Kidder 1981: 365.

---

some authors have provided general guidelines as to what is considered to be ethical research practice. These guidelines generally advocate a **professional approach** to research and focus upon issues such as privacy, confidentiality and anonymity. They suggest that the researcher should weigh carefully the potential benefits of a project against the **negative costs** to individual participants. Such individual costs might include affronts to dignity, anxiety, embarrassment, loss of trust, loss of autonomy and self-determination, and lowered self-esteem (Kidder, 1981). This is clearly a subjective exercise, but one that can be approached in an informed manner. A general rule might be to consider whether the respondent will take part in the study if it had to be repeated again.

It is generally accepted that researchers will be professional in approach and not undertake the ten questionable practices listed in Box 2.7 (depending upon ethical persuasion). Of those practices listed in Box 2.7 the first point, concerning **covert participant observation**, is probably one of the most controversial amongst researchers. Some would argue that covert participant observation is a legitimate and valid research strategy, especially in public or open settings and where anonymity of the person(s) observed is maintained. Observation in **closed systems** where the researcher is deceiving a particular participant raises more difficult ethical questions. In this situation, some would argue that the act of deception is ethically unsound and should not be practised. Those in favour of such covert research argue that 'deep cover' work allows insights that would not be gained if the person observed was guarded because of the researcher's presence. Our

---

Box 2.8   Politically incorrect and correct language

Politically incorrect	Politically correct
Man/mankind	Person, people, human beings
Manmade	Synthetic, artificial, manufactured
Manpower	Workforce, staff, labourpower
Man-hours	Work-hours
Forefathers	Ancestors
Master copy	Top copy, original
He (generic)	They
The disabled	Disabled people, people with disabilities
The blind	Blind people, people with visual impairments

Source: Adapted from Robson 1993: 63.

---

view is that it is for the researcher to decide whether they consider this research strategy unethical, but they should be aware of the inherent risks and legality of deception. For those wishing to read further about research ethics we recommend the discussions in Kidder (1981; Ch. 15) and Miles and Huberman (1992; Ch. 11) which both provide fuller accounts.

2.5.3   Political correctness

In recent years, the term 'political correctness' has been fashionable. Whilst the term itself has come under scrutiny, political correctness does have a place within your research as the concept of **non-offensive language**. Language, from a poststructuralist perspective (Section 1.4), can be extremely important in representing and reinforcing social and cultural life. Therefore, it is important that you try to use language which will not be emotive or cause offence. This is the case regardless of the approach you adopt. Box 2.8 details the difference between politically correct and politically incorrect language. When designing your research project you should also be aware of the language you use both in your proposal and design stage and when you write up your report. A useful source for further guidance on the use of potentially offensive language is The British Sociological Association's guides (1989a, 1989b) to both **sexist** and **racist** language. Similarly, you should be sensitive to bias within your research strategy (see Box 1.15 for gender biases within research).

2.5.4   Conducting research on sensitive issues

We have been discussing research as if all topics can be approached in the same manner. However, you should be particularly cautious when planning and conducting research on topics that might offend some people. Clearly some topics of research are of a more sensitive nature than others. To a large degree the identification of **sensitive topics** is common sense. However there are a number of ways we might classify sensitive research. For some researchers, sensitive research is a study that has social and political implications. For example, Sieber and Stanley (1988: 49) define socially sensitive research as:

studies in which there are potential consequences or implications, either directly for the participants in the research or for the class of individuals represented by the research.

In other words, a sensitive topic is one that might have broad **social implications**. For example, a study which looks at poverty and race would be a sensitive topic because it might lead to a shift in public policy and it might also affect people's attitudes towards a particular group. For example, Lewis (1966) undertook a study of poverty and race in the US and concluded that Black people were poor because they were black (and thus hold certain *natural* attributes such as low self-esteem), not because of society's social arrangement. He thus suggested that a 'culture of poverty' exists whereby Black people are responsible for their poverty. Such a conclusion was used by the State to absolve themselves of responsibility for Black poverty and to underpin the system of social welfare in the US. This finding also reinforced White attitudes towards Black people. Like many sensitive topics, this is clearly a politicised topic as well as being sensitive.

Another sensitive topic is the study of subordinate groups by dominant groups. As such, sensitive research focuses upon groups who are **socially excluded**, and on some aspect of their exclusion. Within these groups there might be a reluctance to talk to members of the dominant group about certain issues for a number of reasons which centre around suspicion and perceived threat. For example, the research might pose an intrusive 'threat' on the subordinate group, dealing with topics that are sacred or private. The research might also reveal information that will lead to the stigmatisation or incrimination of the subordinate group, or feed popular or cultural myths. Sensitive research also concerns '**taboo**' subject matter. Such 'taboo' topics are those about which people feel embarrassed, shy or threatened. For example, it has been noted that men are generally not keen to talk about masculinity or sexuality.

Sensitive research can also be defined as research which raises ethical, moral or legal implications. As such, sensitive research challenges our **social conventions** and the (largely) unwritten rules about how we conduct research and for what purposes. Lee (1993) suggests that sensitive topics like sexuality, race or disability raise a number of questions. For example:

- What kinds of research are deemed permissible by society?
- To what extent may research encroach on people's lives?
- To what extent can we ensure data quality in dealing with certain kinds of topic?
- To what extent is and should research be controlled by powerful groups?

Some research also raises a whole series of ethical and moral questions concerning what the results will be used for.

Sensitivity affects potentially every stage of the research process and there are a number of issues and problems that can arise at each stage (see Box 2.9). You should consider all these aspects before undertaking any study which you think might be sensitive in nature. You should also be aware that sensitive research can also lead to effects upon the personal life, and even personal security, of the researcher. For example, Taylor (1988) has reported that, although welcomed initially into certain communities in Northern Ireland, some researchers have been forced into hiding or have had to leave the Province altogether as fears arose in some communities that confidential research material was finding its way into the hands of the intelligence community and security forces. It is also worth remembering that while we might think that some groups will be sensitive to some research topics, in some cases they are not. For example, we can end up 'tip-toeing' around a topic when the

---

Box 2.9 Issues of sensitivity within the research process

Methodological: Relates to validity and reliability in data generation, e.g., choosing a suitable methodology to gain the confidence of the researched – there is often mistrust between researcher and researched, leading to concealment and suspicion.

Technical: Relates to the mechanics and running of the research process, e.g., gaining access to 'sensitive' government data or to suitable research subjects. Clearly, where previous research has led to a deterioration of opinion in relation to a certain group, this group is going to be reluctant to take part in future research. Other people might be reluctant to take part because they cannot see the usefulness of the research, or fear it might be used against them.

Ethical: Relates to ethical and moral considerations about the research process, e.g., should there be research on a particular topic? Does the subject matter affect human rights? Should a particular methodology such as 'hidden' participant observation be used?

Political: Relates to the political sensitivity of the research, e.g., does the research have an overt or potential political agenda? Will the research be used against those who form its central concern? If the research relates to comparing two groups, will the research be objective and unbiased?

Legal: Relates to the legality of the research, e.g., is the research breaking the law in relation to privacy, confidentiality and human rights? Are the research findings libellous?

---

people we are researching might be open, frank and unoffended by the questions. However, it is advisable to proceed cautiously when you think an issue might be of a sensitive nature.

2.5.5 Practical considerations

As you consider the methods of data generation and analysis you should consider simultaneously some of the practicalities of conducting research. You will need to balance the remit of your study with more **practical constraints** relating to the *size* of the study, *time* available, *resource* availability, and *physical costs* of undertaking the study. These are important considerations. There is no point in conjuring up a worthwhile and brilliantly designed project if it is unresearchable given time, resource, access and cost constraints. Similarly, if the project is too large you might find it difficult to write up the study within any coursework constraints. Before you start the process of generating and analysing your data, you need to assess the viability of your study given these constraints. This means you need to construct a **timetable** and to calculate any costs which are likely to be incurred (see Section 2.9). If you decide that your project is not viable given time, resource, cost or access constraints you may need to redesign elements of your planned research. If your idea is sound you ought to be able to either narrow down the focus further, reduce the size of the project, or reformulate the way in which data is generated and analysed. Physical costs that might be incurred include (Blaxter *et al.*, 1996):

- Travel costs to your research sites.
- The costs of consumables such as paper, tapes and batteries.
- Possible charges for entrance into certain institutions.
- Equipment purchase or hire (e.g., tape recorders, software).
- Book, report or journal purchases.
- Photocopying, printing and publication costs.
- Postage and telephone costs.

Access may be of particular importance, especially if the research subject is of a sensitive nature (see previous section). If you cannot get individuals or institutions to cooperate with your study, then effectively there is no study, regardless of your intentions. Data generation will often require access to one or more of the following (Blaxter *et al.*, 1996):

- **Documents** held in a library or private collection.
- **People**, in their homes, places of work or the wider community.
- **Institutions**, such as private companies, schools or government departments.

In order to ensure access to certain people or documents you will need to approach the relevant 'gatekeeper' and ask permission to carry out the study. You will have to state coherently what the project concerns and why you are interested in gaining access to the resources the gatekeeper controls. Remember, you might need a degree of **forward planning** to schedule a date to visit. Be prepared to offer something back in return, such as a copy of your final report. If access is refused, all is not lost and there are a number of strategies that you might pursue. For example, if your research consists of interviewing people in their own home you might approach other individuals with similar backgrounds. Most sampling

strategies (see Section 3.3) allow for refusal. If your research involves access through gatekeepers, you could try another institution, try to find another route into the same institution, or wait a while and re-approach the same gatekeeper. If you find access is blocked whichever way you approach the issue, then you might need to consider changing your research strategy completely. One way to try to negate some of the problems discussed in this section is to construct a full **research plan** including a timetable, resources and costs before you embark on the project (see Section 2.9).

2.6 Choosing a method to generate data

In conjunction with an exploration of the practicalities of conducting research you have to decide how to generate your data. The first consideration is whether data will be generated by you (termed **primary data**) or whether you are going to use data generated by somebody else (**secondary data**). While the use of secondary data sources is a valid approach to research, and often the only one available in certain circumstances, such as for an historical study, we would suggest that where possible you should consider generating primary data. There are a number of reasons for this. First, the data generated within primary research will be more **context dependent** to your study. Most secondary sources will not have been recorded with your purposes in mind. Second, if you have generated the data you will know *exactly* how the data were produced, whether any problems arose, and any foibles in the encoding. When looking at secondary sources you may have little idea as to why certain things have been recorded; as such, you are reliant upon inserting meaning into your analysis which might not be warranted. Third, if you are undertaking a project as part of an assessed piece of work, many markers expect primary data generation where possible. This is because of the other reservations expressed above, but also because secondary analysis is sometimes seen as an 'easy way out' of data generation. This is not to suggest that secondary analysis, if done well, is not as valid a piece of research. However, while this might not seem fair, it should be realised that data generation is largely seen as central to student projects. A general rule, then, is to undertake primary data generation where possible. Secondary data should only be used to supplement

Table 2.2　The characteristics of quantitative and qualitative methodologies.

Qualitative	Quantitative
Humanistic	Scientific
Subjective	Objective
Data are words, pictures and sounds	Data are numbers
Data gathered personally	Data gathered by technology or prescription
Inquiry from the inside	Inquiry from the outside
Inductive	Deductive
Interpretative	Functionalist
Idealistic	Realistic
Meaning and understanding	Explanation and prediction
Specificity	Generality
Ideographic	Nomothetic
Individuals	Populations
Small sample sizes	Large sample sizes
Concepts and categories	Incidence and frequency
Extrapolation	Generalisation
Natural	Artificial
Micro	Macro
Participants	Subjects/objects
Self	Society

primary data generation or where primary data generation is impossible. The generation of primary data may be considered impossible when the data required are geographically diverse or if for other reasons it is difficult to generate data without help and access to respondents. Sources of secondary data, and their merits and limitations, are discussed in Chapters 3 and 7.

If you are going to generate primary data, your main decision consists of deciding whether to generate qualitative or quantitative data, or a mix of the two. In basic terms, **qualitative data** are generally unstructured and consist of words, pictures and sounds. In contrast, **quantitative data** are generally structured and the data consist of numbers or empirical facts that can easily be 'quantified' and analysed using numeric (statistical) techniques. The characteristics of qualitative and quantitative methods commonly cited in the literature are listed in Table 2.2. Note how the two lists appear to form a set of polar opposites. As we will see, whilst talking about qualitative and quantitative as separate and contrasting methods helps to provide structure to a discussion concerning methodologies, the distinction is quite misleading. Indeed, it is probably best to think of qualitative and quantitative methods as a *continuum* rather than polar opposites, with the two lists in Table 2.2 viewed as the extremities of such a continuum.

The confusion which concerns both the characteristics and use of qualitative and quantitative methods is due, in the main, to two common misconceptions concerning the choice of methods to generate data. The first misconception is that each approach restricts you to a set of designated techniques. These will be either qualitative *or* quantitative in nature. Whilst some approaches do naturally lean towards some methodologies (see Table 1.2), there are no set designations. Therefore, whilst a positivist is most likely to favour quantitative data generation, there are no rules to say that he or she *must*. Similarly, there are no rules to say that a phenomenologist *must* use qualitative methods, and in particular interview respondents within an ethnographic framework. It is sufficient that the method chosen is used within the epistemological and ontological conditions of your underlying approach to provide data that can be interpreted within this approach.

The second misconception is that you generate *either* quantitative *or* qualitative data and that you *cannot* (as with oil and water), or rather *should not* (as with fruit-based and grain-based alcohol), mix them. This misconception is given credence by many of the books on undertaking research in the social sciences which compartmentalise knowledge: 'how to use qualitative methods' and 'how to use quantitative methods'. There is no law that states that qualitative and quantitative methods have to be used in

---

Box 2.10   A proposal for a study combining quantitative and qualitative methods

Synopsis
To describe and analyse the effects of employment status and social attitudes upon patterns of spatial behaviour, using a three-tiered approach.

1. Statistical analysis and mapping
To provide a basis for the primary research, we propose to analyse statistically and map socio-economic data from the census in order to help identify our three areas of study. To provide contextual information we will next map data relating to employment status (and job form) and social attitudes. Consideration will be given to the use of the Labour Force Survey to provide data on employment and the National Social Attitudes Survey to provide data on social attitudes. Data within these databases are often held using different spatial units and a GIS will therefore be used to manipulate the data to utilise different geographies.

2. Questionnaire survey
To gain information on the employment status, social attitudes and spatial behaviour of residents within each of three study areas it is proposed to conduct a large-scale questionnaire survey. A total of 300 questionnaires will be administered in each area, giving a total sample of 900 households. The survey design is based on a stratified sample (e.g., calling at every *n*th house). The

questionnaire will include a selection of closed and open questions. However, it will mainly contain 'closed questions', to help ensure standardisation in collection, given that different interviewers may be used at each location. The results will be analysed using standard statistical procedures.

3. Focus groups
To verify, build on and add depth to the results of the questionnaire survey, we propose to use a system of focus groups to explore the relationship between unemployment and social attitudes upon patterns of spatial behaviour. Each focus group will consist of approximately ten volunteers. Discussions will be taped, transcribed and analysed using standard qualitative techniques (see Chapters 7 and 8). This section of the research is envisaged as having two main parts – area based and group based. Firstly, focus groups will be formed in each of the three areas for in-depth discussions of the most pertinent issues as revealed by the questionnaire. Volunteers will be sought from the questionnaire respondents. Secondly, five focus groups will be formed to deal with thematic issues, to broaden the coverage of the survey. The membership of these groups will be gay people, ethnic minorities, women, adolescent children, and disabled people.

---

isolation from each other. To illustrate this point, Box 2.10 summarises a proposed research strategy combining qualitative and quantitative methods to study the relationship between employment status, social attitudes and spatial behaviour in three areas (e.g., a council estate in a city, a middle-class suburb of a city, and a rural town). As novice researchers you should not perhaps try anything as elaborate as detailed here. However, the proposal does give you an example of the methodological considerations you should be thinking about and the ways in which quantitative and qualitative studies can be combined.

Within the bounds of any epistemological and ontological considerations, and the nature of the question(s) to be answered, the choice of methodology is open, and the selection of which methodology to use to generate data, therefore, is no easy task. You need to weigh up carefully the advantages and disadvantages of all methods, regardless of whether they are qualitative and/or quantitative in nature, and decide on which will provide the most appropriate answer to your question within the constraints of the adopted approach. As we have already stated, whilst the qualitative and quantitative dichotomy is

artificial, it does provide a useful distinction in which to discuss methodologies. As such, we have used this artificial distinction to structure our own discussion. The specific techniques of data generation are detailed in Chapters 3 and 7. We suggest that you read both chapters before coming to any decision about how to generate your data.

2.7   Choosing a method to analyse your data

The next choice you have to make concerns the method of data analysis. Although for clarity we have separated the discussion of data generation from that of data analysis in this and subsequent chapters, this is an artificial division. Data generation and analysis are not divorced and both need to be considered carefully *before* starting your research. For example, you should not just generate your data and then look for a way to analyse them, you should decide on both before you start. This is particularly the case with qualitative research where on-going analysis often occurs whilst data are generated. The material in Chapters 4, 5 and 8 is designed to help you make informed choices about methods of analysis, detailing various common techniques, their advantages and disadvantages, and how to use them. Remember, when choosing your techniques, to keep in mind the research objectives and the theory you are trying to assess.

2.8   Conducting group research

It may be the case that you are involved in a group rather than an individual project. If so, then all of the discussion in this chapter concerning research design, validity, ethics, etc., still applies. However, with **group work** the research plan and design have to be more carefully managed as a new set of key issues arise (see Box 2.11). With group work the group is only *as strong as the weakest link*. If somebody fails to undertake their section of the research then the whole project can fail despite the efforts of the others. It is vital, therefore, that in undertaking a group project you all know what your responsibilities are and you all carry them out to the best of your abilities. Almost inevitably there will be tensions within any research group, especially if one person is not pulling their weight or working towards a common goal. These conflicts must be resolved, if possible. As a group you need to work through the issues in Box 2.11 and each group member needs to be happy that each issue is collectively answered and understood.

Box 2.11   Key issues in group projects

- Does the group need and have a leader?
- Who is responsible for managing each part of the research?
- What are the strengths and weaknesses of each group member?
- How are the different tasks to be distributed?
- Does every group member have a clear idea as to what they should be doing?
- Does each group member respect the abilities of all the other members?
- Are there individuals who do not feel happy about the group structure or individual responsibilities?
- Will the research be written up, individually, collectively or both?

Source: Adapted from Blaxter *et al*. 1996.

Box 2.12   The pros and cons of individual and group research

Individual research:
- Gives you sole ownership of the research.
- Means you are wholly responsible for its progress and success.
- May result in a more focused project.
- Is of an overall quality determined by you alone.
- Means that you have to carry out all elements of the research process.

Group research:
- Enables you to share responsibility.
- Lets you specialise in those aspects of the work to which you are most suited.
- Provides you with useful experience of team working.
- Allows you to take on larger-scale topics than you could otherwise manage.
- Provides you with a ready-made support network.
- May be essential for some kinds of research.

Source: Blaxter *et al*. 1996: 46.

If you have the choice between undertaking individual or group work there are clearly pros and cons for you to assess (see Box 2.12). Both group and individual work can be rewarding but both need careful planning and management. Our suggestion is to explore the possibilities of undertaking a group project, but if you are doubtful as to the success of such a project, to then revert back to an individual endeavour. There is nothing more disheartening than seeing your hard work going to waste at the expense of somebody else's laziness.

2.9   Managing and piloting a research project

As discussed in Section 2.5.5 there are a number of time, resource, access and cost constraints to be considered before you start your research. Once you have planned and designed your project these issues will come to the fore during the realisation of your study. You will need to manage carefully all the various stages of the research process in order to obtain successful results. The best research design in the world will *not* save a project that is badly administered. Managing a project in progress is important, and you should make every effort to make sure that your plan

---

Box 2.13   Example timetable for a project lasting nine months

Month 1	Phase 1	Explore initial ideas
		Visit library
		Read literature
		Plan project
Month 2	Phase 2	Make contacts
		Design data generation and analysis
		Undertake small pilot project
		Evaluate pilot project
Month 3	Phase 3	Generate data
Month 4		
Month 5	Phase 4	Analyse data
Month 6		Interpret results
Month 7	Phase 5	Write up project
Month 8		Submit project
Month 9		

The last month being free allows for mishaps and timetable miscalculations.

---

is put smoothly into practice. As you will no doubt discover, this is not as easy as you might think. A number of problems can occur. For example, data you have been promised might fail to materialise or you might get lower than expected response rates from questionnaires or interviews. As a consequence, you need to try and be **flexible** when conducting the project. In the planning stage you should construct a **detailed timetable** of when you envisage each part of the project to start and end (see Box 2.13 for an example timetable). Do not make this timetable rigid, but rather allow for any mishaps that might happen. As a general rule, you should construct a timetable that will allow you to complete the project comfortably within the overall time. If there are mishaps you can reschedule or repeat the problem event, or increase your work pace accordingly without having to work around the clock to get everything finished by a deadline.

One way to test whether your study is viable is to conduct a small **pilot study**. This will give you an indication of whether your research methodology is suitable, whether you are going to gain access to respondents, and how long the full research project is likely to take. Projects that are undertaken without any pilot study often run into problems which are difficult to solve without scrapping what has been undertaken and starting again. This is often the case, we find, with questionnaire-based research. The pilot study should be a full mini-run of the project, including the processes of data generation and analysis. There is no point discovering that you can generate

the data to then discover later that your mode of analysis is unsuitable. If you are undertaking a group project then the pilot study will also identify whether everyone has been assigned the most suitable research roles and how well the group works together.

2.10   Summary

After reading this chapter you should:

- be able to define, plan and execute your study;
- understand that there also practical considerations to undertaking research;
- be aware of issues of validity and reliability;
- be aware of the ethical implications of your research;
- understand that there are sensitive topics that need to be approached with caution.

You should now have a good idea about not only *what* you are going to study, but also *how* you are going to undertake the research. As we have illustrated, there are a number of issues that you need to consider in advance of the data generation and data analysis steps. You should consider each of these issues carefully and plan your project in detail. This preparation is vital if you are to ensure a smooth-running project, as free as possible of potential problems. Once you are fully prepared, you will then need to put your ideas into practice. The information in the following chapters will guide you through specific issues and techniques, demonstrating how to implement your choices of data generation and data analysis. In the final chapter we discuss the process of finishing off and writing up your project.

2.11 Questions for reflection

- *How focused should your research topic be and why?*
- *Why is a research plan of importance?*
- *How are theory and practice related in relation to your project?*
- *What is the relationship between qualitative and quantitative data?*
- *What sorts of ethical issues might arise in relation to your research project?*
- *Why are validity and reliability important to a study?*
- *How should you approach a topic that is sensitive in nature?*
- *What are the advantages and disadvantages of working in a group?*
- *Why should you undertake a pilot study?*

Further reading

Blaxter, L., Hughes, C. and Tight, M. (1996) *How to Research*. Open University Press, Buckingham.

Homan, R. (1991) *The Ethics of Social Research*. Longman, Harlow.

Kidder, L.H. (1981) *Research Methods in Social Relations*, 4th edition. Harcourt Brace, London.

Kimmel, A. (1988) *Ethics and Values in Applied Social Research*. Sage, London.

Lee, R.M. (1993) *Doing Research on Sensitive Topics*. Sage, London.

Miles, M.B. and Huberman, A.M. (1992) *Qualitative Data Analysis: An Expanded Sourcebook*. Sage, London.

Phillips, E.M. and Pugh, D.S. (1987) *How to Get a Ph.D.* Open University, Milton Keynes.

Robson, C. (1993) *Real World Research: A Resource for Social Scientists and Practitioner-Researchers*. Blackwell, Oxford.

CHAPTER 3

Data generation for quantitative analysis

This chapter covers	**3.1** Introduction
	3.2 Classifying data types and measurement scales
	3.3 Generating primary quantitative data
	3.4 Sampling, estimation and distribution
	3.5 Obtaining and using secondary data
	3.6 Summary
	3.7 Questions for reflection

3.1 Introduction

Within every group of students to whom we have taught basic statistical analysis, at least one-third profess to be 'stato-phobic' and are petrified by the thought of doing anything with numbers. Whilst some statistical tests can be very complicated, those dealt with over the next two chapters are generally basic in nature. At first glance, an equation may look daunting, but as you will see when we work through the examples – particularly the tests in Chapter 5 – calculating a statistic simply involves following a number of steps in a particular sequence. When the computer does all the work for you, using statistics becomes even easier: all you need to do is input the data, run the program and interpret the results. As long as you understand the simple mathematical operators $+$, $-$, $\div$, $\times$, $\sqrt{}$, ^{2} and Σ, using the statistics in this book should be relatively straightforward. We have come to the conclusion that many of our students find statistics difficult to use because they think they ought to be. Many students think there is a catch, that they are missing something obvious. Statistics are, however, relatively straightforward to use. To carry out the analysis you just need to understand the language or 'jargon' being used, and then follow the set of accompanying rules. In our opinion, qualitative data are often much harder to generate and analyse. Ironically, many of our students choose qualitative studies, not because they think that geographic

studies should be qualitative in nature but because they think that qualitative analysis will be easier. As you will see in Chapters 7 and 8, this myth needs to be firmly quashed.

3.2 Classifying data types and measurement scales

The quantitative analysis methods discussed in Chapters 4 and 5 require the collection of **measured data**. Such data represent one type of information which we can collect about phenomena in our environment. Here, measurements relate both to the fundamental properties of objects, where quantities like length extend through space, and to measurement operations performed on individual objects. In the former, measurements are known as **extensive properties** which form the basis of standard measuring systems such as the SI system in use by most scientists today (Chrisman, 1997: 8–9). However, this view of measurement is only appropriate for physical properties, and is not really applicable to the subject matter of most social scientists. The alternative latter view sees the process of measurement as being separate from the object being measured. This is usually referred to as a **representative view** of measurement. It was this view that was adopted in the **levels of measurement** developed by Stevens (1946) and listed in Box 3.1. He identified four levels of measurement: **nominal,**

Box 3.1   Levels of measurement

Nominal   Observations are placed in categories, symbolised by numerals or symbols (e.g., A, B, C, D).

Ordinal   Observations can be placed in a rank order, where certain observations are greater than others. Assigned numerals cannot be taken literally (e.g., first, second, third, fourth).

Interval   Each observation is in the form of a number in relation to a scale which possesses a fixed but arbitrary interval and an arbitrary origin. Addition or multiplication by a constant will not alter the interval nature of the observations (e.g., 1°C, 2°C, 3°C, 4°C).

Ratio   Similar to the interval scale, except that the scale possesses a true zero origin, and only multiplication by a constant will not alter the ratio nature of the observations (e.g., 0%, 5%, 10% as an exam mark).

---

ordinal, **interval** and **ratio**, each of which creates data of a different form. As a consequence, the type of quantitative analysis appropriate to each level of measurement also differs. The choice of a level of measurement is important since it will determine what the generated data will tell you.

Nominal measurement (also known as categorical measurement) is based on mathematical set theory, and is concerned with the assignment of objects into categories often signified by numerals or symbols. Each category is homogeneous and mutually exclusive and the overall categorisation is exhaustive (Reynolds, 1984). For example, we might divide a group of geography students sitting an exam into two groups based on gender, and put these into the categories 'A' and 'B'. We could further subdivide each of these groups into a group who were predominantly interested in physical geography (group 1) and a group predominantly interested in human geography (group 2). We have thus categorised the students into four distinct groups (Figure 3.1).

Although we have assigned those in the physical geography group as '1' and those in the human geography group as '2', we could have chosen any two different codes to differentiate the groups. Although we *seem* to have developed an ordering system – 1A, 1B, 2A, 2B – this is not meaningful, since we could

Table 3.1   Ordinal ranking scale.

---

1   Very interesting
2   Interesting
3   Neither interesting nor boring
4   Boring
5   Mind-bendingly boring

---

just as easily categorise each physical geographer as a 2 and each human geographer as a 1. In addition, we cannot say that 1 + 1 = 2 (i.e., two physical geographers = a human geographer), as arithmetic operations make *no sense* on nominal data.

Ordinal measurement, in contrast, can convey more information since we are able to place objects in some sort of rank order, as the name suggests. For example, the results of an exam will place the above-mentioned students in a rank order. The student with the highest mark will be *first*, the next highest *second*, and so on. In this case we can say that the student who came *first* has a mark which is *greater than* that of the student who came *second*, who has a mark *greater than* that of the student who came *third*, although the actual mark differential may be less between *first* and *second* than between *second* and *third*. Another example of ordinal measurement is the use of ranking scales which might be used in a questionnaire analysis. For example, an ordinal ranking scale similar to that in Table 3.1 might be used to rank responses received from each student about how interesting they find their statistics class.

The results of questioning our six students according to Table 3.1 might produce the results seen in Figure 3.2. Numbers are assigned to each ordinal category only to assist analysis (Weisberg, 1992) and they convey information only about rank order: we cannot assume that the intervals between the ranks are equal, or that the quantities ranked are in any way

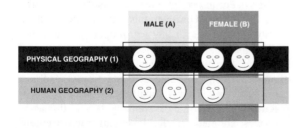

Figure 3.1   Geography students organised into categories.

Figure 3.2 Ordinal ranks of student responses.

☺	☺		☺	☹	☹		☹		
90	**80**		**60**	**40**	**30**		**10**	**0**	**%**
45	**40**		**30**	**20**	**15**		**5**	**0**	**/50**

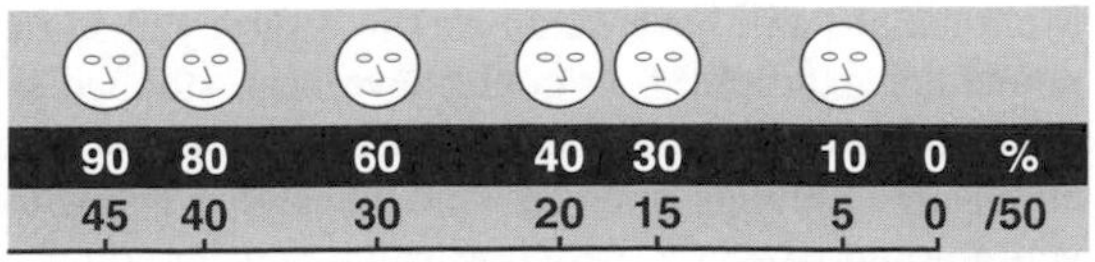

Figure 3.3 Ratio measurement of student exam results.

absolute (Frankfort-Nachmias and Nachmias, 1996). Even though the difference between the codes for the categories 'Very interesting' and 'Interesting' (1 and 2) is 1, the same as the difference between the codes 'Boring' and 'Mind-bendingly boring' (4 and 5), we *cannot* treat this numeric interval as being meaningful. We could have chosen other codes, for example 10, 20, 30, 40 and 50, to replace the existing 1–5 codes.

In contrast to nominal and ordinal measurement, **interval** and **ratio measurement** are inextricably concerned with **quantitative data** (also termed **metric data**). In this case, an observation will usually be in the form of a meaningful number. These can either take on any value within a given range (**continuous variables,** e.g., length in metres) or be restricted to a number of distinct values within a given range (**discrete variables**, e.g., counts of people in a sample category). For an interval measurement this number relates to both an arbitrary interval scale and an arbitrary origin, though we do know *precisely* what the interval is between two measurements. For example, consider the Roman calendar system: 1998 is exactly 100 years greater than 1898, but the origin chosen on the scale is arbitrary – in this case 0 AD. Taking the numbers on such a scale and either multiplying by a constant or adding a constant will not destroy the interval nature of the data (Weisberg, 1992).

For a ratio measurement, which is the highest level of measurement, this number is in relation to a scale of an arbitrary interval, similar to interval data, but with a *true zero origin*. If we return to the example of students in the exam, each student will have obtained a certain number of marks. It is possible for a student to obtain zero marks in the exam, and the ratio of marks between two students might be 3:2, hence the different scores for each student are measured on a ratio scale. The key point here is that we can multiply ratio variables by a constant, but adding a constant will destroy their ratio nature (Weisberg, 1992). For example, our six students may have obtained the set of exam results displayed in Figure 3.3. Two of the

students have scored 30 and 20 respectively out of a total of 50. Multiplying by 2 gives the mark out of 100, i.e., 60 and 40, which maintains the ratio, but adding an extra 5 marks to each score changes the ratio to 7:5. Ratio measurement more commonly applies to metric quantities such as distance and mass, which possess a zero origin.

As noted above, the level of measurement determines the type of analysis we can perform on the data generated. Nominal and ordinal data can *only* be analysed using **non-parametric** tests. Interval and ratio data are generally analysed using **parametric** statistics, although it is possible to convert data from an interval/ratio measurement scale to an ordinal or nominal scale if required. As a rule, properties measured at a higher level of measurement can be converted (measured) at lower levels (Frankfort-Nachmias and Nachmias, 1996). For example, we can convert the student grades displayed in Figure 3.3 to an ordinal scale, simply by placing them in rank order. This is particularly useful if we wish to use a non-parametric statistical technique for analysis of data which, although interval or ratio in nature, do not fit the assumptions of an appropriate test (see Section 4.7). We will return to describe further parametric and non-parametric methods of analysing data, and the differences between them, in Chapter 5.

3.3 Generating primary quantitative data

Whatever the specific research questions you have established from Chapter 2, you may wish to generate/collect your own data: a process we have termed **primary data collection**. Primary data generated for quantitative data analysis, as discussed above, do not need to be interval in format. However, such data need to be generated in such a way that they can easily be described and analysed using quantitative techniques. We often use **surveys** to obtain such data. A survey is a study which seeks to generate and analyse data on a specific subject from a particular sample population.

Most people will have taken part in a survey at some point. For example, you could have been stopped in the street, questioned on the doorstep, or phoned at home about things such as who you were going to vote for and why? Or, which washing powder you use or chocolate you eat? Consumer and market research rely almost exclusively upon large-scale surveys to generate data about all aspects of our daily lives. As indicated, surveys can take a number of forms ranging from postal surveys to telephone surveys, and can use a number of different sampling strategies. In general, surveys use **questionnaires** to generate quantitative data from which they can calculate statistical information (e.g., 39% of males aged 21–30 smoke regularly), and it is this type of survey we will concentrate on below.

3.3.1 Designing a questionnaire survey

When designing any survey there a number of factors you need to consider (see Box 3.2). How you resolve each of these will determine the nature of the data you generate and what they will tell you. You will need some sort of purpose or objective, usually generated by your research problem, in order to motivate your survey. This could be a simple question, for example: 'What motivates people to make intra-urban residential moves?'. The **sample population** are the respondents taking part in the study, which you will have sampled to try and answer your question. You need to decide upon your sample population carefully. There is no point in including people whose opinions and thoughts are not relevant to what you are researching. The characteristics of your sample are usually delineated by factors such as age, occupation and economic status. For example, a survey concerning voting patterns will usually include only people of voting age (18+). Similarly, if you are interested in the behavioural processes underlying intra-urban residential mobility (mobility within a city), there is no point in including people in the survey who are making an inter-urban move (mobility between cities).

After deciding upon who you want to survey, you need to decide how your data will be generated. There are two options. First, you can **interview** your respondents using a discussion situation aimed at generating qualitative data (see Chapter 7). Second, you can interview your respondents using a more formal, structured **questionnaire** aimed at generating quantitative data. These two data generation methods are not mutually exclusive and many surveys mix discussion-style **open questions** with more rigid **closed questions**, which we will examine in the next section.

Before designing the questionnaire, you will need to decide on the medium in which your questionnaire will be conducted. There are a number of options open to you, each with merits and limitations. The most common questionnaire medium is **face-to-face meeting**. Unlike an interview, where the discussion may need an informal, comfortable setting due to its length and nature, a questionnaire can be conducted in a variety of settings, for example on the doorstep or in the street. This is because a questionnaire usually takes less time to complete and the relationship between researcher and researched is more formal. Face-to-face meetings have the advantage of personal contact and a higher response rate (especially when administered on the doorstep). However, they can also be quite time-consuming. A second option is to use a **postal questionnaire**. Here, the questionnaire is sent to all the potential respondents for self-completion. The postal questionnaire increases the potential sample size and can reach those respondents who are geographically dispersed, but has some distinct disadvantages: the response rate can be much lower (especially if return postage is not included), the costs of administration are higher (postal costs), questionnaires are often returned incomplete, and the method provides no opportunity to clear up misunderstandings. It is essential, with a postal questionnaire, that you send a covering letter stating explicitly the reasons for the survey and thanking the respondents in advance. A third option might be to administer the questionnaire using the telephone. **Telephone questionnaires**

Box 3.2 Designing a survey

Purpose/objective	Why do you want to survey – what is your objective?
Sample population	Who is being surveyed?
Survey methodology	How the data are generated (e.g., questionnaire; interview – see Section 7.3)
Survey medium	How the survey is undertaken (e.g., face-to-face, postal, telephone)
Survey design	How the survey is constructed (e.g., elements of questionnaire design)
Sampling strategy	How the survey is administered to your sample population (see Section 3.4)

tend to be quicker to complete and increase the potential sample size. They can also limit the sample as respondents must have a telephone. Refusal rates are generally much higher and phone bills can be expensive. A final option might be to distribute the questionnaire via **e-mail.** Several studies have utilised e-mail as a questionnaire medium, posting their forms to members of a relevant mailing list. The sample population for these questionnaires usually have specialised interests. As with postal administered questionnaires, the response rate can be variable, especially if the questionnaire has been posted to people who have no relevant interest in the topic.

Designing a questionnaire

There is more to designing a questionnaire than just assembling a series of questions. The type, wording and order of the questions need to be carefully planned. The questions you ask will be determined by what you want to do with the data. There are generally four types of variables that need to be generated by questions so that they can be statistically analysed (see Box 3.3). When constructing your questionnaire you must be careful to ask questions that will generate these variables to enable subsequent statistical analysis.

There are nine basic types of questions which can be used in a questionnaire, and examples of these are given in Box 3.4. In general, questionnaires usually seek a mix of **descriptive** and **analytical** answers. Descriptive questions tell us 'what' and analytical questions tell us 'why'. These aim to generate both factual and subjective data relating to people and their circumstances, the behaviour of people, and attitudes, opinions and beliefs. Questionnaire data to be analysed quantitatively are usually generated using what are termed **closed questions** (also known as *closed-ended* questions).

A closed question is one where the respondent is given a set number of answers, one of which they must choose as the most representative of their facts/views. In Box 3.4 questions 2–8 are all different forms of closed questions. **Categories** are useful mechanisms to obtain factual information, but the options offered to the interviewee must reflect the likely pattern of variation in the sample. For example, if we asked question 2 to a sample of households who lived in new suburban subdivisions, we might find that the bulk of the responses occur in categories 1 or 2. This characteristic, termed *heaping* (Bourque and Clark, 1992), may occur at either end of a category

---

Box 3.3 Questionnaire variables for analysis

- **Experimental or independent variables**: factors which you think might influence the subject of your study. For example, factors you think might influence the decision to buy a particular house include price, investment and location.
- **Dependent variables**: the factor(s) you are interested in explaining. This is achieved by examining the influence of the independent variable upon the dependent variable. For example, is the reason for moving house (dependent variable) contingent upon price (independent variable)?
- **Controlled variables**: these are independent variables that are used as constants to try and determine the full effect of another independent variable. It might not be unreasonable to expect people's house-buying decisions to vary with wealth. If we wanted to know whether people who have the same income make the same choices we could use wealth as a control variable. Wealth would then be held constant and the other independent variables (price, investment, etc.) within the wealth constraints could be examined for their effect upon house choice.
- **Uncontrolled variables**: these are other independent factors that could also influence house choice which have been omitted from the questionnaire and therefore are not included in the analysis. Because of the complexity of the world and the limited space of a questionnaire there are always factors that are omitted. You should be aware that these factors exist.

Source: Adapted from Oppenheim 1992.

---

scale. Again, if we sampled households comprised of retired people, we might find that the upper limit '5+' heaps respondents in this category. **Lists, scales** and **ranks** are also useful devices for closed question construction. A scale allows respondents to qualify their feelings or thoughts on a fixed scale ranging from negative to positive reactions. Questions 4, 5 and 7 are all types of scale. The bipolar scale used in Question 5 is usually referred to as a **semantic differential scale** (Frankfort-Nachmias and Nachmias, 1996). Ranking questions require the respondent to place a series of choices into an order of importance/ preference. Which of these question types you choose to adopt depends upon exactly *what* you want to know, determined ultimately by your research problem. There is no point asking respondents to rank all choices if all you want to know is which variable is

Box 3.4   Types of questionnaire questions

1   Quantity or information

In which year did you move to your current address? _____

2   Category

How long have you lived at this address? (circle number):
```
0–1 years ................................................. 1
1–2 years ................................................. 2
2–5 years ................................................. 3
5+ years .................................................. 4
```

3   List or multiple choice

Which was the main reason for buying your current house? (circle one number):
```
Price ....................................................... 1
More suitable residence ........................... 2
An investment ......................................... 3
Better location ........................................ 4
Children's schooling ................................ 5
Nearer work/relatives ............................. 6
Other ....................................................... 7
   specify _____
```

4   Scaling

Please rate on the scale below how important price was when buying your house (tick one box)

Not important	Relatively unimportant	Of consideration	Quite important	Very important	Not sure
☐	☐	☐	☐	☐	☐

5   Semantic differential scaling

Indicate on the scale below how important price was when buying your house (tick one box)

	Very	Fairly	Slightly	Neither	Slightly	Fairly	Very	
Unimportant	☐	☐	☐	☐	☐	☐	☐	Important

6   Ranking

Please rank the reasons for buying your current house (please rank all relevant categories from 1 (most important) to 6 (least important))

Price	☐	More suitable residence	☐	An investment	☐
Better location	☐	Children's schooling	☐	Nearer work/relatives	☐

7   Complex grid and table

Please detail the relative importance of each factor listed in the table (tick one box per line)

	Not important	Relatively important	Of consideration	Quite important	Very important	Not sure
Price	☐	☐	☐	☐	☐	☐
Size	☐	☐	☐	☐	☐	☐
Investment	☐	☐	☐	☐	☐	☐
Location	☐	☐	☐	☐	☐	☐
Schooling	☐	☐	☐	☐	☐	☐
Distance	☐	☐	☐	☐	☐	☐

8   Contingency

If in Question 4, Price = 'Very Important', then how much did you pay for your current house? (tick approx range in £)

40–60,000	60–80,000	80–100,000	100–120,000	120–140,000	Over 140,000
☐	☐	☐	☐	☐	☐

9   Open-ended

Do you have any further comments?

_____

_____

_____

most important. Finally, contingency questions (e.g., if answered 'yes' to Question 5) are asked only to respondents who gave specific answers. They act as filters to probe a particular issue further. These types of question create **skip patterns** within a questionnaire, where certain questions are skipped if a respondent gives a certain answer.

The other main type of question is the **open question** (or *open-ended* question) where the respondent is given no set of possible answers. Although this type of question is generally easier to put into a questionnaire, and avoids the problem of suggesting potential answers to the respondent, it is harder to analyse quantitatively, requiring some form of **content analysis**.

Once you have thought about the types of question, and the variables which you need, it is worth considering the various options available for the design of your questionnaire so that the surveyed data can be used for computer-based quantitative analysis with a minimum of difficulty. Bourque and Clark (1992) give many helpful suggestions in this regard, and recommend the use of both closed and **pre-coded** questions where possible. In Box 3.4 we have deliberately used different styles for several of the closed-ended questions. For example, in questions 2 and 3 we have pre-coded the appropriate number to circle guided by dotted lines, but for questions 4–8 we have chosen boxes to tick or number. Both are acceptable styles to use, but information entered into the boxes will require subsequent coding, and the pre-coded questions are considerably easier for taking the data and entering into a computer. Bourque and Clark (1992) advise that when designing questionnaires with closed and pre-coded questions, the researcher should be very careful to ensure that the categories offered in each question are *exhaustive* (or if not, an 'other' category is included to catch other answers, as in question 3 in Box 3.4), are *mutually exclusive*, and use *consistent coding* across different questions for common replies (e.g., no = 1, yes = 2, don't know = 3, missing data = 999). Finally, some thought should be given as to whether alternative answers are read aloud to the respondent. Doing so may suggest potential answers which the respondent may not have thought of.

Taking care over **question wording** is crucial. Questions should be concise and clear, so that they cannot be misinterpreted or misunderstood. Some general rules concerning question construction are detailed in Box 3.5. Questions should be divided into appropriate sections and ordered in a progressive manner. Respondents should feel that the question-

naire has a logical order so that there is a purpose to their answers. A general guide is to start by providing a set of instructions and a preamble. The instructions should tell the respondent how long the questionnaire will take to complete, and if appropriate, provide a return date and details of how to return the completed questionnaire. The preamble should provide a brief explanatory statement of the purpose of the research. Keep the statement short but as informative as possible. Remember that you are relying on the goodwill of the respondents to complete your questionnaire. You should also explain whether the responses will be treated confidentially. After the preamble, it is usual to seek factual background information (e.g., age, sex, etc.). This is followed by the main sections of the questionnaire. At the end always thank the respondent for their answers and time.

However the questionnaire is worded and ordered, it should not be too long. Respondents will soon become bored when answering closed questions as there is little discussion and interaction. We recommend that you ensure that your questionnaire takes less than ten minutes to complete and consists of only a few pages. If your questionnaire consists of reams of paper then many people will be put off, especially if it is a postal questionnaire. As well as length, presentation is important. Your questionnaire should be presented in a professional manner with appropriate spacing and boxing. A poorly presented questionnaire does not place confidence in the respondent and can be difficult to complete and code. Finally, you may well encounter problems with respect to administering the questionnaire in person. Sometimes people can overreact to someone knocking at their door. Indeed, the father of one of the authors recalls an instance whilst surveying where he was greeted by a man with a revolver who was determined he wasn't going to come into that particular house! This is an extreme example, and is unlikely to happen to you. However, if you are administering the questionnaire in person, it is useful to have an official covering letter – ideally on headed notepaper from your department – which describes who you are and the purpose of your survey.

As with all methodologies your questionnaire should be piloted to determine whether the questions work well and produce the data that you require for your research. Even if this is just a case of trying out the intended questionnaire on your family and friends, feedback received at this stage can be very useful in clarifying the questions you are trying to ask. Some of the key things to look for in inspecting the results of a pilot test are listed in Box 3.6.

Box 3.5   General guidelines for asking questions

- **Make sure the question is relevant.** There is no point asking a question if you are not interested in the response.
- **Choose the correct question type** to gain the information you require.
- **Place the questions in a progressive and logical order.** Questions placed in a random order can confuse respondents.
- **Keep the wording simple and use plain English.** Respondents should be able to understand the question without having to ask for clarification.
- **Do not begin with difficult/sensitive questions.** Put these near the end so that if the respondent decides that they do not want to answer them it will not affect responses to earlier questions.
- **Do not make respondents feel they ought to know the answers.** Help them by saying 'Perhaps you have not had time to give this much consideration?'. Maybe they can 'find' the answer later.
- **Remember that respondents may not have the answer.** Even though they might like to cooperate, respondents may not have the answer. Perhaps they do not know, cannot remember, cannot express the answer well in words, or have no strong opinion.
- **Decide carefully whether you should avoid emotional or sensitive words.** Using words like 'greedy', 'oppressed' or 'immoral' may seem to imply a judgement. Such words can cause bias in the answers. However, depending on the situation, you may need to use such words.
- **Avoid making assumptions.** For example, do not ask 'What sort of degree did you obtain?' until you know whether the person attended university.
- **Do not use leading questions.** For example, 'Do you enjoy living in Belfast?' should be rephrased to 'What is living in Belfast like?'. Do not start questions with phrases such as 'Do you not think that. . . .' or 'Is it not likely that . . .'.
- **Do not use biased questions.** Questions should be neutral in nature and not seem to favour one position or another. For example, 'Do you favour the Nationalist solution to marches in Northern Ireland?' should be rephrased to 'What solution to "The Marches" do you favour?'.
- **Do not use confusing questions.** Avoid questions like 'Would you prefer your child not to be vaccinated?'. Keep it simple and positive. Ask 'Do you wish your child to be vaccinated?'.
- **Do not use long questions.** The interviewee might remember only part of the question and respond only to that part.
- **Do not use double- or multi-barrelled questions.** For example, 'What do you feel about the welfare system now compared with that ten years ago?'. The solution here is to break the question down into simpler questions. For example, 'What do you feel about the welfare system?'; 'Can you recall the welfare system ten years ago?'; 'How do you feel welfare ten years ago compares with today?'.
- **Be prepared to be able to rephrase the question.** There may be a need to rephrase a question to tailor it to the situation or individual or to explain more fully what information the question is seeking.
- **Try to avoid hypothetical questions.** Respondents may not have direct experience and answers are often unreliable.
- **Randomise possible alternatives on questions that involve category choices.** The sequence of choices should not overtly influence the answer given.

Sources: Barrat and Cole 1991;
Mikkelsen 1995; Robson 1993.

When you administer your questionnaire, it is possible to introduce error due to non-response, which we term **non-response error**. As a rule, it is hoped that all respondents in a selected sample will answer all questions. In this case, there will be no non-response error. However, in most surveys there is a degree of non-response because of reasons such as refusal to answer questions, absences or lost forms. The concern is that those who do not participate in a survey may well differ from those that do. In general, the larger the non-response error the larger the biasing effects are within the data caused by certain portions of the sample population being missing. One of the largest non-response errors recorded was by the *Literary Digest* in the poll for the 1936 US election between Roosevelt and Landon (Frankfort-Nachmias and Nachmias, 1996). The *Digest* conducted a poll of 2.4 million individuals (then the largest in history). The method they used was to mail 10 million questionnaires to people whose names and addresses were taken from sources such as telephone directories and club membership lists. From their sample, they predicted a victory for Landon with 57% of the votes. Roosevelt in fact won the election with 62% of the vote! Despite the large sample size, the error was enormous. This occurred because the sampling scheme failed to recognise that many poor people did not have a telephone or belong to clubs, so they were

---

Box 3.6   Pilot test evaluation

Variability	Is the pattern of variability what you expected? For example, with Question 8 in Box 3.4, you may have underestimated the cost of houses in your sample area such that all responses are in the top category (over £140,000).
Interpretation	Are some of your questions ambiguous or confusing?
Difficulty	Some questions may never have occurred to the interviewee, or simply be too complex and difficult to answer.
Interviewee interest/boredom	Too many questions will send even the most interested interviewee to sleep.
Question flow/order	Did your pilot test interviewees think that the question order seemed logical?
Skip patterns	Do these seem to be logical with no questions inadvertently missed out?
Timing	How long does it take to administer the questionnaire? This will be a crucial factor as it will determine how long it will take for you to question all those in your sample.

Source: Adapted from Converse and Presser 1986.

---

missing from the sample framework. In general, poor people voted for Roosevelt. A common non-response error occurs if you try to sample your data between 9 am and 5 pm, as many people are at work and therefore unable to answer your questions.

We can calculate the proportion of responses to non-responses (R) as follows:

$$R = 1 - \frac{n - r}{n}$$

where n is the original sample size and r is the response rate. For example, if the original sample was 500, and 400 responses are obtained, the response rate is $1 - (500 - 400)/500 = 0.80$, and the non-response rate is 0.20, or 20%.

The proportion of responses to non-responses depends upon such factors as the nature of the population, the method of data generation, the kinds of questions being asked, and the skill of the interviewer, and if a postal questionnaire is used, whether return postage has been paid for, and the number of call-backs (Frankfort-Nachmias and Nachmias, 1996). To try to minimise non-response error we suggest that you design your survey carefully and sensitively; thoroughly pilot your survey; use a strategy of calling back to survey respondents who may have been out or busy when you first called; check the response rate for certain groups against what you might expect (e.g., if you have a response rate of only 20% for women, when an equal proportion of men and women were meant to be sampled, then there may be a problem that needs to be addressed/considered); and consider drawing a sample larger than needed to ensure that a certain sample size is achieved. The

final point should be used with care as non-response usually affects certain sections of the population more than others (e.g., older people, immigrants, disabled people), so increasing the sample size might not mean that they will necessarily be included. We will now turn our attention to taking a sample from a population.

3.4 Sampling, estimation and distribution

Once you have designed and piloted your questionnaire, the next step is to determine how you will sample your respondents. Why sample? The answer to this is straightforward and will be illustrated with a simple example. Suppose you are interested in the decision-making process underlying intra-urban residential mobility, and have decided to interview households to investigate the motivation behind their decision to move. You will normally have a variety of candidate towns and cities in the part of the world in which you live. Even if you could identify all those people who are currently moving house, constraints on time and resources would make it impractical for you to interview all candidate households in all cities. You will clearly have to select a subset of households, perhaps restricting your focus to one urban area, or even a part of one urban area. We term the total of all possible people who display the characteristic we are interested in, a **population**. The size/extent of a population is determined by the size of the group that we wish to make generalisations about. In the above case the population could be the total of all people in UK cities who are desiring to move within the urban area of their city, or it could be restricted to all the people in one city or part of a city. Once we have identified a population we are now in a position to select a subset of that population. This subset is termed a **sample**, and in the example above this would constitute households selected from a list of those in the process of moving or wishing to make such a move. Whilst nearly everyone can identify a population and select a sample, it should be noted that different methods of selection exist, some of which may be more valid tham others. As a rule, we want our sample to be **free of bias** and as **representative** of the larger population as possible.

There are many different **sampling methods** (also known as **sampling designs**) available for the collection of data, and these are described in Table 3.2, with examples given in the context of selecting a sample of households moving house within Belfast to investigate the decision-making process underlying intra-urban residential mobility. The choice of which sampling method to use depends on a variety of questions which we are usually faced with when designing a sampling strategy:

- Do you want the sample to be representative of a larger population?
- Do you want to employ probability-based statistical methods?
- Does a convenient sampling frame exist?
- How big do you want your sample to be?
- Are you able to afford the time and money to carry out the sample collection?

We shall consider each of these questions in turn, and provide some answers which will guide you towards the setting up and execution of a sampling strategy in your own research project.

3.4.1 Do you want the sample to be representative of a larger population?

We often sample for the purpose of generalising about the larger population from which the sample was drawn. The need for some sort of **representative sample** is most clear in the polling process prior to elections, where polling organisations such as *Gallup* and *Harris* want to obtain a sample about voting intentions which is as representative as possible of the larger voting population. As an example of an unrepresentative sample, consider the following case (taken from Chatfield, 1995). Several years ago,

Puerto Rico was hit by a hurricane and suffered considerable damage, resulting in 10,000 claims for hurricane damage. The US government initially decided to base its total grant aid on the first 100 claims received, and then multiply this by 100. Clearly, the first 100 applications need not necessarily constitute a representative sample (i.e., small claims requiring less work would be likely to come in first). A more representative sampling scheme was needed, and was eventually applied. There are a variety of methods by which we can produce a representative sample. These include non-probability-based methods such as **quota sampling** and probability-based methods based on **random sampling** (see Section 4.6 for details on probability concepts).

The **judgemental sample** (also known as a purposive sample) is the most subjective sampling method. Here, sample elements are selected based on judgement derived from prior experience. In an interview situation, individuals may be selected on the basis of the sort of response that they are likely to give, and the responses the interviewer is looking for. The **quota sample** selects those elements to be included in the sample on the basis of satisfying a predefined quota. For example, an interviewer might have a quota to fill of 30 men in the age-group 20–25, who are in an income bracket of high income earners, or as in the example in Table 3.2 homeowners aged 20–40 years. The interviewer has the choice of selecting certain people who might be included in the quota, and therefore introduces bias into the sample. Quota sampling has advantages in that it is easier, generally costs less to perform and does not require a sampling frame, but the key point to emphasise is that it should be used only in those limited situations where experience has shown it to work with minimal interviewer/ sample bias (Ehrenberg, 1978). This is usually only when we know enough about the population being sampled to say that any sample bias introduced by the method or interviewer is in fact minimal. Another non-random sampling method which is used with the selection of people, is the **snowball sample**. This is based on a number of initial contacts who are asked for the names and addresses of any other people who might fulfil the sampling requirements.

3.4.2 Do you want to employ probability-based statistical methods?

The sampling designs considered so far are only really appropriate if you wish to use very simple non-probability-based descriptive techniques such

Table 3.2 Sampling methods

	Representative	Probability	Random	Description	Example
Judgemental	✗	✗	✗	Sampling elements are selected based on the interviewer's experience that they are likely to produce the required results.	Several houses for sale in Belfast, perhaps with families known to the interviewer, are chosen subjectively.
Quota	✔	✗	✗	Sampling elements are selected subject to a predefined quota control.	The quota is the first 30 homeowners selling their houses in Belfast who are also making an intra-urban move, and are aged between 20–40 years.
Systematic	✔	✔	✗ (N.B. first unit selected at random)	Sampling elements in the sampling frame are numbered. First sampling unit is selected using random number tables. All other units are selected systematically k units away from previous unit.	Sampling frame of 600 homeowners selling their houses in Belfast. These houses for sale are ordered and numbered. A random number is selected for a start point, from which every tenth property is selected for inclusion in the sample.
Simple random	✔	✔	✔	Sample size of n elements selected from a sampling frame without replacement, such that every possible member of the population has an equal chance of being selected.	All 600 houses for sale in the sampling frame are numbered 1–600. A sample of 30 units is selected using a random number table, excluding those numbers outside the range 1–600.
Stratified random	✔	✔	✔	Sampling frame divided into sub-groups (strata) which are then each sampled using the simple random method.	All 600 houses for sale come from lists provided by six estate agents. These are each randomly sampled for houses to include in the sample.
Multi-stage random	✔	✔	✔	Sampling frame divided into hierarchical levels (stages). Each level is sampled using a simple random method which selects the elements to be included at the next level.	All 600 houses for sale are distributed to enumeration districts within several wards. A random sample of these wards is selected and of these random samples of both enumeration districts and finally houses for sale are selected.
Clustered random	✔	✔	✔	Sampling frame divided into hierarchical levels (stages). Levels are selected using random sampling similar to the multi-stage random method. However, all elements are selected at the final level.	Similar to the above method, except that all the houses for sale in a given enumeration district are selected.

Table 3.3   Random numbers.

71118	41798	34541	76432	40522	51521	74382	06305	11956	30611
53253	23100	03743	48999	37736	92186	19108	69017	21661	17175
12206	24205	37372	46438	67981	53226	24943	68659	91924	69555

Source: Silk 1979, taken from Neave 1976.

as raw counts and proportions. If you want to use a probability-based statistical method such as those considered in Chapter 5, then designs where we can estimate the probability of each unit being included in the sample are required. The simplest is the **systematic sample**. Such a sample involves the systematic selection of cases from a sampling frame. Although the initial start point in the sampling frame is random, the subsequent selection of points is usually selected with a regular interval, and can be subject to selection bias. This aside, since we can determine the probability of each unit being included in the sample, the method is probability-based. However, usually those methods more firmly based on a **random sample** are more useful, as they avoid any interviewer/sample bias. At the simplest level, the **simple random sample** occurs when each unit of the population is given the *same probability* of independent selection. There are a variety of other designs based on the simple random sample, which include the **stratified random sample,** the **multistage random sample** and the **clustered random sample**, which are described in Table 3.2. All these methods possess several key elements in common (Ehrenberg, 1978):

1   The chances of obtaining an unrepresentative sample are *small*.
2   This chance *decreases* as the size of the sample increases.
3   This chance can be *calculated*.

Other sampling designs can be obtained from combinations of those described above, such as combining quota sampling with a random route, or modifying the systematic sample by implementing a randomly varying interval. All the random sampling methods are based on the use of random numbers. These can be read off from a convenient table of random numbers, or generated using the random number generator of a spreadsheet package such as *Microsoft Excel*, or a statistical package such as MINITAB. Table 3.3 displays an extract from a random number table – in this case the last three lines from the table which appears in the appendices of Silk (1979).

There are various methods for obtaining numbers from such a table, but the simplest is to choose a start point somewhere on the table, and then to move along either the rows or columns, selecting groups of numbers which fall within the range of numbers in which we are interested. For example, if we wanted a sample of ten numbers coded from 1 to 450, and we had selected the middle '1' of the 71118 as our start point, we could then extract the following three-digit numbers moving along the first row:

118	417	983	454	176	432
✔	✔	✗	✗	✔	✔
405	225	152	174	382	063
✔	✔	✔	✔	✔	✔

We omit the 983 and 454 since they exceed our range (which has an upper limit of 450), and stop once we have obtained ten numbers. If we were to use instead the example given for the simple random sample in Table 3.2, since our codes range from 1 to 600, we would have included all the numbers listed above (except 983), with the addition of a further 18 numbers to bring the sample up to the desired size of 30. For further details on the sampling methods discussed above, you should consult a text such as Cochran (1977) or Dixon and Leach (1978) for a more geographical perspective.

If we take several random samples of the same size from the same population, we would not expect the statistics of each sample, such as the mean, to be the same. This is most clearly demonstrated in Table 3.4 and Figure 3.4 which displays data taken from Ehrenberg (1978). Although not geographic, this example is most helpful in the understanding of some of the characteristics of sampling distributions.

The data in Table 3.4 detail the consumption of Corn Flakes packets in a period of six months by a set of 491 households, where the known mean of this population was 3.4 packets. This population is sampled repeatedly using sample sizes of 1, 2, 10 and 40 to produce an empirical distribution of sample means which are displayed in histogram form in Figure 3.4. The distribution of means at each sample size is termed a **sampling distribution**. We can make four useful observations from this table (Ehrenberg, 1978):

Table 3.4 Sampling distributions (%) of Corn Flakes consumption.

Sample size	Value of sample mean												Mean	s
	0.0–0.99	1.0–1.99	2.0–2.99	3.0–3.99	4.0–4.99	5.0–5.99	6.0–6.99	7.0–7.99	8.0–8.99	9.0–9.99	10.0–10.99	11.0+		
1	40	17	11	4	4	5	3	1	4	0	2	9	3.4	4.7
2	29	21	12	8	5	4	5	4	1	1	2	8	3.4	3.9
10	5	5	30	15	20	15	10	0	0	0	0	0	3.4	1.5

Sample size	Value of sample mean									Mean	s
	<1.8	1.8–2.19	2.2–2.59	2.6–2.99	3.0–3.39	3.4–3.79	3.8–4.19	4.2–4.59	4.6+		
40	0	10	10	15	25	10	20	10	0	3.4	0.8

Source: Ehrenberg 1975: 300–301.

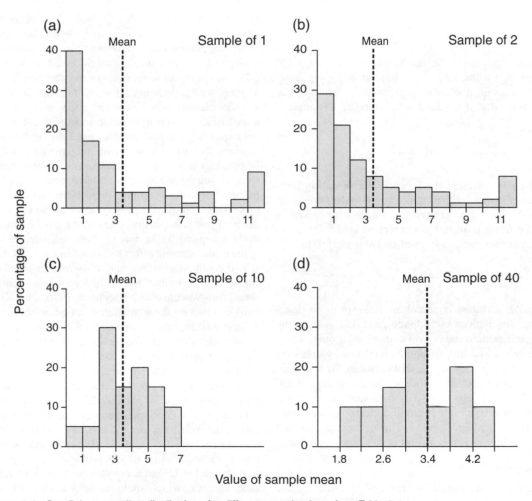

Figure 3.4 Cornflakes sampling distributions for different sample sizes, from Table 3.4.

1   *Large* samples are more accurate in estimating the population mean than *small* samples.
2   If we average the individual sample means for each sample size we obtain a mean of 3.4 packets, which coincides with the population mean.
3   The standard deviation *s* of the individual sample means for each sample size *decreases* with the size of the sample.
4   The form of the sampling distribution is skewed/non-normal for small samples but approximates a normal distribution with larger sample sizes.

In general, the form of the sampling distribution follows the **Student's *t*-distribution** for small samples, but for samples greater than 30 follows a **normal distribution** (see Section 4.6). The standard deviation of the sampling distribution is termed the **standard error of the mean** (*SE*) and is found by:

$$SE = \frac{\sigma}{\sqrt{n}}$$

where sigma (σ) is the standard deviation of the population and *n* the size of the sample. Since we rarely know enough about our population to define sigma, we make use of *s*, which is the standard deviation of the sample, to give:

$$SE = \frac{s}{\sqrt{n}}$$

If the population is small, where the **sampling fraction** – the ratio of the sample size *n* to the population size *N* – is 0.2 or greater, a correction factor known as the **finite population correction** is usually incorporated into the above equation (Malec, 1993):

$$SE = \frac{s}{\sqrt{n}}\sqrt{1 - \frac{n}{N}}$$

This *SE* estimate is useful as it allows us to determine, for a given **confidence level**, the distance by which a sample mean is within a stated range of its true value, i.e., how *accurate* an estimate such as the sample mean is of the population mean. We will consider the subject of confidence levels in more detail in Chapter 5. For now, it is sufficient to say that a confidence level represents the chance of our sampled estimate being correct. This distance *c* is termed the **confidence interval**, and is calculated as follows:

$$c = \pm z \times SE$$

where *z* is the standard error unit measure for the desired confidence level (Dixon and Leach, 1978).

For example, if we adopt a 95% confidence level (i.e., we allow a 95% chance of being correct that the true population mean is within this range), then for a normal distribution we know that 95% of the values lie within $\pm 1.96 SE$ of the mean. The above equation therefore becomes:

$$c = \pm 1.96 \times SE$$

So, the advantage of using a simple random sample in experimental design is that sampling distributions for a variety of summary statistics such as the mean, variance and correlation coefficients can be estimated from the sample data, and corresponding confidence intervals calculated around such statistics.

3.4.3   Does a convenient sampling frame exist?

Once you have identified a method of sampling, the next step is to consider whether or not a convenient **sampling frame** exists. An example of a sampling frame is a list of names, as might be found in a register of voters, an organisation's membership list, or a membership directory. If such a frame does not exist, or cannot easily be created, it may be costly and impractical or even impossible to implement a random sampling method. In this case, it might be more appropriate to use a quota or judgemental method which does not require access to a sampling frame. Even if you have access to an appropriate sampling frame, there are a variety of problems which may exist. The sampling frame may be incomplete, particularly if compiled several years in the past. Elements in the sampling frame may be clustered together and require considerable effort to extract individuals. Finally, the sampling frame may also contain elements which are ineligible for inclusion in the sample (Frankfort-Nachmias and Nachmias, 1996, after Kish, 1965). However, these problems can be overcome or at least ameliorated. With respect to the omission of eligible elements, and the inclusion of ineligible elements, supplementary lists can be used to make the sampling frame more robust and up to date.

In some cases it may be necessary to construct a sampling frame. This is particularly so for questionnaire surveys in less developed parts of the world where there may well be a lack of accurate and/or official lists. Dixon and Leach (1978, 1984) provide useful introductions to some of the problems and suitable techniques for researchers attempting to undertake such surveys, which includes advice on sampling frame construction that is appropriate in all survey research contexts. It is important that care is taken in this part of the sampling design. Dixon and Leach (1984: 25) note that:

The nature of the sampling frame, how it was compiled . . . its discovery or possible defects are essential components of the research report; no frame is perfect, and the validity of the findings will be strengthened rather than undermined by such an appraisal.

3.4.4 How big do you want your sample to be?

With regard to sample size, it is often thought that 'more is better'. We have seen from the example displayed above in Figure 3.4 that larger samples will more accurately estimate the population mean than small samples. Indeed, as a general rule of thumb we can say that the larger the sample, the more confident we can be that the statistics derived from it will be similar to the population parameters. However, a large sample with a poor sampling design will probably contain less information than a smaller but more carefully designed sample (Chatfield, 1995). We have seen from the equations above that the standard error of the mean will clearly decrease as the sample size n increases. With measured variables, a sample size of between 20 and 30 is usually recommended as a minimum to conform to a normal distribution. However, sample size also depends on the variability of the population you are sampling. If you already know something about this variability, you can estimate the size of sample needed to estimate population values with a certain degree of confidence. Rearranging the equation used above to calculate the standard error of the mean gives us:

$$n = \frac{s^2}{SE^2}$$

which can be adjusted if n is too large relative to N by incorporating the finite population correction:

$$n' = \frac{n}{1 + (n/N)}$$

The important point from looking at calculations of n' using the SE in this fashion, is that it takes a *considerable increase* in sample size to make a significant *decrease* in the SE. It is salutary to look at an example. If we take $s = 10$ and $SE = 2$, then substituting in the last-but-one equation above gives us:

$$n = \frac{10^2}{2^2} = \frac{100}{4} = 25$$

Table 3.5 Sample sizes needed to estimate population values with given levels of confidence assuming a variability of 50%, and a very large population.

Confidence limit (±% of mean)	Confidence level	
	99%	95%
1	16587	9604
2	4147	2401
3	1843	1067
4	1037	600
5	663	384
6	461	267
7	339	196
8	259	150
9	205	119
10	166	96
15	74	43
20	41	24

Notes:
1. Table assumes a variability of 50%, which for a continuous variable is where the standard deviation is 50% of the mean.
2. Confidence limit is measured as a percentage. For a continuous variable, it is given as the percentage of the mean.
Source: Adapted from Dixon and Leach 1978: 10.

If we wanted to decrease the SE by 50%, then the required sample size would be:

$$n = \frac{10^2}{1^2} = \frac{100}{1} = 100$$

In other words, we need to *quadruple* the size of n. If we incorporate desired confidence limits at a selected confidence level, we can make a more precise estimate of n. Table 3.5 is taken from Dixon and Leach (1978) and gives the sample size needed to estimate the population values for continuous variables to within a certain percentage, with a probability of being right, when the sample standard deviation is 50% of the mean.

3.4.5 Are you able to afford the time and money to carry out the sample collection?

It is worth stressing that the most elegant sample design may be impractical due to logistical considerations. This is particularly so for research projects undertaken by students. In the context of the example given above, the collection of a simple random sample of 600 households in Belfast may be impractical with the time and resources at your disposal. All too often logistical considerations are cited in student research as being the main constraint on sampling and the collection of a data set, when very often the data collection task which has been set by the student (or indeed the supervisor) is over-ambitious. In these situations direct experience of the costs in both time and money for sampling the data is hard to beat. You may have neither of these if you are undertaking an analysis for the first time, and in this situation the only answer is to seek advice, either from the published literature, or from your supervisor or other staff/faculty in your department.

3.5 Obtaining and using secondary data

The quantitative data which you might collect in a questionnaire survey is a form of **primary data collection**. However, there are a number of secondary sources of data which can be used for what is termed **secondary data analysis**. These can consist of survey data collected by governments in the form of national censuses which provide data at a variety of different spatial scales, or other archives of socio-economic data which may be available to the researcher. Some of these are produced by government agencies, others by independent surveys. We will consider some of the sources of secondary data of interest to the human geographer, concentrating on digital data available from archives, and census data. However, as a useful preliminary step, it is useful to consider why you might want to use such data in the first place, and some of the pitfalls of so doing.

There are three main **justifications** for the use of secondary data: *conceptual*, *methodological* and *economic*. These points are largely taken from Frankfort-Nachmias and Nachmias (1996) and Kiecolt and Nathan (1985). From the conceptual perspective, the data which you are wanting to use may simply not be available in any other form. For example, historical geographers rely almost exclusively on the secondary data available in record offices and other archives to examine historical trends. Methodologically, secondary data can enable the replication of analysis, allowing different researchers to corroborate analytical findings, allowing the possibility of longitudinal and trend analysis, broadening both the scope and dynamism of variables, and the size of the data sets used in a piece of research. The economic justification is perhaps the easiest to appreciate: collecting raw data is a time-consuming and costly process, particularly for the undergraduate or postgraduate student! The availability of suitable secondary data represents a considerable saving on the expenses which would otherwise be incurred to do your research. In some cases, the data collection task would be impossible for one researcher alone: data surveyed at the national scale, for example. The costs that you are likely to face with secondary data are likely to be restricted to data purchase, preparation and costs incurred in undertaking the analysis. However, even these costs, particularly for data purchase, can be prohibitive for student research.

Although there might be considerable savings through, in effect, getting someone else to collect your data, there are many drawbacks. All secondary data are cultural products produced by organisations and individuals with views of the world which may be incompatible with the view adopted in your research (Clark, 1997). Other limitations are summarised in Box 3.7.

Box 3.7   Limitations of secondary data

Dated	The data were collected some years/decades previously (e.g., census data, which are collected in many countries only every ten years).
Too specific	The data were unlikely to have been collected with the goals of your own research project in mind (e.g., census data) and may have been subject to changes in the collection procedures if collected over time.
Inaccessible	The data may be available in a digital form, perhaps over the Internet, but may only be held in a restricted access government archive in a remote geographic location, and any data, if supplied at all, may take time and cost money to be accessed.
Inaccurate	The data originally collected may be inaccurate, sampled using a non-random method, and of generally poor quality.
Poorly documented	Existing data sets may have little or poor documentation to describe the nature and quality of the data or any pre-processing which may have taken place.

Of course, the degree to which these problems are important in your research will be determined by the nature of your research problem. The availability of census data collected in 1981 and 1991 may be a problem if you wish to analyse *contemporary* geodemographic characteristics, though the dated nature of these data is an essential component of an historical perspective on geodemography.

Of all the problems listed above, the most critical concern those of **data quality**, and adequate data documentation, which is increasingly being referred to by the term **metadata**, particularly in relation to digital databases and data archives. Some statement or feel for the quality of the data in secondary sources is essential, since errors which might have been made in the original data collection exercise are effectively hidden. Errors due to inappropriate sampling methods, or mistakes in the administration of the sample, may be magnified by the use that you might intend for the data. Indeed, data sets generated from nationally representative samples, with carefully designed questionnaires and well-designed coding and pre-processing, do not always exist (Kiecolt and Nathan, 1985). It is therefore wise to be cautious whenever you are using published, secondary data. Indeed, Jacob (1984: 45) recommends that:

Perhaps the most important attribute for the user of published data is a large dose of scepticism. Whether data are found in libraries or data archives, they should not be viewed simply as providing grand opportunities for cheap analyses: they should be seen as problematic. In every case the analyst should ask: Are these data valid? In what ways might they have been contaminated so that they are unreliable?

Metadata is usually defined as 'data about data':

'All the things that surround the actual content of the data to give a person an understanding of how it was created and how it is maintained'.

(Rubenstrank 1996: 1)

This includes not only the description of the data, including quality, but also information about the location of the data. The term has come to subsume what we might previously have referred to as 'data documentation'. The increasing quantity of information in digital databases has led to efforts to develop standards of metadata description. This is particularly so for digital geographic data, with the establishment of a metadata standard for **geospatial** data by the Federal Geographic Data Committee of the United States in 1994. This requires all federal agencies in the US to document all new and existing geospatial data sets, and make such documentation available to a National Geospatial Data Clearinghouse (NGDC) (FGDC, 1994). In this form, metadata contains information relating to data availability, fitness for use, data access and data transfer (FGDC, 1994). Efforts to improve the quality of metadata are also being made by those involved in developing a National Geospatial Data Framework for the UK. The practical implication for your research is that the increasing standards of data documentation will make it considerably easier for you to obtain secondary data with higher levels of supporting documentation for your research.

3.5.1 Sources of data

In this section we will identify some general sources of data for geographic research. We would like to stress that it is impossible to give an exhaustive overview of all the various types of secondary data which you can obtain for your research. Some of the data sets available in digital form that have been derived from official government statistics and other independent surveys are listed in the section on data archives below. General sources of data on various aspects of human geography in a UK context are listed in Box 3.8.

Box 3.8 Secondary data sources in the UK

Census data	Decennial Census from the Office of National Statistics (ONS)
Agricultural data	Annual Census from the UK Agriculture Department
Manufacturing data	Annual Census of Production and Business Monitors from ONS
Employment data	Labour Force Survey, and New Earnings Survey by the Department of Employment
Public attitude data	British Social Attitudes Survey
Tourism data	British Tourism Authority's Digest of Tourism Statistics
Personal expenditure dada	Family Expenditure Survey and General Household Survey from ONS
Crime data	British Crime Survey from the Home Office

Source: Adapted from Clark 1997.

In addition to those data listed in Box 3.8, it may be possible to obtain various types of health data at a county/state or local level from a city or local authority. Many regional, national and international agencies collect epidemiological data which might be used in the context of your research project. Details of other official and unofficial data sets can also be obtained by searching publications which list available data sets. Clark (1997) provides a useful starting point for such material, and suggests reference to various publications which include *Guide to Official Statistics* (Office for National Statistics, 1996), *Sources of Unofficial UK Statistics* (Mort, 1990), *UK Statistics: A Guide for Business Users* (Mort, 1992), *European Statistics: Official Sources* (EUROSTAT, 1993) and *ECO Directory of Environmental Databases in the UK 1995/6* (Barlow and Button, 1995). With the advent of the Internet and the World Wide Web (WWW) it is now possible to use one of the standard Web search engines to look online for useful data for your research.

3.5.2   Data archives

There are an increasing number of national social science data archives available to the human geography researcher which can supply all types of quantitative and non-quantitative data. We will restrict our concern for the moment to quantitative data, and leave the consideration of non-quantitative archival data (e.g., letters, diaries and autobiographies) to Chapter 7. Many of these archives either offer data sets which are restricted to certain categories such as health or crime, or encompass a wide variety of different data sets amenable to quantitative analysis. Archives will generally supply data files either free or by purchase – but often in a digital form by CD-ROM, or increasingly over the Internet via the WWW. Some of the larger archives will offer assistance in the search for appropriate data sets, and some may well be able to help with customised data extraction (Kiecolt and Nathan, 1985). There are a variety of **national archives** which can supply data, and a useful survey of these has been provided by Kiecolt and Nathan (1985). A current listing, a WWW-clickable map access to a variety of European and North American archives (Figure 3.5) and an Integrated Data Catalogue (IDC) covering ten international archives are provided on the WWW site of CESSDA (Council of European Social Science Data Archives) at http://www.nsd.uib.no/cessda.

We will use two examples, one from the UK and one from the United States, to illustrate some of the

data products which can be obtained for quantitative analysis in the context of your research project. Many of the larger data archives, including the examples below, can be found on the WWW. We include the WWW addresses below which contain further information and links to other national and international archives.

The Data Archive at the University of Essex (http://dawww.essex.ac.uk/)

The Data Archive at the University of Essex (formerly known as the ESRC Data Archive) was set up in 1967 to act as a repository of quantitative information for use by social scientists, and consists mainly of data collected in the UK. There are a wide number of different categories of data available in this archive, with well over 7,000 data sets in total. Some of the largest data sets are listed in Box 3.9. The data holdings of The Data Archive can be searched using a WWW interface to BIRON: 'Bibliographic Information Retrieval Online', or as part of the CESSDA's IDC. The Data Archive also provides access to quantitative data sets held further afield within Europe: the Data Archive is a member of CESSDA and the International Federation of Data Organisations (IFDO). Of particular note is the service the archive provides to postgraduate students (see Figure 3.6). Data sets are disseminated to authorised users in a variety of formats and media including CD-ROM, network file transfer, tape (8 mm and DAT), and floppy disk. In the higher education community in the UK, data are charged at the cost of the media.

Some of the larger data sets can be accessed at the University of Manchester's Information Datasets and Associated Services (MIDAS) service, which is a JISC-funded national data centre. This service provides online access for authorised users in the UK higher education community across the Internet using Telnet and Web-based interfaces (http://midas.ac.uk/) to data for teaching and research purposes. The current data holdings of MIDAS (as of 1998) are listed in Box 3.10 (on p. 65).

ICPSR data archive (http://www.icpsr.umich.edu)

The Inter-University Consortium for Political and Social Research (ICPSR) is located at the University of Michigan, in the United States. Formed in 1962, this archive is the largest repository of social science data accessible by computer. A selection of some of

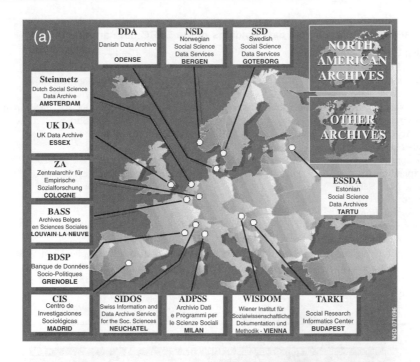

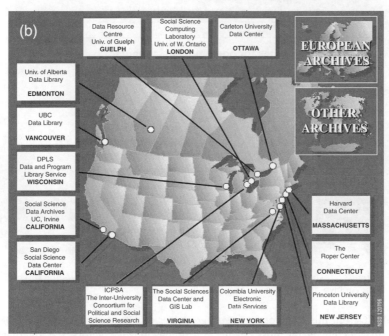

Figure 3.5 CESSDA clickable maps for data archives in (a) Europe and (b) North America (source: CESSDA 1998).

Box 3.9   A selection of the larger data holdings of the ESRC data archive

- Agricultural Census (Great Britain)
- British Crime Survey
- British Election Studies
- British Household Panel Study
- British and Northern Ireland Social Attitudes Surveys
- Census of Great Britain
- Company Accounts Analysis Data
- Continuous Household Survey
- European Community Studies and Eurobarometers
- Family Expenditure Survey
- Family Resources Survey
- Farm Business Survey
- Financial Expectations
- Gallup Political Polls
- General Household Survey
- Health Survey for England
- International Passenger Survey
- Juvos Unemployment Statistics
- Labour Force Survey
- Mori Political Polls

- Mortgage Lenders Survey
- NOP National Political Surveys
- National Child Development Study
- National Food Survey
- National Health Service Patient Re-registrations
- National Readership Survey
- National Travel Survey
- New Earnings Survey
- ONS Databank
- OPCS Omnibus Survey
- Postcode Address File
- Postzon File
- Road Accident Data
- Scottish Young People's Survey
- Survey of Personal Incomes
- Vital Statistics for England and Wales
- Workplace Industrial Relations Survey
- Youth Cohort Study of England and Wales

Source: The Data Archive 1998a.

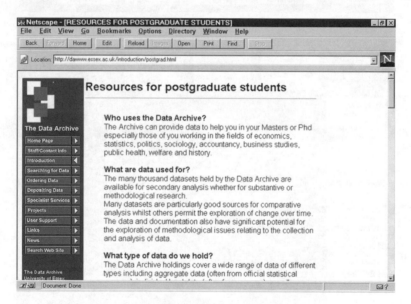

Figure 3.6   The Data Archive postgraduate resources Web page (source: The Data Archive 1998b).

the data sets available from this archive are listed in Box 3.11. Although your university must belong to the consortium to obtain the full range of services and data sets, it is possible to obtain access to data sets for an appropriate fee. In the UK, data from the ICPSR can be obtained via The Data Archive covered above. Similar to the UK's Data Archive, the main routes of data supply are by CD-ROM, by floppy disk or downloaded across the Internet by FTP or the WWW. Of particular use for research students is the ICPSR Summer Program in Quantitative Methods of Social Research. This offers a wide set of courses in research design, statistics and data analysis of relevance to the social scientist.

Box 3.10 Data sets available at MIDAS

- **Census and related data sets:**
  Census Information Gateway
  1991 Local Base and Small Area Statistics
  Special Workplace and Migration Statistics
  1991 Samples of Anonymised Records
  1981 Small Area Statistics
  1991 Census Digitised Boundary Data
  1981 Digitised Boundary Data
  Table 100
  Census Monitor county/district tables
  Topic Statistics
  Population Surface Models
  Estimating with Confidence data
  ONS ward and district level classifications
  GB Profiler
  The Longitudinal Study
  1971/81 Change File
  Postcode to ED/OA Directories
  Central Postcode Directory POSTZON File
  ONS Vital Statistics for Wards

- **Government and other continuous surveys:**
  General Household Survey
  Labour Force Survey
  Quarterly Labour Force Survey
  Family Expenditure Survey
  Family Resources Survey
  Farm Business Survey

National Child Development Study
1970 British Cohort Study
British Household Panel Study
Health Survey for England
Macro-Economic Time Series Databanks
ONS Time Series Databank
OECD Main Economic Indicators
UNIDO Industrial Statistics 3 digit level of ISIC code
UNIDO Industrial Statistics 4 digit level of ISIC code
UNIDO Commodity Balance Statistics Database
IMF International Financial Statistics
IMF Direction of Trade Statistics
IMF Balance of Payments Statistics
IMF Government Finance Statistics Yearbook

- **Spatial data sets:**
  Landsat Satellite Data
  SPOT Satellite Data
  Bartholomew Digital Map Data
  1991 Census Digitised Boundary Data
  1981 Census Digitised Boundary Data
  OS Digital Map Data (Sample data sets)
  The JANUS Visualisation Gateway Project

- **Miscellaneous:**
  Several scientific data sets and electronic journals

Source: MIDAS 1998.

Box 3.11 Main data holdings of the ICPSR

- American national election studies
- Roll call voting records
- Election returns
- National crime survey
- Panel study of income dynamics
- The general social survey
- Census data
- Education data (IAED)
- Ageing data (NACDA)
- Criminal Justice data (NACJD)
- Substance Abuse and Mental Health (SAMHDA)

Source: Adapted from Kiecolt and Nathan 1985 and ICPSR 1998.

3.5.3 Population census data

The source of secondary data you are most likely to encounter is that based on the **census of population** which is carried out regularly in most industrialised countries. For example, in the United States, decennial (every 10 years) censuses of population have taken place since 1790, with the last in 1990 and the next planned for 2000. In the UK, similar decennial censuses have taken place since 1841, with the last in 1991 and the next planned for 2001. In Canada, a national decennial census was in place by 1851, which by 1971 had become quinquennial (every five years), with the next planned for 2001.

Confidentiality requirements usually result in the individual census returns being unavailable to the public. For example, in the UK the individual census return information is confidential for 100 years. Census information is usually available in the form of statistics aggregated to various geographic areas which vary in size from roughly 100 people to several thousands and larger. Neither is all the information collected by the questionnaires included in these aggregated statistics. For both the US and UK census, although many variables are sampled at 100%, some variables reflect only a certain percentage (usually 10%) of houses sampled; these are listed in Box 3.12.

Box 3.12   UK 1991 census variables

100%

Sex and date of birth
Marital status
Whereabouts on census night
Usual address
Term-time address of students
Usual address one year ago
Country of birth
Ethnic group (England, Wales and Scotland)
Long-term illness
Welsh/Gaelic/Irish language
Economic activity previous week
Type of accommodation/sharing
Number of rooms
Tenure of household
Household amenities
Availability of cars/vans
Lowest floor level (Scotland)
Religious adherence (Northern Ireland)

Sample (10%*)

Relationship in household
Hours worked
Occupation
Name and business of employer
Workplace
Journey to work
Higher qualifications

* All variables in Northern Ireland were sampled at 100%.

Source: Martin 1993.

Box 3.13   1991 census geography for the UK

England and Wales	**Scotland**	**Northern Ireland***
Basic units for census data:		
Counties	Regions	Regions (SAS)
Districts	Districts	Districts (SAS)
Wards	Postcode Sectors	Wards (SAS)
Enumeration Districts (SAS)	Output Areas (SAS)	Enumeration Districts (SAS)
Other areas for which data are available:		
Regional/District Health Authorities	Wards (SAS)	Postcode Sectors (SAS)
SAR areas	Regional Electoral Divisions (SAS)	Parliamentary Constituencies (SAS)
Standard Regions of England	Localities (SAS)	European Constituencies (SAS)
Postcode sectors (SAS)	Inhabited Islands (SAS)	Education Board Areas (SAS)
1991 SAS in 1981 wards	New Towns (SAS)	Belfast Urban Area (SAS)
Parliamentary constituencies	Health Board Areas	Health Board Areas (SAS)
European constituencies		
English Civil Parishes (SAS)		
Welsh Communities (SAS)		

* Note: SAS are also available for Northern Ireland in 100 m grid squares (see http://www.nics.gov.uk/nisra/ for details).

Source: Modified from MIDAS 1998 and Martin 1993.

The basic census geography for the UK is given at the top of Box 3.13 and is depicted in Figure 3.7. The *Enumeration District (ED)/Output Area (OA)* is at the lowest level of aggregation, containing in the region of 400 people, although census data can be obtained aggregated to a variety of other (usually larger) zones, including postcode sectors. In the UK, two main sets of census statistics are available: **Local Base Statistics**

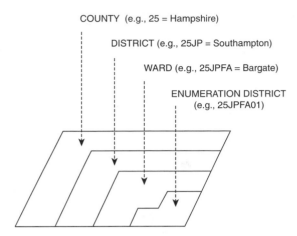

COUNTY (e.g., 25 = Hampshire)

DISTRICT (e.g., 25JP = Southampton)

WARD (e.g., 25JPFA = Bargate)

ENUMERATION DISTRICT
(e.g., 25JPFA01)

Figure 3.7 UK census geography (source: redrawn from Martin 1996: 74).

(LBS) which for the 1991 census consisted of 99 tables of information, and **Small Area Statistics** (SAS) which consist of some 86 tables of information. The LBS are unavailable for the lowest levels of aggregation. Both the LBS and SAS data sets are available to authorised users over the Internet from the Census Dissemination Unit at MIDAS, or via CD-ROM from the Data Archive. For access to these data over the Internet, users must make use of either the SASPAC91 software package to extract and output the required datasets as they relate to the census geography outlined above, or the SURPOP (V.2) software package to output selected SAS variables at a 200 m grid resolution.

In the UK, although census SAS are available for small areal units such as unit postcodes and EDs, users of these data should be aware of the 'suppression procedures' which operate if an ED contains fewer than 50 people, to avoid the possibility of disclosing individual information. The counts from such zones are frequently merged with those from adjacent zones, and a zero count will be coded to the zone in question. Similar procedures are followed for wards with the LBS, which are suppressed if they contain fewer than 1,000 people or 320 households. Various other data sets are also available, including census microdata in the form of the *Sample of Anonymised Records* (SAR), *Special Migration Statistics* (SMS), *Special Workplace Statistics* (SWS) and *Longitudinal Study* (LS) data. Further details of the 1991 UK census can be found in the volumes edited by Dale and Marsh (1993) and Openshaw (1995 – particularly the chapter by Rees) and the monograph by Martin

(1993) which also contains an overview of the census geography in the UK. Details of the 1981 census can be found in Rhind (1983) and Dewdney (1985).

In the US, the 1990 census of population provides information on a set of sample and 100% categories similar to the UK census, which are listed in Box 3.14.

The census geography for the US is displayed in Box 3.15 and Figure 3.8, with the lowest level of aggregation being the census *block* (US Census Bureau, 1996).

In metropolitan areas, blocks contain information on approximately 100 people; block groups on about 800 people; and tracts on about 4,000 people. There are about six million block records, about 300,000 block group records, and about 60,000 tract records. The 1990 US Census data are primarily released as Summary Tape Files (STF) in the form of printed reports, microfiche and digital data files that may be obtained online from the Internet, on tape, on CD-ROMs, or in diskette form. US census data can be supplied aggregated by ZIP code (post code) areas of various sizes, although often by third-party commercial data suppliers.

For both the UK and US census there are an increasing number of locations providing both Internet and Web-based access to census products. For the US census, of particular note is the archive of several past censuses held at the Lawrence Berkeley National Laboratory (http://parep2.lbl.gov/mdocs/LBL_census.html). If you are interested in using US census data in your research you should look at the Web home page for the census, which contains much useful information (http://www.census.gov/). The Census Dissemination Unit at MIDAS in Manchester is probably the most useful means of obtaining census data over the Internet/WWW in the UK, and contains many online help documents as well as census data in a variety of formats (see Box 3.10 above).

3.6 Summary

After reading this chapter you should:

- understand data types and measurement scales;
- understand how to design a survey and questionnaire;
- understand the different sampling frameworks and their merits and limitations;
- be able to generate quantitative data using surveys and questionnaires;
- be able to obtain secondary quantitative data and assess its merits and limitations.

Box 3.14   US 1990 census variables

100%	Sample (15.9%)
Household relationship	Social characteristics
Sex	Veteran status
Race	Education – enrolment and attainment
Age	Place of birth, citizenship and year of entry to US
Marital status	Ancestry
Hispanic origin	Language spoken at home
Number of units in structure	Migration (residence since 1985)
Number of rooms in unit	Disability
Tenure (owned or rented)	Fertility
Value of home or monthly rent	Income in 1989
Congregate housing (room and board)	Labour force
Vacancy characteristics	Occupation, industry and class of worker
	Place or work and journey to work
	Work experience in 1989
	Year last worked
	Year moved into residence
	Number of bedrooms
	Plumbing and kitchen facilities
	Telephone in unit
	Vehicles available
	Heating fuel
	Source of waste and method of sewage disposal
	Year structure built
	Condominium status
	Farm residence
	Shelter costs including utilities

Source: Calkins 1990.

Box 3.15   1990 census geography for the US*

- **Statistical areas**
  Block
  Block group
  Census tract/block numbering area
  Urbanised areas
  Census designated places
  Census county division
  Metropolitan areas
  Division
  Region

- **Government areas**
  States
  Counties
  Congressional districts
  Voting districts
  American Indian reservations

* Note: many of these zones are used *only* for data obtained in the decennial population census.

Source: US Census Bureau 1996.

You should now have a basic understanding of quantitative data and their generation. As we have illustrated, there are a number of issues that you need to consider when designing and implementing a study which aims to generate primary quantitative data. You should consider each of these issues carefully to ensure a valid piece of research. Of particular note is how you construct your questions and the sampling frame

(a)

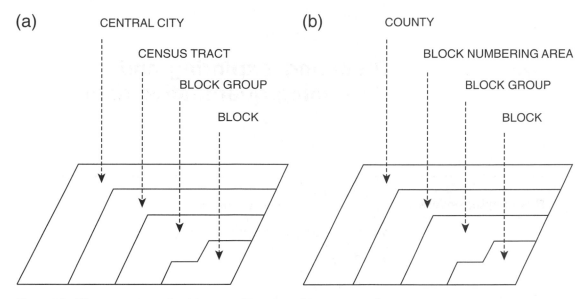

Figure 3.8 US census geography: (a) metropolitan areas; (b) non-metropolitan areas.

adopted. These are important decisions which should not be taken lightly. There are many different sources of secondary quantitative data, a few of which have been detailed above. These data sets can provide a useful source of information for your study, but you must be careful to assess their merits and limitations prior to use to ensure that they are suitable for your purpose. In the following chapter, we introduce the subject of how to prepare quantitative data for analysis and the initial steps of such analysis.

3.7   Questions for reflection

- *What characteristics of interval and ratio data make them appropriate for quantitative analysis?*
- *Summarise the various advantages and disadvantages of closed and open questions for use in a questionnaire.*
- *What is the point of piloting a questionnaire?*
- *What is the main limitation of a judgemental sampling strategy?*
- *Is there any such thing as an 'ideal' sample size?*
- *Outline the advantages and limitations of secondary data for use in a research project.*
- *Are all questions sampled at a 100% rate in the US and/or UK Census of population?*

Further reading

Bourque, L.B. and Clark, V.A. (1992) *Processing Data: The Survey Example.* Sage University Papers Series on Quantitative Applications in the Social Sciences 07-085. Sage, Newbury Park, CA.

Clark, G. (1997) Secondary data sources. In Flowerdew, R. and Martin, D. (eds), *Methods in Human Geography: A Guide for Students Doing a Research Project.* Longman, Harlow, pp. 57–69.

Dixon, C. and Leach, B. (1978) *Sampling Methods for Geographical Research.* Concepts and Techniques in Modern Geography 17, Invicta Press, London.

Ehrenberg, A.S.C. (1975) *Data Reduction: Analysing and Interpreting Statistical Data.* Wiley, Chichester.

Frankfort-Nachmias, C. and Nachmias, D. (1996) *Research Methods in the Social Sciences.* Edward Arnold, London.

Martin, D. (1993) *The UK Census of Population 1991.* Concepts and Techniques in Modern Geography 56, Geo Books, Norwich.

Openshaw, S. (1995) *Census Users' Handbook.* Geo-Information International, Cambridge.

CHAPTER 4

Preparing, exploring and describing quantitative data

This chapter covers

4.1 Introduction
4.2 Data pre-processing and checking
4.3 Computer programming
4.4 Using the MINITAB statistical package
4.5 Initial data analysis
4.6 Probability
4.7 Transforming data
4.8 Summary
4.9 Questions for reflection

4.1 Introduction

From Chapter 3, you should now be familiar with the methods of sampling and collecting both primary and secondary data. If we refer to Chatfield's (1995) schema of the stages of a statistical investigation (Box 4.1), these stages correspond to points 1 and 2. This chapter is concerned with the next two stages in analysis, summarised by points 3 and 4, as well as introducing the use of MINITAB, a computer package which performs statistical analysis, which we shall use both in this chapter and in Chapter 5.

Box 4.1 Stages of a statistical investigation

1 Understand the problem and its objectives.
2 Collect the data.
3 *Look at the quality of the data and pre-process the data.*
4 *Perform an initial examination of the data including description.*
5 Select and carry out appropriate statistical analysis, often suggested by the results of 4.
6 Compare the findings with any previous results.
7 Interpret and communicate the results.

Source: Adapted from Chatfield 1995.

The process of initial examination or analysis of data is referred to as Exploratory Data Analysis (EDA) or Initial Data Analysis (IDA) (Chatfield, 1995). The preliminary part of IDA is concerned with the pre-processing of the data, checking on data quality, errors and missing observations, with the latter part more concerned with the description of data using visualisation techniques (e.g., graphs) and descriptive statistics.

4.2 Data pre-processing and checking

Data pre-processing consists of a number of components including *coding*, *data entry* (usually entering values into a file for use by a computer package), *error checking* and any subsequent *data editing*.

4.2.1 Coding

Prior to any formal analysis, your data will need to be coded into meaningful values. If your data are in the form of digital data, or are already pre-coded (see Section 3.3), you can probably skip this section and move to the section considering *data entry* below. If, however, your data exist as a pile of completed questionnaires, full of ticked answers to closed questions or textual answers to open questions, the task will be to assign numbers to the range of responses you have obtained for each question. Where you have used closed questions (see Box 3.4) the task is relatively straightforward. For example, if your question was of the form:

Which was the main reason for buying your current house? (tick box):

Price....................................☐	➔	1
More suitable residence.......☐	➔	2
An investment ....☐	➔	3
Better location....................☐	➔	4
Children's schooling.............☐	➔	5
Nearer work/relatives ....☐	➔	6
Other...................................☐	➔	7
specify _____		

we can assign numbers 1 to 7 for the seven options listed (although you might want to add extra codes to cover specific reasons revealed in the 'other' category returns). Where you have used open questions within the structure of your questionnaire, the coding task is more difficult. Here, you have to first identify appropriate categories within the data and then classify each response prior to numerical assignment. In this case, you will need to construct both a **coding frame** of the codes or classification used for each open question in the questionnaire, and then an accompanying **code book** which lists all the codes used in your study. The number of categories used, and the form of your classification, will be determined by the objectives of your research (i.e., the problem that you are trying to solve), the variation in your responses, and the size of your sample. For example, we could have asked the above question using an open form:

Which was the main reason for buying your current house?

_____

_____

_____

If the objective of your research was to examine the various types of economic factors underlying the selection of a house, then you may wish to focus on categories such as *price, as an investment, to be near to job*, etc., consigning any response which does not fall into one of these categories into a *non-economic* category. Alternatively, you may be more concerned

with personal factors, such as *more suitable residence, children's school, better location*, etc., with other categories classed as *non-personal*. Of course, it is important to take into consideration the variability displayed in your responses: if you are getting a wide range of different personal factors for buying a house, you may wish to increase the number of corresponding categories. This explains why it is generally more useful to design your coding frame *after* collecting your data. For example, with a question such as:

How long have you lived at this address? _____

you might find that the majority of responses are falling well outside your anticipated range, and beyond the magnitude of any pre-sample coding frame. The number of categories will also be dictated by the overall size of your sample: too many categories with a small data set will allow small numbers for each category, which may make the data inappropriate for statistical analysis. Oppenheim (1992) notes that 12–15 categories will usually be sufficient for most purposes, allowing for the inclusion of codes such as *don't know* or *other*. Once you have designed a coding frame for each open question, you should test it on a further subsample of your returns, and then assemble the codes into a code book which you then use as the basis to code all remaining questionnaire returns.

4.2.2 Data entry

Once your data have been coded, if you are going to perform your analysis using a computer package such as MINITAB (see Section 4.4), you will need to enter them into the computer. There are three main possibilities for entering such data: typing the values into an ASCII file; typing the values into a spreadsheet; and using an optical mark reader. Of these options, an **optical mark reader** is the quickest. This consists of a machine that will scan your questionnaires, automatically coding checked boxes. If you do not have access to such a machine, then you will have to type your data into the computer. The first option is to enter your data into an **ASCII text file** using the text editor on your computer. If you are using a PC, Windows 95 supports either the *Notepad* editor, or the older DOS *Edit* editor. If you are running on an Apple Mac, then you can use the *SimpleText* editor. UNIX workstations have a variety of text editors depending on the hardware/operating system, ranging from fairly simple (e.g., Solaris Openwindows *textedit*) to quite specialised (e.g., *VI* and *GNU EMACS*).

Table 4.1 Data file records and fields.

Respondent no.	Q1	Q2	Q3	Q4	Q5	Q6	...	Q20
1	62	7	6	1	5	3	...	5
2	32	6	6	3	2	1	...	8
3	28	1	3	4	2	1	...	2
...	...	...	...	...	...	...	...	...
...	...	...	...	...	...	...	...	...
...	...	...	...	...	...	...	...	...
63	23	3	6	1	1	7	...	7

The advantage of using a text editor is that you can enter data in a simple format which can be read by most statistical packages.

When typing in your data you will need to adopt a standard format. Common data formats include the *free format* (or list), and the *space delimited* (values separated by spaces), *comma delimited* (values separated by commas), and *tab delimited* (values separated by tabs) formats. Entering data in free format is very simple, but perhaps less useful in that we can enter the **records** (row) only for one **field** (column) at a time. More useful are the space and comma delimited formats, since we are able to enter several fields against each record. Here, each record might be a questionnaire respondent, or some other sample unit such as a ward or city. The fields correspond to the variables

we have for each record. This might be the count of people in each ward or city, or the coded response the person gave to a particular question.

Consider the sample of hypothetical data in Table 4.1. Each record (row) corresponds to a respondent to whom you have administered a questionnaire, whereas each field (column) corresponds to the responses to different questions. In this case we have 63 respondents and 20 questions, of which only a portion are reproduced in the table.

If we were to enter the first three lines of these data codes into a file using a text editor in the space-delimited, comma-delimited, tab and free formats, we might end up with the displays in Figure 4.1.

The alternative to generating your own delimited ASCII text file is to input your data into a **spreadsheet**.

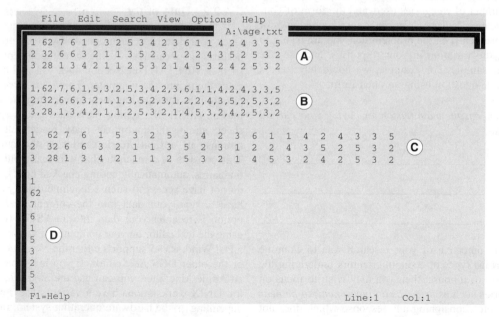

Figure 4.1 Different formats of ASCII text: space delimited (A), comma delimited (B), tab delimited (C), free format (D).

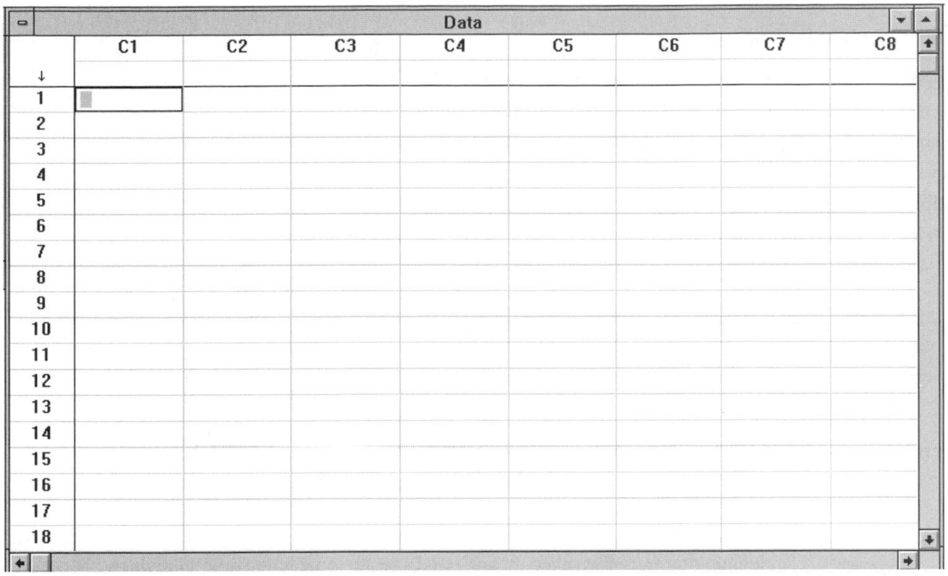

Figure 4.2 MINITAB spreadsheet.

A spreadsheet consists of a set of boxes in rows and columns, in which we can enter numeric data. Both the rows and columns are numbered conveniently, and will allow the entry of text headings so we can end up with a display very similar to that in Table 4.1. There are many varieties of spreadsheet ranging from *Microsoft Excel* to the simple spreadsheet available in statistical packages such as MINITAB. An example of MINITAB's internal spreadsheet can be seen in Figure 4.2.

Usually, statistical packages such as MINITAB will allow you to import data in a variety of ASCII text and spreadsheet formats, though it is probably advisable to check what formats you can import before choosing to use a particular spreadsheet package. Spreadsheets (and for that matter some PC text editors) often impose an upper limit for the maximum number of records they will allow in a file. If you are intending to use very large data sets with thousands of records (e.g., census data) you may well find that you can hold these only in a text editor or other software program on a powerful computer such as a UNIX workstation. Generally, such data sets can be entered into a computer package with any degree of efficiency only if they are already in a digital form. Whether your data set is big or small, if it is *not* in a digital form, you must consider carefully the amount of time needed to input the data into the computer. There is no point in collecting a large and complex data

set if it is impossible to enter the data physically into a computer package, given time and labour constraints.

4.2.3 Error checking and safeguarding your data

The input of data into a computer is a relatively straightforward process, but errors can and do occur, particularly if you are dealing with large and complex data sets. There are a variety of reasons why you might have errors in your data. These are summarised in Box 4.2.

Some of these errors can be spotted fairly easily during checking and preliminary analysis: these are the more mechanical **typing, transcription, inversion** and **repetition** errors. However, **recording** and **deliberate** errors can be nearly impossible to detect. Although we can spot errors such as data outliers (extreme values) using some of the initial data analysis methods (IDA) considered in Section 4.5 below, there are a variety of checks we can perform at the entry stage. These steps are often known collectively as *data cleaning* and are summarised in Box 4.3.

The majority of the steps in Box 4.3 will be useful only for spotting gross errors, and will not catch those errors where there is only a slight change (e.g., a '10' entered when a '19' should have been). To perform more rigorous checking, you will have to duplicate the data by typing it into the computer a second time,

---

Box 4.2   Types of error

Recording	The wrong questionnaire category was recorded.
Typing	An observation has been typed into the computer incorrectly (e.g., 2.369 rather than 23.69).
Transcription	Copying data from one medium to another can cause errors.
Inversion	Two successive digits are transposed (e.g., 23.69, rather than 23.96).
Repetition	A number is mistakenly typed in twice.
Deliberate	A questionnaire respondent has deliberately not told the truth!

Source: Adapted from Chatfield 1995.

Box 4.3   Data cleaning checks

Case numbers	Try to introduce some simple redundancy. Include the case number or count with each data row. This will make it easier to check for omissions (e.g., a '1' against the first record, '2' against the second, etc.).
Consistency checks	For example, if you know the total number of cases which should be in your data set, try to calculate this by some other means (e.g., use intermediate subtotals which should add up to the required total). Alternatively, check the number of lines in your data file if you have one case per line.
Magnitude checks	If your codes range from 1 to 20, check that you don't have any numbers included outside this range.
Variable consistency	Check pairs of variables for approximate consistency, if this is to be expected. For example, you might expect that income and house size might be related, or those who answered 'no' to Q. 10 should *not* have answered Q. 11 but skipped to Q. 12.
Duplicate your data	Enter your data set into the computer twice, and check that both sets correspond.

Sources: Adapted from Beasley 1988 and Bourque and Clark 1992.

---

and verify that both sets are the same. In this case, a simple method to check for errors is to subtract one data set from the other, which should of course lead to a set of zero values (Bourque and Clark, 1992). Alternatively you can use a computer system utility to compare files. If you are using a computer that operates using Windows 95 you will have access to an old DOS utility 'FC.EXE' which will detect differences between two files. For example, if we have two data files, ONE.DAT and TWO.DAT, which should be identical, we can check for differences by issuing the command at the DOS prompt (C:\windows>):

C:\windows>FC ONE.DAT TWO.DAT

Corresponding parts of each file which are different will then be flagged on the computer screen. On UNIX systems, there is a more powerful utility '*diff*', which allows the detection and listing of differences between two files. To run this, issue the following command at the UNIX prompt (%):

% diff one.dat two.dat

Any corresponding rows in each file which contain different values will be flagged on the screen. Such utilities are invaluable during any subsequent processing of your data, since you can usually output the data that the computer program is using as part of your analysis. With large data sets, and complex processing, it is always worth checking that the data you *think* the computer is processing are the same as the data which are *actually* being processed, particularly if you make use of your own purpose-written programs in a higher language (see Section 4.3).

Once you have entered your data, do not stop checking for errors! Sometimes errors come to light only after you have started your analysis, forcing you to scrutinise the data you have typed in. The methods of IDA which we will consider below are particularly useful for spotting certain data errors. However, as a preliminary, there are several simple checks you can make. Depending on your computer platform (e.g., *PC* or *UNIX box*), there are a variety of operating systems tools which you can use to interrogate any ASCII text file you have created. This could be simply a matter of typing the data to the screen or to the printer, or performing some simple checks on the size of the data file. For example, in a UNIX environment we can check the number of records (rows) in a file by using the word count command '*wc -l*', and we can also check the number of characters and size in bytes using the '*wc -c*' and '*wc -m*' commands respectively. The command '*wc*' with no qualifiers will give the number of lines, characters and bytes. For example, for a file 'three.dat':

```
% wc three.dat
%    734      1153      17862    three.dat
         ↑         ↑          ↑
      lines   characters   bytes
```

If you obtain the data in a digital form, it is useful to compare the size of the data file in terms of the number of bytes before and after you have transferred the file to your computer. This will ensure that you have copied the entire data set.

Once you have entered and checked your data, it is always advisable to do two things: **print out** a copy, and **save** a copy to a floppy disk or other storage medium such as a DAT tape or ZIP/JAZZ drive disk (which you keep in a safe place). It is often difficult to see some errors on screen, but these are much more apparent when you print out your file. Accidents leading to data and disk corruption do happen; it is always wise to take precautions – particularly with irreplaceable data! Hard disks can (and do) fail, floppy disks can attract all sorts of physical errors, particularly if you have been carrying them around loose in a pocket, and a fire will take care of even the most carefully stored data. When one of the authors was a research student, he had the stressful experience of a computer tape containing his data being overwritten by a member of the university's computer centre who had mistakenly confused two tapes! Fortunately in this case the bulk of the data was soon replaced. But the moral is that even if you are relying on someone else to look after your computer data, you should make it *your business* to keep regular backup copies, preferably at home or in a fireproof safe.

As with all statistical analysis, the confidence in any results will be increased if the data have been carefully checked, but for large data sets this may require a considerable amount of time. Perhaps the best advice here, particularly if you are doing a research project for the first time, is not to underestimate the amount of time needed to error-check your data. Chatfield (1995) estimates that the process of collecting, examining, processing and cleaning data prior to analysis can take anywhere up to 70% of total effort expended in the project!

4.2.4 Coping with missing data

In contrast to the examples displayed in textbooks on statistics, you may find that your data sample contains missing entries. You will have to decide on an approach to deal with this problem, which will largely depend on the nature of the missing values, the overall size of your data set and the types of statistical analysis you wish to perform. The first question which you must try to answer is: *Why is a particular observation missing*? If your data sample is composed of interview results, you may find that there are gaps in an otherwise complete record for a respondent. This could be due to the respondent (or interviewer) having forgotten to enter certain data or some other random effect. If missing values are distributed randomly across all respondents, then the data are **missing completely at random** (MCAR) (Bourque and Clark, 1992). If missing data are not randomly distributed across all respondents, but randomly distributed within a group or several groups, then such data are **missing at random** (MAR). Alternatively (and more importantly), missing data could be non-random, owing to some **correlated bias** in the data set (Oppenheim, 1992) particularly arising from non-response (see Section 3.3). Is the reason for non-response due to the nature of the questions or the characteristics of the individual being polled? If so, then this is a serious problem which could undermine the validity of the data sample.

The deletion of the record containing the missing data is one approach to dealing with the problem. We can delete all records which contain missing data, so that the only data used for analysis consist of records with no missing values. For example, if we have sampled 100 respondents, where the data for each respondent form a row (record), we might find that there are one or more missing fields for 35 respondents. We would then delete all 35 from our sample. The alternative is to delete only the record for those pairs of variables which have missing values. For example, from the same 100 respondents we find that fields 1 and 2 have only 14 rows with missing values. We would therefore delete only the corresponding 14 records from any bivariate analysis between these

---

Box 4.4   Methods for the imputation of data

Mean substitution	MCAR	The arithmetic mean of non-missing data is substituted into the missing case(s) usually if the variable is uncorrelated.
Least squares	MCAR	Missing data are estimated using least-squares regression if two variables are correlated.
Least squares + residual	MCAR	Residual component is added to the least-squares method (can also be applied to the mean substitution method).
Hot deck	MAR	Data from a similar case are substituted into the missing case(s).
Maximum likelihood	MAR	A likelihood function is derived from a model for the missing data set using a maximum likelihood technique, but this is beyond the scope of this book. The reader is referred to Chatfield (1995) and Little and Rubin (1987) for details.

Sources: Bourque and Clark 1992; Chatfield 1995.

---

fields. These methods are known as **listwise deletion** and **pairwise deletion** respectively (Oppenheim, 1992). The former is usually the preferred strategy unless the data set is very small (Bourque and Clark, 1992).

An alternative to deleting cases is to try to 'guesstimate' or **impute** the missing data values from the data (Chatfield, 1995). There are a variety of methods which can be used depending on the structure of the missing data (MCAR or MAR), which are listed in Box 4.4.

Missing data are also a problem at the practical analysis stage. If you have collected your own data, you will need to insert an appropriate **missing data code** prior to using a computer package (e.g., a '*' for MINITAB). If you are using a secondary data source you should have information on whether any missing data has been re-coded and what codes have been used. Occasionally this is not the case, and in such a situation you should be suspicious of any repetitive or very large or very small values in the data set. For example, Chatfield (1995) notes a number of examples where data sets had already been coded with a missing data code of 999 or 0 before the statistical analysis stage, but were not documented as such. In this context it is important that you look at the distribution of the data using an IDA, and be suspicious of any very large or very small data values.

4.3   Computer programming

It is almost inevitable that you will require some exposure to computers if you are going to perform quantitative analysis of your data. This may vary from the simple use of built-in functions in spreadsheets such as *Microsoft Excel*, and the use of statistical packages such as MINITAB, SPSS and SAS, to programming your own analysis. Where quantitative analysis is concerned, there is no doubt that the skill of being able to write computer programs can be extremely valuable (although as our step-by-step guides illustrate, not essential). This can range from the use of a structured high-level language such as BASIC, FORTRAN, C or C++, to the use of UNIX shell-scripting. Although increasingly you can make do with statistical packages (we will consider the MINITAB package below), the process of writing a computer program to analyse data does convey certain advantages (as well as disadvantages!). These are listed in Box 4.5.

Box 4.5   Pros and cons of computer programming your statistical analysis

Pros	*Cons*
Programming is a useful skill to learn.	**Difficult**: you need to be able to program!
Analysis tailored to your needs.	**Commercial software** is almost as good.
Improves your understanding of the statistics.	**Program** will need thorough testing.
Time saved in automating processing.	**Time consumed** to write program.

It is beyond the scope of this book to cover in detail the programming languages and the details for both compiling and debugging such programmes. However, there are an ever-increasing number of simple guides to programming, which range from the simple 'how to' guides (e.g., Aitken and Jones, 1997) to the more specific programming of statistics and numerical analysis (e.g., Press *et al.*, 1989; Cooke *et al.*, 1992).

As an illustration, the C code for the calculation of standard deviation is listed in Box 4.6. This will look complicated – particularly if you don't understand the C programming language! However, armed with an introductory text, a computer, a C compiler (and a lot of time!) you should soon be able to understand the elements of Box 4.6 and produce your own programs.

4.4 Using the MINITAB statistical package

The MINITAB statistical package was first released in 1972, and was designed to help teach students a variety of basic statistics. There are a variety of versions of the software available, on several different computer hardware platforms. The version that we will use is version 10.1, running on a PC with a *Windows* operating system. The original versions of MINITAB were based on entering commands at a command line. Version 10.1 can be used either by entering commands or by accessing pull-down menus. In the examples throughout this book we will use both

---

Box 4.6 C code for the calculation of standard deviation

```
/*      Program to illustrate how to calculate    */
/*      standard deviation using C     */

/*      main calls a function called 'standard_d' to    */
/*      calculate the standard deviation. It takes an    */
/*      array of integers as its argument and returns    */
/*      a float as the standard deviation    */

/* included header files */
#include <stdlib.h>
#include <stdio.h>
#include <math.h>

/* function prototypes   */
float standard_d(int *data);

/* main (which calls standard_d) */
main()
{
    int data[100];
    int count = 0;
    float standard_deviation = 0.0;
    FILE *inputfile;
    /* open File pointer */
    source("shell");
    inputfile = fopen('data.txt','r+');
    /* check inputfile exists */
    /* if it does not exist print error and quit */
    if(inputfile == NULL) {
        printf("\n File does not exist");
        goto quit;    }

    /* set data values to -999 (or any unused value) */
    while(count <100)  {
        data[count] = -1;
        ++count;      }
```

Box 4.6   (cont'd)

```
    /* read in the data from the file */
    /* assumes you want all the data */
    count = 0;
    while(!feof(inputfile)) {
          fscanf(inputfile,"%d",&data[count]);
          ++count;        }

    /* call standard_d() */
    standard_deviation = standard_d(data);

    /* print out the result */
    printf("\n standard deviation = %f\n",standard_deviation);

quit:
/* main should return an int */
return(0);
}

    /* standard_d function */

    float standard_d(int data[])
    {

    int count = 0;
    int sum = 0;
    float mean = 0.0;
    float var_sum = 0.0;
    float variance = 0.0;
      float sd = 0.0;

    while((count <100) & (data[count]!=-999)) {
          sum = sum+data[count];
          ++count;  }
    mean = sum/count;

    count = 0;
    while((count <100) & (data[count]!=-999)) {
          var_sum = var_sum+((data[count]-mean)*(data[count]-mean));
          ++count;  }
    variance = var_sum/count;
    sd = sqrt(variance);

    return(sd);
}
(Code courtesy John Spain)
```

methods, and leave the choice of which to use to you. When you start up MINITAB on a PC, you will be presented with a screen similar to that in Figure 4.3, which comprises two windows and a **command bar**.

The command bar consists of a row of keywords which when selected display one or more pull-down menus. To select a keyword, you can either use the mouse to simply point and click (with the left mouse button), or press the *ALT* key on the keyboard, and keeping this pressed, press the letter which is underlined in the keyword. For example, to pull down the *File* menu, you would need to type the *ALT* and *F* keys.

If you prefer to use the command line interface, the **Session** window acts as the location in which you type your commands and provides a command line prompt in the form of MTB >. This window also lists

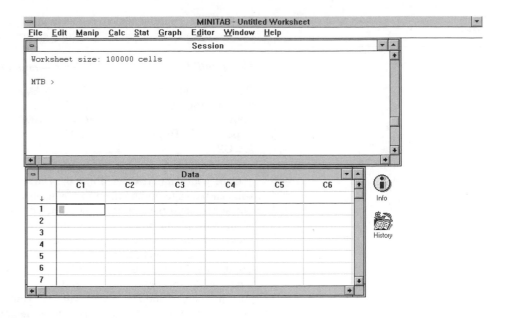

Figure 4.3 MINITAB initial screen.

the results of any MINITAB commands which you execute on your data. This could be a simple operation such as loading data from, or saving to, a floppy disk. Alternatively, this could be the result of a statistical operation. The **Data** window is where you enter your data. This consists of a set of columns labelled C1–C100 across the top and a set of rows labelled 1–n (where n is the number of possible data entries) on the left-hand side. If you are familiar with the use of spreadsheets, you will note that the data window is a simple spreadsheet. Each variable can be entered in a separate column (e.g., C1), with each row corresponding to an individual record.

4.4.1 Input and output of data in MINITAB

The simplest method to input data into MINITAB is to enter a number into the appropriate row/column in the **Data** window. However, this can be time consuming, particularly if you have a large data set with many variables and/or records. In this case, it is more efficient if you can obtain your data already in a digital form. This could be either a simple ASCII text file as discussed above in Section 4.2.2, or data in some other proprietary file format. MINITAB will allow the import of data files saved by a variety of other

computer packages. For example, in version 10.1 import filters include *Microsoft Excel, Quattro Pro, Lotus 1-2-3* and *dBase*. As an example, we will assume that our data are in two columns in a space-delimited ASCII text file with the name 'INTERNET.DAT' stored on the C:\ drive. You can use either the pull-down menu (under *File*) or the *READ* command at the command line:

Using pull-down menus:

➡ File
➡ Open Worksheet
➡ List Files of Type [ ]
Select text (*.txt; *.dat)
Highlight INTERNET.DAT
➡ Options
➡ Variable Names [ ]
Select none
➡ Field definition [ ]
Select single character separator, and space
➡ OK
➡ OK

Using command line:

MTB > READ
'C:\INTERNET.DAT'
C1-C2.

It will be useful to explain some of the elements of the shorthand written under the pull-down menu

Table 4.2 Number of Internet networks in 13 countries in 1992 and 1995.

	C1 1992	C2 1995
USA	2112	15920
Canada	174	1578
Sweden	33	230
Australia	111	569
UK	85	769
Germany	170	884
France	123	1078
Japan	93	868
Netherlands	61	251
Spain	12	128
Greece	5	46
Italy	52	306
Ireland	7	86

Source: ftp://nic.merit.edu/nsfnet/statistics

section, as these will be used below and in subsequent sections of the book. We use the symbol ➥ to indicate either a selection from a menu or a button, which needs to be clicked on with the mouse cursor. Where this is followed by a name and two square brackets, i.e., ➥Variable Names [ ], this means that you should click on a box/section with the name 'Variable Names'. Other information to be entered not in the form of menus, boxes and buttons is requested explicitly (e.g., 'Click on none' or 'Type in C4') at the appropriate place in the sequence of instructions. For example, in the above list you are required to click on the box called 'Variable Names', and then select the 'none' option to indicate that you have not included the names of your variables in your data file. You should always end commands entered on the command line with a full stop '.' or a semi-colon ';'. A full stop signifies the end of a command, whereas a semi-colon indicates an optional sub-command. If you end with a semi-colon MINITAB will issue a new prompt SUBC > at which you should enter a command, and then end with a full stop.

The above example loads two columns of data from a text file INTERNET.DAT (see Table 4.2) which are then entered into two corresponding columns in the MINITAB **Data** window. You will see that the columns are referred to in MINITAB simply as C1 and C2. Although you can maintain this simple naming strategy, it is probably more efficient to give each column the name of the variable, particularly where

the variables are more numerous than in this example. In this case we can enter the following commands:

Using pull-down menus:	*Using command line:*
Select box in Data window immediately below 'C1'. Type '1992'. Repeat for 'C2'	MTB > NAME C1 '1992'. MTB > NAME C2 '1995'.

Once your data are entered into the **Data** window, you can navigate around them in a number of ways. The slider bar at the extreme right of the window can be moved up and down to move either forwards or backwards through the data. Alternatively, you can use the *Page Up* and *Page Down* keys on your keyboard. Pressing the *Ctrl* and *End* keys together will send you to the bottom of the **Data** window, and pressing the *Ctrl* and *Home* keys will send you to the top. You can also use the arrow keys to move around the worksheet in a cell-by-cell manner.

You may also want to print out your data and then save it to a floppy disk. You can print out your data either from the **Data** window or as a text file from MINITAB. The former is the simpler, and requires you to first select the **Data** window with the mouse. Then pressing *Ctrl P* will bring up the MINITAB **Print** window, to which you should select the defaults and click on the OK button. For saving data, there are again a variety of options which range from simple ASCII text files to files in a MINITAB worksheet format. In this case we will save the data as a standard MINITAB worksheet, although MINITAB will allow you to save data in a portable format to transfer to other versions of MINITAB on different types of computers. To save the data we have used above, you should use either the menu command *Save Worksheet As* (found under *File* on the command bar) or the appropriate *SAVE* command at the command line:

Using pull-down menus:	*Using command line:*
➥ File ➥ Save Worksheet As ... ➥ Save File as Type: [ ] Select Minitab ➥ File name [ ] Type in c:\internet.mtw ➥ OK	MTB > SAVE 'C:\INTERNET.MTW'; SUBC > REPLACE.

4.4.2 Checking and manipulating data in MINITAB

Once you have entered your data, it is a good idea to perform some basic checks to ensure that you have both the number of cases and the number of variables you expected. The command *INFO* can be typed in at the prompt, or alternatively you can select the command *Get Worksheet Info* (found under *File*) on the command bar.

Using pull-down menus:	*Using command line:*
➡ File	MTB > INFO.
➡ Get Worksheet Info ...	
➡ OK	

This will produce a list of the variables in your data set, with the name of each variable, the number of cases and the number of missing values (if any). For example, for the data set from Table 4.2 we would obtain the following output:

```
MTB  >  INFO.

Column    Name     Count
C1        1992     14
C2        1995     14
```

This confirms the correct name and count for each variable. MINITAB denotes each missing value with an '*' symbol. If the data value for USA in 1995 was missing, we would insert an '*' in the appropriate cell for C2, and the results of the INFO command would be as follows:

```
MTB  >  INFO.

Column    Name     Count    Missing
C1        1992     14       0
C2        1995     14       1
```

Note that we now have one missing value. If you were not expecting any missing data values and have some listed, or you know you have missing data but none have been listed, then you will need to check and edit your data in the **Data** window.

Each column of data can be managed within the MINITAB environment. For example, the *COPY* command will copy a column of data and put the replica in a new column.

Using pull-down menus:	*Using command line:*
➡ Manip	MTB > COPY C1 C4.
➡ Copy Columns ...	
Click on C1	
➡ Select	
➡ To Columns [ ]	
Type in C4	
➡ OK	

If you now return to the **Data** window you will see that the data in C1 are now replicated in C4. Alternatively, you can remove unwanted columns. For example, to remove C4:

Using pull-down menus:	*Using command line:*
➡ Manip	MTB > ERASE C4.
➡ Erase variables ...	
Click on C4	
➡ Select	
➡ OK	

In some of the statistical calculations considered below and in Chapter 5, you may find it useful to stack (add) two columns of data into one column. You can do this as follows:

Using pull-down menus:	*Using command line:*
➡ Manip	MTB > STACK C1
➡ Stack ...	C2 INTO C4.
Click on C1	
➡ Select	
➡ Second space [ ]	
Click on C2	
➡ Select	
➡ Store Results In [ ]	
Type in C4	
➡ OK	

If you go to the **Data** window you will see that the data in C1 and C2 are now in the same column (C4). You can also manipulate the data using mathematical formulae. For example, if we wanted to sum the number of networks in one year (e.g., 1995) (say, as a preliminary step to calculating the mean number of networks), then we might do the following:

Using pull-down menus:	*Using command line:*
➡ Calc	MTB > SUM C2.
➡ Column Statistics	
➡ Input Variable [ ]	
➡ Click on C1	
➡ OK	

We can also manipulate whole columns. For example, if you want to add 10 to each network estimate in 1992, and store the results in C3, you can use the *Let* command:

Using pull-down menus:	*Using command line:*
⇨ Calc	MTB > LET
⇨ Mathematical	C3 = C1 + 10.
Expressions ...	
⇨ Variable (new	
or modified) [ ]	
Type in C3	
⇨ Expression: [ ]	
Type in C1 + 10	
⇨ OK	

A useful feature of MINITAB is that you can save all your work automatically to a text file which you can then load into a word processor to edit and print out. To do this, you create an *outfile* with the name 'INTERNET.LIS', and save to the floppy disk (A:\):

Using pull-down menus:	*Using command line:*
⇨ File	MTB > OUTFILE
⇨ Other files	'A:\INTERNET.LIS'.
⇨ Start Recording	
Session	
⇨ Record output in	
file and display in	
session window	
⇨ Select file	
⇨ File Name [ ]	
Type in age.lis	
⇨ OK	

Everything you do in the MINITAB session is recorded to this file. Once you decide that you want to stop the recording session then:

Using pull-down menus:	*Using command line:*
⇨ File	MTB > NOOUTFILE.
⇨ Other files	
⇨ Stop recording	
session	

We have only scratched the surface of some of the general data manipulation features available in MIN-ITAB. For further details you should refer to one of the various texts available for general use (e.g., Ryan and Joiner, 1985) or more directly in the context of

social science data (e.g., Cramer, 1997). We will return to some of the statistical features of MINITAB in the section below and in Chapter 5.

4.5   Initial data analysis

Once you have your data coded, performed preliminary checks, and if appropriate entered the data into a computer, it is useful to begin exploring your data through the process of Initial Data Analysis (IDA) (Chatfield, 1995). Throughout the section below, we will describe the mechanics of some of the main methods of IDA using worked examples by hand, and also using the MINITAB statistical package. We adopt a standardised format:

Name of method	*e.g., mean*
Level of data	*e.g., interval or ratio*
Function of method	*e.g., measure of location*
What the method does	*A simple description of what the method is used for*
Explanation and procedure	*Explanation of why and what you are doing, with step-by-step details of how to calculate the statistic*
Example data	*A real geographical data set*
Handworked example	*Step-by-step using a calculator and the example data*
MINITAB example	*Step-by-step using MINITAB and the example data*

The 'Explanation and procedure' section will describe the statistic, and then explain the procedure of calculation referring to a number of steps (e.g., **Step 1**) which are described in the 'Handworked example' box. Although for the purposes of your research you are more likely to use the 'MINITAB example' (or equivalent in a different computer package), we advise that you read through the 'Handworked example' to familiarise yourself with the mechanics of each statistic.

We will concern ourselves here only with IDA statistics for *one variable*. Although we have deliberately adopted a 'cook book' approach, we would like to stress that there is *no substitute* for a thorough

---

Box 4.7   Statistics for one variable

- Mean, mode and median
- Histogram
- Stem-and-leaf
- Range, variance and standard deviation
- Box-and-whisker plot

---

understanding of the method, including some appreciation of any limitations of use. In particular, it is important to consider whether a certain method is appropriate in the context of your research. The treatment within this chapter is designed to give you a basic overview. The book by Erickson and Nosanchuk (1992) provides further detail on some of the exploratory methods considered below. For additional details, you are referred to the textbooks listed in the further reading section at the end of this chapter.

There are two main characteristics of a set of data which we can describe: a summary **measure of location** (central tendency), and a **measure of variability** (spread) of the data values. Box 4.7 summarises the statistics described in this section.

4.5.1   Mean, mode and median

Level of data: Interval or ratio.

Function of methods: Measures of central tendency in a data set.

What the methods do: The mean, mode and median all calculate the 'centre' of the data set, but use different strategies to achieve such an end.

Explanation and procedure: The mean, mode and median are measures of the **central tendency** of a distribution. There are a variety of means, which include the arithmetic mean, the geometric mean and the harmonic mean. We will concern ourselves only with the **arithmetic mean**. This is the *average* of the values in a data set. The **median** is the number exactly in the middle of the data set, i.e., there are an equal number of values before and after it in an ordered sequence. The **mode** is the data value which occurs most frequently in the data set, or alternatively the category, or class in a histogram, with the greatest frequency of observations. Where the distribution of values in a data set is symmetric (e.g., a normal

distribution: see Section 3.4), the mean, median and mode will coincide. However, where the distributions are not symmetric (skewed) they will differ. The mean is often calculated for use in statistical tests, though it is not resistant to extreme data values (outliers) which can influence its value greatly. Some statistical packages allow the calculation of a **trimmed mean**, which excludes the largest and smallest values from the calculation. In contrast to the mean, the median is resistant to extreme data values and will give a 'truer' measurement of central tendency in data containing outliers. The mode merely reports the most popular data value. If possible, calculation of the 95% confidence interval around the mean (see Section 3.4.2) is a useful accompanying statistic, although we will not attempt this here. You should refer to Gardner and Altman (1989) for details.

To calculate an **arithmetic mean**, place all the data in a table (**Step 1**) and then add all your data values together ($\sum x_1, \ldots, x_n$) to obtain the overall sum ($\sum x_i$) (**Step 2**). Next, divide the overall sum by the number of data values n (**Step 3**):

$$\bar{x} = \frac{\sum x}{n} = \frac{\sum_{i=1}^{n} x_i}{n}$$

The procedure for calculating the **median** will depend on whether the number of observations is even or odd. First, rank your data in sequence from the largest to the smallest value (**Step 4**). Where n is an odd number, the median is the single data value in the middle of the data set. Where n is an even number, take the sum of the two data values in the middle of the data set and divide by two to obtain the midpoint between them. This value is taken as the median (**Step 5**). To calculate the **mode** determine which value appears most often in your data set (**Step 6**). We will make use of data on sheep numbers in County Down to illustrate the calculations (Table 4.3) which are set out in Boxes 4.8 and 4.9.

4.5.2   Histogram

Level of data: Interval or ratio.

Function of method: Frequency description.

What the method does: A histogram communicates information on the frequency of occurrence of values within a data set. These are displayed in a block diagram consisting of a number of equal sized groups.

Table 4.3   Sheep in the parishes of County Down, Northern Ireland, in 1803.

Aghaderg	532	Drumbeg	30	Loughinisland	676
Annaclone	195	Drumbo	235	Magheradrool	746
Ardglass	183	Drumgath	170	Magheralin	304
Bright	326	Hillsborough	511	Moira	104
Clonallan	645	Hollywood	228	Rathmullan	192
Donaghcloney	146	Kilbroney	2379	Saintfield	666
Donaghmore	268	Kilclief	177	Saul	524
Dromara	787	Kilmegan	764	Tullylish	416

Source: Turner 1984.

Box 4.8   Handworked example: mean, mode and median

532	30	676
195	235	746
183	170	304
326	511	104
645	228	192
146	2379	666
268	177	524
787	764	416

Step 1   Place the data in a table as above.

Step 2   *Mean:* Sum of the data values $\sum x_i = 11204$.

Step 3   Number of data values $n = 24$, therefore:

$$\frac{\sum\limits_{i=1}^{n} x_i}{n} = \frac{11204}{24} = 466.83 = 467 \text{ sheep}$$

Step 4   *Median:* Rank the data from smallest to largest as follows (where x is a data value, $r(x)$ = rank):

x	r(x)	x	r(x)	x	r(x)
2379	24	524	16	195	8
787	23	511	15	192	7
764	22	416	14	183	6
746	21	326	13	177	5
676	20	304	12	170	4
666	19	268	11	146	3
645	18	235	10	104	2
532	17	228	9	30	1

Step 5   Here, $n = 24$ (even number), therefore the median = $(326 + 304) \div 2 = 315$.

Step 6   *Mode:* There is no mode for this data set: all observations occur with equal frequency (one).

Box 4.9   MINITAB example: mean, mode and median

Place the numbers of sheep in a single column (C1).

There are two methods of calculating descriptive statistics in MINITAB. (1) The command 'describe' can be used to obtain a summary of simple descriptive statistics which include the mean, median, trimmed mean (which excludes the largest 5% and smallest 5%) and standard deviation, which we consider below. (2) Individual commands can be used.

(1)   Describe

Using pull-down menus: *Using command line:*

➥ Stat   MTB > Describe C1.
➥ Basic Statistics ➤
➥ Descriptive Statistics...
Click on C1
➥ Select
➥ OK

MINITAB results:

```
MTB > describe c1.

        N    MEAN  MEDIAN  TRMEAN  STDEV  SEMEAN
C1     24   466.8   315.0   399.8  469.8    95.9

       MIN     MAX     Q1      Q3
C1    30.0  2379.0  185.3   660.8
```

(2)   Individual commands

Using pull-down menus: *Using command line:*

Mean ➥ Calc   MTB > Mean C1.
        ➥ Column Statistics...
        ➥ Input variable [ ]
        Click on C1
        ➥ Select
        Select Mean
        ➥ OK

Median ➥ Calc   MTB > Median C1.
        ➥ Column Statistics...
        ➥ Input variable [ ]
        Click on C1
        ➥ Select
        Select Median
        ➥ OK

Examination of the example data set will reveal that there is no mode: you do not really need to use MINITAB in this instance, though we will include the commands and output for the sake of completeness:

Mode ➥ Stat   MTB > Tally C1;
        ➥ Tables ➤   SUBC > Counts.
        ➥ Tally...
        ➥ Variables [ ]
        Click on C1
        ➥ Select
        Select counts
        ➥ OK

Box 4.9   (cont'd)

MINITAB results:

```
MTB > mean c1.

Column Mean

Mean of C1 = 466.83
MTB > median c1.

Column Median

Median of C1 = 315.00

MTB > Tally C1;
SUBC > Counts.
       C1   COUNT
       30     1
      104     1
      146     1
      170     1
      177     1
      183     1
      192     1
      195     1
      228     1
      235     1
      268     1
      304     1
      326     1
      416     1
      511     1
      524     1
      532     1
      645     1
      666     1
      676     1
      746     1
      764     1
      787     1
     2379     1
       N=    24
```

Explanation and procedure: The construction of a **histogram** provides us with a graphical representation of the frequency distribution of our data. It does this by simply displaying the data, grouped into equal sized classes. There are other graphical methods available to view the frequency distribution which are closely related to the histogram; these include the *dotplot* (not described here), the *stem-and-leaf plot* and *box-and-whisker plot* which we describe below. The mechanics of constructing a histogram are quite simple and straightforward. The only areas of difficulty involve the selection of the number of classes and the selection of the class intervals themselves.

First assemble your data values into a table, making a note of the maximum and minimum values (**Step 1**). Next, decide on the number of equal interval classes in your histogram. There are a variety of ways to select the number of classes. For example, you can select this number subjectively or by using a simple calculation such as $5(\log_{10} n)$, where $n =$ the

Table 4.4   Number of milch cows per 1000 acres of crop and pasture in Ireland.

Carlow	67	Kerry	194	Louth	69	Tipperary	107
Cavan	99	Kildare	52	Mayo	97	Waterford	122
Clare	110	Kilkenny	91	Meath	55	Westmeath	43
Cork	146	Laoighis	69	Monaghan	85	Wexford	64
Donegal	102	Leitrim	102	Offaly	55	Wicklow	79
Dublin	108	Limerick	181	Roscommon	66		
Galway	69	Longford	74	Sligo	92		

Source: Cliff and Ord 1973.

number of data values (**Step 2**). The $5(\log_{10} n)$ calculation is meant only as a guide, to avoid the selection of either too large or too small a number of classes. The classes chosen must be **exhaustive** (i.e., include the full range of values in the data set) and **mutually exclusive** (values cannot belong to more than one class). We find that the latter point can often be missed by students, who might select limits for the example in Boxes 4.10 and 4.11 as 0–30, 30–60, 60–90, etc. In which class do we place a value of 30? We have a choice of two – the classes used in this case are *not* mutually exclusive, as this class choice would be incorrect. A correct set of class intervals is shown below (**Step 3**). Next, record the class in which each observation falls, and sum for each class to obtain the class frequency (**Step 4**). Construct the histogram, plotting class frequency on the vertical axis, and class magnitude on the horizontal axis (**Step 5**).

It is important to be aware that changing the number of classes, and the class intervals used, can have an important influence on the way the histogram appears. We can influence the message that a histogram communicates, by the way in which it is drawn. The book by Tufte (1983) is worth referring to as a summary of the ways that the drawing of a graph can emphasise, confuse or completely change the message from a data set. We will make use of data on the number of milch cows in Ireland to illustrate the calculations (Table 4.4).

4.5.3 Stem-and-leaf plot

Level of data: Interval or ratio.

Function of method: Frequency description.

What the method does: A stem-and-leaf plot, similar to a histogram, communicates information on the frequency of occurrence of values within a data set.

Unlike a histogram, the plot uses all the data values to create the display.

Explanation and procedure: A stem-and-leaf plot is a visual representation of tally counts, and is constructed on the basis of determining a base – termed a **stem** value – and tallying other values called **leaves** in relation to this. In many ways, the stem-and-leaf works similarly to the categorisation process for a histogram, but uses the data structure to determine the stem values (equivalent to the bar in a histogram). This negates the problem of determining category classes required for histogram construction. The easiest way to understand the use of a stem-and-leaf plot is to examine its construction. First, assemble your data values into a table, making a note of the maximum and minimum values (**Step 1**). Next, divide up each data value into a stem and leaf, where a stem might be in tens of units, and a leaf might be in individual units. If there are extreme values (outliers) in the data set, consider transforming the data (see Section 4.7) (**Step 2**) and trimming any very large or very small values (**Step 3**). Make a clear note of the units used for both stem and leaf (**Step 4**) and draw the stem and leaf as follows (with the corresponding numbers included here for comparison) (**Step 5**):

Numbers		Stem-and-leaf	
10, 11, 12, 15, 15, 16		1	012556
20, 22, 22, 23, 24, 26, 28, 28		2	02234688 ◣
34, 37, 39, 39	◤	3	4799 leaves
48	stem	4	8

To illustrate the construction of a stem-and-leaf plot, we will make use of a data set on the population of Prince Edward Island settlements (Table 4.5). The calculations are set out in Boxes 4.12 and 4.13 (pp. 90–92).

Box 4.10   Handworked example: histogram

67	91	55
99	69	66
110	102	92
146	181	107
102	74	122
108	69	**43**
69	97	64
194	55	79
52	85	

Step 1   Place your data into a table as above. Maximum = 194, minimum = 43, $n = 26$.

Step 2   Calculate the number of classes in your histogram:

$$5 \times (\log_{10} n) = 5 \times (\log_{10} 26) = 5 \times (1.4149) = 7 \text{ classes}$$

Step 3   With the data set above, our class intervals are:

1	30–54	XX	2
2	55–79	XXXXXXXXXX	10
3	80–104	XXXXXXX	7
4	105–129	XXXX	4
5	130–154	X	1
6	155–179		0
7	180–204	XX	2

Step 4   Record the class in which each observation falls (X), and sum for each class to obtain the class frequency (see above), checking that the total number of frequencies equals the size of the data set.

Step 5   Construct the histogram (Figure 4.4), plotting class frequency on the vertical axis and class magnitude on the horizontal axis. Label each class with the mid-point of that class.

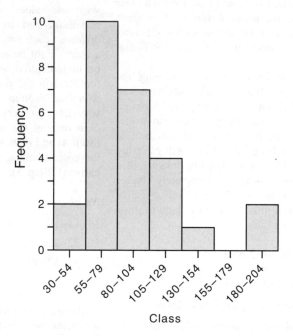

Figure 4.4   Histogram of milch cows in Ireland.

Box 4.11   MINITAB example: histogram

Place the numbers of cows in a single column (C1).

You can construct a histogram in MINITAB in two ways. The first (1) displays the histogram in graphical form, and the second (2) presents the findings in character form in the **Session** window. When using the graphical form, you must be careful to select the classes you require yourself, as MINITAB's default settings often class the data into too few classes to be meaningful.

(1)   Graphical histogram

Using pull-down menus:

➡ Graph
➡ Histogram…
Click on C1
➡ Select
➡ Options
➡ Define interval values using [ ]
Type in classes
➡ OK
➡ OK

Using command line:

MTB > Histogram C1;
SUBC > MidPoint;
SUBC > Bar.

MINITAB results (Figure 4.5):

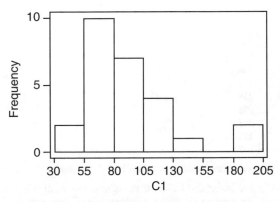

Figure 4.5   MINITAB output for milch cows in Ireland.

(2)   Character histogram

Using pull-down menus:

➡ Graph
➡ Character graphs▶
➡ Histogram…
Click on C1
➡ Select
➡ OK

Using command line:

MTB > Histogram C1;
SUBC > MidPoint;
SUBC > Bar.

MINITAB results:

```
Histogram of C1   N = 26

Midpoint  Count
      40      1   *
      60      9   *********
      80      3   ***
     100      8   ********
     120      2   **
     140      1   *
     160      0
     180      1   *
     200      1   *
```

Table 4.5   1996 census population in Prince Edward Island (Canada) settlements.

Abram Village	328	Hunter River	354	Murray Harbour	356	Union Road	218
Alberton	1084	Kensington	1383	Murray River	420	Victoria	158
Bedeque	148	Kings Royalty	246	North Rustico	650	Warrengrove	295
Borden-Carelton	829	Kinkora	321	O'Leary	877	Wellington	427
Brackley	367	Lennox Island	222	Cavendish/N.Rustico	255	Winsloe South	207
Breadalbane	171	Linkletter	304	Sherbrooke	160		
Cardigan	371	Meadowbank	354	Souris	1293		
Central Bedeque	182	Miltonvale Park	1242	Stratford	5869		
Charlottetown	32531	Miminegash	210	St Louis	100		
Clyde River	601	Miscouche	679	St Peters Bay	283		
Cornwall	4291	Montague	1995	Summerside	14525		
Crepaud	378	Morell	336	Tignish	839		
Georgetown	732	Mount Stewart	310	Tyne Valley	231		

Source: Government of PEI 1997.

Box 4.12   Handworked example: stem-and-leaf plot

328	378	679	5869
1084	732	1995	**100**
148	354	336	283
829	1383	310	14525
367	246	356	839
171	321	420	231
371	222	650	218
182	304	877	158
32531	354	255	295
601	1242	160	427
4291	210	1293	207

Step 1   Assemble your data values into a table as above. In this case the maximum is 32531 and the minimum is 100.

Step 2   Divide up the data into stems and leaves. If the stem is too small this produces either a very large plot or, if too large, a very skewed data set, hiding much of the variation within one stem. In this case, the step unit is 100, and the leaf unit is 10.

Step 3   Trim five values that are much larger than the others and are skewing the data set (e.g., 1995, 4291, 5869, 14525, 32531). Once these values have been removed, the stems now range only from 1 to 13.

Step 4   Make a clear note of the units used for both stem (integers representing hundreds) and leaf (integers representing tens).

Step 5   Draw the stem and leaf as follows, rounding up or down where appropriate:

```
 1 | 055678
 2 | 011223589
 3 | 001234556778
 4 | 23
 5 |
 6 | 058              Stem unit = 100
 7 | 3                Leaf unit = 10
 8 | 348
 9 |
10 | 8
11 |
12 | 49
13 | 8
```

Box 4.13   MINITAB example: stem-and-leaf plot

Place the population values into a single column (C1).

MINITAB provides an additional column of summary data to accompany the stem-and-leaf plot. This is the first column displayed, which provides a summary of the distribution around the median, and is known as the **depth** (McCloskey *et al.,* 1997). Where the number is contained in brackets, e.g., (41) or (11), this is the group which contains the median value. The numbers above and below detail the number of entries above or below that value. For example, in the second plot below, (12) identifies the class containing the median, the '16' indicates 16 values at this level and above, and the '6' indicates 6 values at this level and above.

Using pull-down menus:

➥ Graph
➥ Character graphs➤
➥ Stem-and-Leaf...
Click on C1
➥ Select
➥ OK

Using command line:

MTB > Stem-and-Leaf C1.

MINITAB results:

If we adopt the MINITAB defaults, the stem is defined in units of 10,000 which will give us four stems – 0, 1, 2, 3 – and leaf units of 1000, i.e.:

```
MTB > Stem-and-Leaf C1.
```

Character Stem-and-Leaf Display

```
Stem-and-leaf of C1   N = 44
Leaf Unit = 1000

(41)  0  00000000000000000000000000000000000111114
 3    0  5
 2    1  4
 1    1
 1    2
 1    2
 1    3  2
```

This is unsatisfactory, given that the stem-and-leaf plot of C1 has an extreme positive skew (41 in one category). If we reduce the size of the stem unit by specifying an increment (100), and trim the extreme values, we obtain a more satisfactory plot:

Using pull-down menus:

➥ Graph
➥ Character graphs➤
➥ Stem-and-Leaf...
Click on C1
➥ Select
➥ Increment [ ]
Type in 100
➥ OK

Using command line:

MTB > Stem-and-Leaf C1.
SUBC > Increment 100;
SUBC > Trim.

MINITAB results:

Character Stem-and-Leaf Display

```
Stem-and-leaf of C1   N = 44
Leaf Unit = 10
```

Box 4.13   (cont'd)

```
    6    1  056678
   14    2  11223568
  (12)   3  001234556778
   18    4  23
   16    5
   16    6  058
   13    7  3
   12    8  348
    9    9
    9   10  8
    8   11
    8   12  49
    6   13  8
        HI  200,  429,  587,  1453,  3253,
```

The 'HI' values are those that have been trimmed from the plot.

4.5.4 Range, variance and standard deviation

Level of data: Interval or ratio.

Function of method: Determine data spread.

What the method does: The range, variance and standard deviation examine the variability of data, which for variance and standard deviation is expressed as a spread around the mean.

Explanation and procedure: The **range** is a simple measure of variability, calculated by subtracting the smallest value from the largest value. Although simple to calculate, the range tells us about only two values which may be atypical of the rest of the data set. The variance and standard deviation in a data set are both measures of dispersion of all the data around the mean. The base of both statistics is the **mean deviation** defined as:

$$\frac{\sum_{i=1}^{n} |x_i - \bar{x}|}{n}$$

where $\bar{x}$ is the mean, and $|x_i - \bar{x}|$ refers to the absolute value of $x_i - \bar{x}$. The **variance** is defined as the sum of the squared deviations $(x_i - \bar{x})^2$ divided by the number of data values, i.e.:

$$\text{variance} = \frac{\sum_{i=1}^{n} (x_i - \bar{x})^2}{n}$$

Variance as calculated in this equation is usually represented by the symbol σ^2. This is used with a large number of data values usually representing a population. However, when the variance is calculated for a small number of values, as we would have in a sample, the denominator of the above equation is modified to $n - 1$, and the variance is represented by s^2. Unfortunately, reporting the variance of a data set leaves the units squared. Taking the square root of the variance removes this problem, and produces what is known as the **standard deviation**, represented by σ or s depending on whether our data is a sample or not:

$$\text{standard deviation} = \sqrt{\frac{\sum_{i=1}^{n} (x_i - \bar{x})^2}{n}}$$

The variance/standard deviation are both measures of the *average* distance the data values are from the mean. The larger the value of σ^2 or σ (or s^2 or s), the more spread out is the set of data values. The standard deviation is often reported with a $\pm$ symbol, e.g., ±3.2. This signifies that the spread is the same on either side ($\pm$) of the mean. We can often visualise the spread in a data set by constructing a histogram or a stem-and-leaf plot in addition to calculating the variance or standard deviation.

Since the range is easy to calculate, and the standard deviation is only the square root of the variance, we will illustrate all three together. To obtain the range, we subtract the smallest value from the largest (**Step 1**). To obtain the standard deviation and variance, we will make use of a slightly different computational form of the equation used above:

Box 4.14   Handworked example: range, variance and standard deviation

x	x^2	x	x^2
67	4489	74	5476
99	9801	69	4761
110	12100	97	9409
146	21316	55	3025
102	10404	85	7225
108	11664	55	3025
69	4761	66	4356
194	37636	92	8464
52	2704	107	11449
91	8281	122	14884
69	4761	**43**	1849
102	10404	64	4096
181	32761	79	6241

$\sum x_i = 2398$ $\sum x_i^2 = 255342$ $(\sum x_i)^2 = 5750404$

Step 1   Tabulate your data as above. Range $= 194 - 43 = 151$.

Step 2   Obtain the sum of the observations $\sum x_i$ and then the mean $\bar{x}$ by dividing by the number of observations n:

$$\bar{x} = \frac{\sum\limits_{i=1}^{n} x_i}{n} = \frac{2398}{26} = 92.23$$

Step 3   Calculate the square x_i^2 of each observation, and sum the squares together to obtain $\sum x_i^2 = 255342$.

Step 4   Square the sum of the observations $\sum x_i$ to obtain $(\sum x_i)^2 = 5750404$.

Step 5   Substitute into the equation:

$$s = \sqrt{\frac{255342 - \dfrac{5750404}{26}}{26 - 1}} = \sqrt{\frac{255342 - 221169.38}{25}} = \sqrt{1366.90} = 36.97$$

The variance s^2 is 1366.90, and the standard deviation s is 36.97.

$$s = \sqrt{\frac{\sum\limits_{i=1}^{n} x_i^2 - \dfrac{\left(\sum\limits_{i=1}^{n} x_i\right)^2}{n}}{n - 1}}$$

This equation is used to avoid rounding error arising from repeated mean subtractions. First, obtain the sum of the observations $\sum x_i$ and then the mean $\bar{x}$ by dividing by the number of observations n (**Step 2**). Next, calculate the square of each observation, x_i^2, and sum the squares together to give $\sum x_i^2$ (**Step 3**). Square the sum of the observations $\sum x_i$ to obtain $(\sum x_i)^2$ (**Step 4**). Substitute into the above equation (**Step 5**) to obtain the variance and then take the square root to obtain the standard deviation.

We will again make use of the data on numbers of milch cows in Ireland (Table 4.4) to illustrate the calculation of the range, variance and standard deviation (see Boxes 4.14 and 4.15).

4.5.5   Box-and-whisker plot

Level of data: Interval or ratio.

Function of method: Visualisation of variance around the median.

What the method does: A box-and-whisker plot provides a summary plot of the variance of the data. In contrast to the standard deviation, it does this by using the median and the data quartiles (a quartile is the median of the data either below or above the median of the whole data set) and plotting them.

Box 4.15   MINITAB example: range, variance and standard deviation

Place the numbers of cows in a single column (C1).

Using pull-down menus: *Using command line:*

Range

    ↦ Calc   MTB > Range C1.
    ↦ Column Statistics…
    ↦ Input variable [ ]
    Click on C1
    ↦ Select
    Select Range
    ↦ OK

Standard deviation

    ↦ Calc   MTB > StDev C1.
    ↦ Column Statistics…
    ↦ Input variable [ ]
    Click on C1
    ↦ Select
    Select Standard Deviation
    ↦ OK

Variance

MINITAB does not perform a direct calculation of the variance similar to the above. There are various methods which can be used; the simplest is obtained by taking the square of the standard deviation calculated above. You can get MINITAB to do this as follows:

Using pull-down menus: *Using command line:*

    ↦ Calc   MTB > Let C3 = stdev(C1)*stdev(C1).
    ↦ Mathematical Expressions…
    ↦ Variable (new or modified) [ ]
    Type in C3
    ↦ Expression [ ]
    Type in stdev(C1)*stdev(C1)

MINITAB results:

Range

```
MTB > Range C1.
 RANGE = 151.00
```

Standard Deviation

```
MTB > StDev C1.
 ST.DEV. = 36.972
```

MINITAB will place the calculation for the variance of C1 in the first cell in C3.

Explanation and procedure: A box-and-whisker plot examines the variance in the data through the median and quartile values. It therefore examines the variance in a data set accounting for data values that deviate strongly from the rest of the data. These **out-** **lier** values would have a significant effect upon both the mean and standard deviation. Moreover, the plot allows us to determine the amount of variance either side of the median, whereas a standard deviation returns one value representing variance on both sides

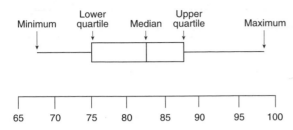

Figure 4.6 Box-and-whisker plot.

of the mean. In this manner we can determine the **skewness** of a data set. It is therefore a useful summary statistic, especially when we have a data set with extreme values. The construction of a box-and-whisker plot is based on five numbers: the median, the upper and lower quartiles, and the maximum and minimum (Figure 4.6).

A disadvantage of the box-and-whisker plot is that it is based only upon these five numbers, unlike the histogram or stem and leaf which are based on all the values in a data set.

To construct a box-and-whisker plot first rank all your data (**Step 1**). Next, determine the median (as described above) (**Step 2**). Now calculate the **quartile** values. The upper quartile (also known as the *upper hinge*) is obtained from the value which is midway between the median and the highest value, and the lower quartile (*lower hinge*) is obtained from the value which is midway between the median and the lowest value. Similar to the calculation of the median, this may not coincide with an actual data value, but may be in between two values (**Step 3**). Using these values, we can define the **interquartile range** (also known as the *midspread*) as the difference between the upper and lower quartiles. This is often used to identify the farthest points of the whiskers in the plot, and any extreme values (termed outliers) beyond these points (Hartwig and Dearing, 1979). Then, multiplication of the midspread by 1.5 produces a quantity known as the **step** (Erickson and Nosanchuk, 1992) (**Step 4**). Observations larger than the upper quartile plus the step, or smaller than the lower quartile minus the step, are identified as outliers. The largest and smallest values in the data set which are *not* beyond these limits are the end-points for the whiskers in the plot (**Step 5**). The final stage is to draw the box plot. Each end of the box is marked by the upper and lower quartiles, with the median drawn as a line across the box. The whiskers extend to the smallest and largest values identified in Step 5, with any outliers identified by a symbol. This

is often a dot, or as in the MINITAB software, an 'O' instead. Different data sets can be compared by constructing a series of adjacent box-and-whisker plots, and this often provides a useful comparison prior to a statistical test to test for differences between data sets such as the analysis of variance or *t*-test, which we will encounter in Chapter 5. As an example we will plot the sheep from Table 4.3: see Boxes 4.16 and 4.17.

4.6 Probability

Probability is concerned with the likelihood or **chance of occurrence** of a certain event. An **event** can be defined as an outcome of sampling a set of data. Events may be very simple, such as tossing a coin with two possible outcomes (heads or tails), or more complex, such as a sample of questionnaire respondents with certain characteristics defined in terms of a variety of variables. In both cases, by using simple concepts of probability, we can estimate the chances of these events occurring. Probability forms the basis of inferential statistics and hypothesis testing which we will consider in detail in Chapter 5. We will outline only the basics of probability here: the reader is directed to Callender and Jackson (1995) for further details. There are a variety of definitions of probability in use in geographical contexts, but all make the distinction between **objective** (or *a priori*) **probabilities**, which are based on the true probabilities in the state of complete information, and **subjective** (or *a posteriori*) **probabilities** derived from experimental or empirical study in a state of limited/incomplete information. It is the latter which is of most use in geographical research, since we usually have access only to a sample of limited size, and simply do not know the characteristics/size of the population we wish to study. However, the various probability laws are best described using simple objective probabilities.

4.6.1 Set theory

Before we talk about probabilities and events, it will be useful to outline some very simple elements of **set theory**. As an example, we have a class of eight geography students who are divided into groups depending on their gender, and whether they are more interested in physical or human geography. If we consider group A to contain all female students of whom there are four, we can define this as set A:

Box 4.16   Handworked example: box-and-whisker plot

x	$R(x)$	
532	30	
195	104	
183	146	
326	170	
645	177	Quartile = (183 + 177)/2 = 180
146	183	
268	192	
30	195	
235	228	
170	235	
511	268	Median
228	304	
2379	326	
177	511	
676	524	
746	532	Quartile = (645 + 532)/2 = 588.5
304	645	
104	666	
192	676	
666	746	
524	2379	

Step 1 Rank all the data values to create a new column $R(x)$.

Step 2 Determine the median. We have an odd number of data values (21) so the median is the eleventh value after ranking (268).

Step 3 Calculate the quartile values by calculating the median of values above and below the median. We obtain 180 for the lower quartile and 588.5 for the upper quartile.

Step 4 Calculate the interquartile range (588.5 − 180 = 408.5), and then obtain the step:

$$\text{step} = 1.5 \times 408.5 = 612.75$$

Step 5 Add the step to the upper quartile (588.5 + 612.75 = 1201.25), and subtract the step from the lower quartile (180 − 612.75 = −432.75). The largest value in the data set which does *not* exceed 1201 is 746, which is the point to which the upper whisker is drawn. Since the lower limit is negative, the lower whisker is drawn to the smallest value in the data set (30).

Step 6 Draw the box plot (Figure 4.7) using these values.

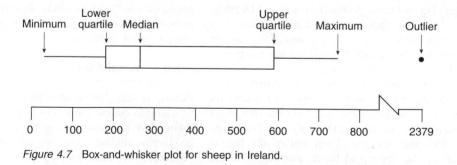

Figure 4.7   Box-and-whisker plot for sheep in Ireland.

Box 4.17   MINITAB example: box-and-whisker plot

Place the numbers of sheep in a single column (C1). In this instance, we are plotting the box-and-whisker plot across the page (Transpose). Omission of this command will orient the plot from top to bottom of the page. Several variables may be selected and plotted side by side to compare distributions between groups (e.g., prior to an ANOVA or Kruskal–Wallis test for differences).

Using pull-down menus:

➥ Graph
➥ Boxplot...
Click on C1
➥ Select
➥ Options
Click on 'Transpose X and Y'
➥ OK
➥ OK

Using command line:

MTB > Boxplot C1;
SUBC > Transpose;
SUBC > Box;
SUBC > Symbol;
SUBC > Outlier.

MINITAB results (Figure 4.8):

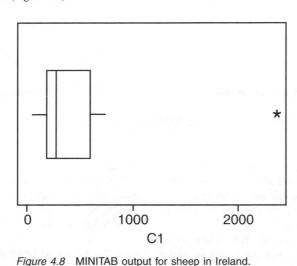

Figure 4.8   MINITAB output for sheep in Ireland.

A = {Mary, Heather, Nic, Carol}

We can also define another set B to contain all those students (male or female) who are more interested in human geography:

B = {Heather, Carol, Steve, Chris}

Two important operations we can perform on these sets is to add them together or take the product (multiply). For the former, termed a **union** operation, which we represent by the symbol '∪':

A ∪ B = {Mary, Heather, Nic, Carol, Steve, Chris}

which consists of students in set A OR in set B, i.e., female *or* more interested in human geography (or both). If we take the product of set A and set B, termed an **intersection** and represented by the symbol '∩', we obtain those students who are in both set A AND set B:

A ∩ B = {Heather, Carol}

i.e., just those female students who are more interested in human geography. In the examples above, there are elements of set A and set B which are not **mutually exclusive**. The concept of mutual exclusivity is an important one for probability as we will see later. If we add a set C of the form:

C = {Mike, Jim, Joe, Pete}

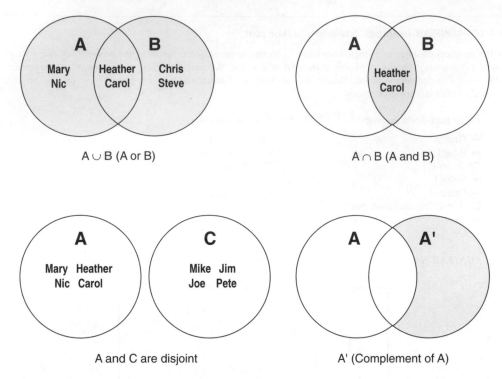

Figure 4.9   Venn–Euler diagrams of geography students.

then sets A and C would not contain any common elements. In set theory, sets A and C would be termed **disjoint**. We can visualise the exact elements these include in terms of shading the appropriate parts of **Venn–Euler diagrams** (Figure 4.9).

We can also think of events in terms of sets. If we have a group of ten students in a class, with six males and four females, and we are intending to sample students on the basis of gender, we will create two subsets, and therefore two possible events/outcomes:

E_1: the student is male
E_2: the student is female

In this case the events would form disjoint subsets; they are mutually exclusive: i.e., if E_1 occurs then E_2 cannot occur. However, if we included a third outcome, such as 'tall', with three tall male students and two tall female students out of the group of ten:

E_3: the student is tall

then this would create events which are not mutually exclusive, since clearly E_1 and E_3 can occur together, as can E_2 and E_3. The two most common logical com-

binations of these events are OR, i.e., the probability of E_1 OR E_2 ($E_1 \cup E_2$) occurring, and AND, i.e., the probability of E_1 AND E_3 ($E_1 \cap E_3$) occurring. The visualisation of such sets in probability space is similar to the shaded areas of the diagrams in Figure 4.9.

4.6.2   Laws of probability

If we take a random sample of one student from the group of ten discussed above, we can calculate the probability $p(E_1)$ as the fraction of males in the total group, i.e., $6/10 = 0.6$. In other words, we are basing our probability estimate on the **relative frequency** with which males occur in our group. For females, $p(E_2) = 4/10 = 0.4$. In this case, both E_1 and E_2 are mutually exclusive and exhaustive; the **complement** of E_1, denoted by E_1', refers to the alternative of event E_1, and here

$$p(E_1') = 1 - p(E_1) = p(E_2)$$

In fact, we can say that

$$p(E_1') + p(E_1) = 0.4 + 0.6 = 1$$

and

Box 4.18 Probability laws

Addition law for mutually exclusive events:	$p(E_1 \cup E_2) = p(E_1) + p(E_2)$
Addition law for non-mutually exclusive events:	$p(E_1 \cup E_3) = p(E_1) + p(E_3) - p(E_1 \cap E_3)$
Multiplication rule for independent events:	$p(E_1 \cap E_3) = p(E_1) \times p(E_3)$
Multiplication rule for non-independent events:	$p(E_1 \cap E_3) = p(E_1) \times p(E_3 \mid E_1)$

$$p(E_1 \cup E_2) = p(E_1) + p(E_2) = 0.4 + 0.6 = 1$$

This introduces the first probability law: the **addition law for mutually exclusive events** (Box 4.18).

If we consider the probability of the student in our sample being male *or* tall, i.e., $p(E_1 \cup E_3)$, then since $p(E_3) = 5/10 = 0.5$:

$$p(E_1 \cup E_3) = p(E_1) + p(E_3) = 0.6 + 0.5 = 1.1$$

Probabilities cannot exceed 1, so we cannot apply the simple addition formula in this case. This has to be adjusted for those people who are *both* tall and male. These events are not mutually exclusive, since clearly a student could be both, and would be counted twice in the simple addition formula. We have to subtract those who are both tall and male. In fact, the adjusted formula becomes the **addition law for non-mutually exclusive events**:

$$p(E_1 \cup E_3) = p(E_1) + p(E_3) - p(E_1 \cap E_3)$$

We have introduced the **multiplication rule for independent events** into this formula:

$$p(E_1 \cap E_3) = p(E_1) \times p(E_3)$$

This is saying that $p(E_1 \text{ AND } E_3) = p(E_1) \times p(E_3)$. Therefore, since being tall and being male are independent of each other:

$$p(E_1 \cap E_3) = 0.6 \times 0.5 = 0.3$$

which we can check since there are a total of three tall male students out of a total of ten, i.e., a probability of $3/10 = 0.3$. If our two events were not independent, we would say that the probability of the second event was conditional upon the first, and the **multiplication rule for non-independent events** would be used instead.

4.6.3 Factorials and probability trees

So far, the calculation of the relative frequencies and probabilities of events has been relatively straightfor-

ward since we have been selecting a sample of only a single individual. Since any of our ten students could be selected, we have a total of ten choices. We can calculate the total number of choices using **factorials**. For example, selecting a single student from a group of ten:

$$\frac{10!}{1! \times 9!} = \frac{3628800}{1 \times 362880} = 10$$

where $10! = 10 \times 9 \times 8 \times 7 \times 6 \times 5 \times 4 \times 3 \times 2 \times 1$. Similarly, the total number of males which can be sampled is:

$$\frac{6!}{1! \times 5!} = \frac{6}{1} = 6$$

The probability of selecting one male from this group, i.e., $p(\text{one male}) = 6/10 = 0.6$. This is quite simple. However, if we were to select two students from this group, there are many more choices available. For any two students out of ten we would have 45 combinations:

$$\frac{10!}{2! \times 8!} = \frac{3628800}{2 \times 40320} = 45$$

So, if we wanted to calculate the probability of selecting some combination of one male and one female out of our group of students, we first calculate the choices of one male:

$$\frac{6!}{1! \times 5!} = \frac{6}{1} = 6$$

and similarly with one female:

$$\frac{4!}{1! \times 3!} = \frac{4}{1} = 4$$

Therefore, $p(\text{one male and one female}) = 6 \times 4/45 = 24/45 = 8/15$. We can visualise the possible outcomes of choosing two students in terms of a **probability tree**: (Callender and Jackson, 1995)

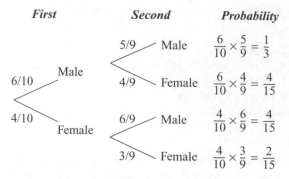

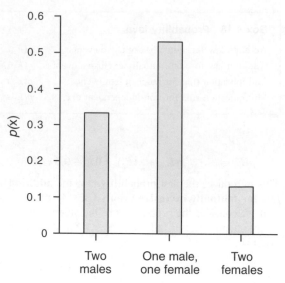

Figure 4.10   Probability distribution of male/female student selection.

In this case, we can estimate the probability of any given path, by applying the multiplication law of probability, multiplying the probability of the individual branches. If more than one path leads to the same result, the probabilities of each of the paths are added. In the above example, we find that the second and third paths lead to the result of selecting one male and one female. Therefore, for the second path:

$$\frac{6}{10} \times \frac{4}{9} = \frac{4}{15}$$

and similarly for the third path:

$$\frac{4}{10} \times \frac{6}{9} = \frac{4}{15}$$

Therefore, p(one male and one female) is:

$$\frac{4}{15} + \frac{4}{15} = \frac{8}{15}$$

which is the same as the result using factorials above. Note that the probabilities in the probability tree above sum to 1. Calculating probabilities in this fashion for each combination allows us to plot the probability for each outcome (Figure 4.10).

4.6.4 Probability distributions

The above is a very simplified example of a discrete probability distribution with very few outcomes. Probability distributions are based on the concept of a random variable (RV), which takes on numerical values according to the outcome of an experiment (Chatfield, 1995) and which may be discrete or continuous. There are a variety of probability distributions, of which the most common and useful in a geographic context are listed in Table 4.6. Discrete distributions are defined from a **point probability**

function represented by $p(x = X)$, whereas continuous distributions are defined from a **probability density function** (pdf) represented by $f(x)$. The positive **binomial distribution** (often referred to simply as the binomial distribution) is a discrete distribution used for events which have two outcomes: heads or tails, yes or no, male or female.

The formula for the binomial distribution looks intimidating, but contains some familiar components. The first term is similar to the above factorial-based combinations formula to determine the possible combinations in the data set. Parameter n is the number of the outcome in which we are interested (e.g., heads in a coin toss), x is the total number of trials (e.g., the number of times an individual is sampled, or a coin is tossed), and p is the probability associated with the outcome we are interested in. For example, if we toss a coin eight times, we can calculate the probability of obtaining three heads:

$$p(x = 3) = \frac{8!}{3!(8 - 3)!} \times 0.5^3(1 - 0.5)^{8-3}$$

$$p(x = 3) = \frac{8 \times 7 \times 6}{6} \times 0.125(0.03125) = 0.21875$$

If we calculate the probabilities for all other combinations of heads (i.e., 0, 1, 2, 4, 5, 6, 7, 8) we can graph the results to produce a histogram of the discrete probability distribution (Figure 4.11).

Table 4.6 Common probability distributions

Binomial	$p(x = X) = \binom{n}{x} \times p^x (1-p)^{n-x}$	where $\binom{n}{x} = \dfrac{n!}{x!(n-x)!}$
Poisson	$p(x = X) = \dfrac{\mu^x \times e^{-\mu}}{x!}$	
Normal	$f(x) = \dfrac{1}{\sigma\sqrt{2\pi}} \times e^{-\frac{1}{2}\left(\frac{x-\mu}{\sigma}\right)^2}$	

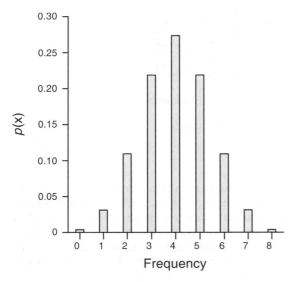

Figure 4.11 Binomial distribution of coin toss.

Note that the probabilities sum to 1, and are discrete. For $p = 0.5$, the binomial probability distribution, as in Figure 4.11, is symmetrical, but where $p \neq 0.5$ the distribution becomes skewed.

The **Poisson distribution** is also a discrete distribution, and occurs when events occur independently, but with a constant average frequency. The classic example of a Poisson distribution is the numbers of

soldiers in the Prussian army kicked to death by a horse over a 20-year period! In this case, we might expect each incidence of a horse-kick death to be rare and to occur independently. For 10 army corps over a 20-year period, 122 deaths were observed, with a population mean (μ) of 0.61. The data are presented in Table 4.7 and come from Ehrenberg (1975).

To fit the Poisson distribution to the observed data in Table 4.7, we substitute into the formula for the Poisson distribution above for each observation. For example:

$$p(x = 0) = \frac{0.61^0 \times e^{-0.61}}{0!} = \frac{0.54335}{1} = 0.54335$$

We have a period over which 200 observations are possible (10 corps × 20 years), therefore to obtain the number of horse-kick deaths predicted by the Poisson distribution, we multiply 200 × 0.54335 = 109 (rounding up). We can graph the results to produce a histogram of the probability distribution in Figure 4.12(a). Note that the shape of the Poisson distribution will change depending on the magnitude of the mean. For example, if we change the mean in the above example from 0.61 to 1.61, the shape of the distribution changes to that in Figure 4.12(b).

The last probability distribution we shall consider is the **normal** (or Gaussian) **distribution**. This is an example of a **continuous distribution**. The probability distribution of a continuous random variable is in the form of a normal distribution, and is referred to

Table 4.7 Horse-kick deaths in the Prussian Army, and their theoretical Poisson distribution

	Frequency of deaths per year for 10 Corps for 20 years					
	0	**1**	**2**	**3**	**4**	**5**
Observed	109	65	22	3	1	–
Poisson	109	66	20	4	1	0.08

Source: Ehrenberg 1975: 192.

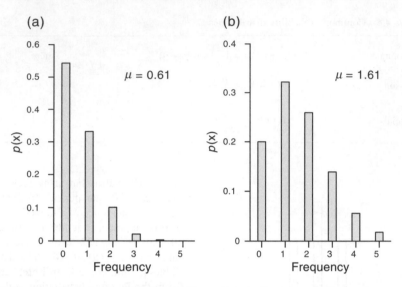

Figure 4.12   Poisson distributions of horse-kick deaths (μ = mean).

as a probability density function (pdf). The normal distribution is the most important and useful continuous distribution, particularly since the distribution of many continuous variables is similar to this form. Where data are sampled from a non-normal distribution, the **central limit theorem** states that the sampling distribution of the mean (see Section 3.3) becomes more like the normal distribution as the size of the sample increases (Pett, 1997: 14). The normal distribution pdf always has the same symmetrical bell-shaped form, which depends only on the value mean (μ) and standard deviation (σ). Example normal distribution pdfs are displayed in Figure 4.13.

For a given normal distribution pdf, we can calculate the probability that a value occurs within certain limits by calculating the area under the curve between the limits (Figure 4.14).

In fact, because of its symmetrical shape, the mean, mode and median are identical, and we can say that:

- **68.26%** of observations lie within **±1 standard deviation** from the mean
- **95.44%** of observations lie within **±2 standard deviations** from the mean
- **99.72%** of observations lie within **±3 standard deviations** from the mean

The calculation of areas and associated probabilities from a normal distribution pdf can be achieved by the use of a computer. An alternative is to make use of standard normal probability tables, which list the areas for different intervals under a certain type of normal distribution pdf termed a **standard normal distribution**, which has a mean of zero and a standard deviation of one. We can transform a normal distribution to a standard normal distribution by calculating the units of standard deviation, termed **Z-scores** (Table A.7):

$$Z = \frac{x - \mu}{\sigma}$$

Where the Z-score = 1, this is equivalent to one standard deviation. For example, as part of a medical geography study we may be interested in the pollutants emitted from a hospital incinerator located in a city. In this case, we are interested in the quantity of polychlorinated biphenyls (PCBs). If the average daily emission rate is 1.5 ppm (parts per million) with a standard deviation of 0.75 ppm, we can determine (Figure 4.15) the number of days per year on which emissions are:

1   Above average
2   Less than 2.25 ppm
3   Between 1.75 and 2.5 ppm.

Case 1   In this case, we do not need to estimate the Z-score, since we know that the total area under the normal distribution pdf must correspond to a probability of 1. The probability of the emission being above average is equivalent to the shaded portion of

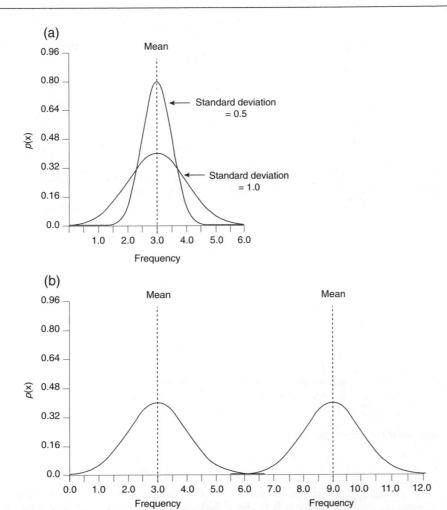

Figure 4.13 Normal distribution pdfs for different standard deviations and means:
(a) same means, different standard deviations; (b) same standard deviations, different
means.

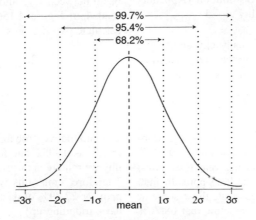

Figure 4.14 Normal distribution pdf and spread
of observations around the mean.

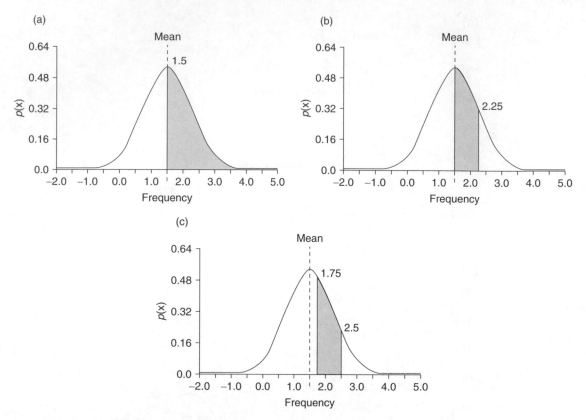

Figure 4.15 Area under the normal distribution.

Figure 4.15(a): this covers 50% of the normal distribution pdf, and therefore the probability is 0.5. The number of days per year would therefore be 182.5 (or 183 if we round up). In fact the Z-score in this case equals 0, which if we look at a table of cumulative probabilities for the normal distribution (Appendix A, Table A.7) gives a probability of 0.5.

Case 2 The probability of a daily emission being less than 2.25 ppm is given by the area shaded in Figure 4.15(b). If we calculate the Z-score we obtain:

$$Z = \frac{2.25 - 1.5}{0.75} = \frac{0.75}{0.75} = 1$$

From the table of cumulative probabilities, the probability associated with a Z-score of one is 0.8413. Therefore, for 84.13% of days per year (307 days) PCB emissions will be less than 2.25 ppm. In fact, again we don't *really* need to calculate the Z-score or look up the probability, since we know that 68.26%

of observations lie within ±1 standard deviation from the mean. Hence, the shaded area in Figure 4.15(b) is $0.5 \times (1 - 0.6826) + 0.6826 = 0.8413$.

Case 3 Lastly, the probability of daily emission being between 1.75 and 2.5 ppm is given by the area shaded in Figure 4.15(c). In this case we need to calculate two Z-scores:

$$Z = \frac{2.5 - 1.5}{0.75} = \frac{1.0}{0.75} = 1.33$$

$$Z = \frac{1.75 - 1.5}{0.75} = \frac{0.25}{0.75} = 0.33$$

The cumulative probability associated with a Z-score of 1.33 is 0.9082, and for 0.33 is 0.6293. The probability we require is therefore $0.9082 - 0.6293 = 0.2789$. Therefore, emissions will be between 1.75 and 2.5 ppm for 27.89% of days per year, or 101.7 days (102 days, rounding up).

4.7 Transforming data

Often, the tests which we will detail in Chapter 5 require the data to be normally distributed. If your data are not normally distributed, for example positively or negatively skewed, you can use a process of **data transformation** to convert your data into a form suitable for other types of analysis. You may also want to transform your data so that attention can be focused on features of particular interest, or to convert the data units into more relevant units (e.g., transforming data collected in miles to kilometres). As a rule you can transform data only within and down the measurement scale (see Section 3.2). For example, ratio data can be transformed to other ratio data and to interval, ordinal or nominal data, but ordinal data can be transformed only into other ordinal classes or nominal data. Nominal data can be transformed only by merging classes together to form other nominal classes.

Ratio and interval data can be transformed using a large number of methods – Robson (1994) reports on 35. We will concern ourselves with five.

1   **Transforming the data units** of a data set is simply a case of multiplying the data values by an appropriate constant. For example, to convert data generated in miles to kilometres you simply multiply each data value by the constant 1.6094. To reverse the transformation, and convert kilometres to miles, multiply by 0.6214.

2   **Transforming the data** towards that of a normal distribution. This can often be achieved by a non-linear transform such as the **square root**, **logarithm** or **inverse** ($1/x$). Figure 4.16(a) displays data for the percentage of the population aged 0–4 years (Column C3 in Table 5.6). The data are positively skewed. As Figure 4.16(b) and (c) demonstrate, transforming the data by calculating the square root or log has the effect of removing the skew from the data to produce a more normal distribution.

A variable that is negatively skewed might be transformed towards a normal distribution with a square or antilog, and a distribution that is flat (platykurtic) can be made more normal with a $1/x$ transformation (Pett, 1997).

3   **Transforming the range** of a data set involves truncation of the data set, assigning all values above

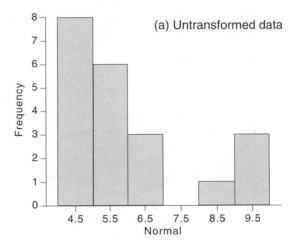

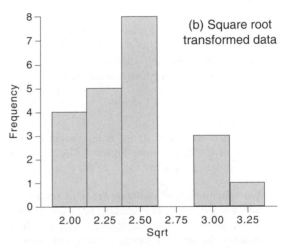

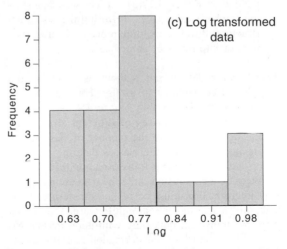

Figure 4.16   Transforming data using the square root or natural log.

or below a certain value to a set value or removing them completely. This is particularly useful when the data set contains a few residual values (data that are unusual by being much greater or less than the rest of the data set). These values tend to exert a great influence on calculations such as the mean. Assigning them to a set value or removing them from the data set removes their excessive influence on the analysis. For example, Figure 4.17(a) displays adult literacy rates (%) for 17 of the world's less developed countries. Two residual data values (Western Somoa 91, and Maldives 78) are clearly visible. The mean of these data is 27.19 with a standard deviation of 20.06. When these values are removed (Figure 4.17(b)) the mean becomes 21.47 with a standard deviation of 8.64, and the data are more normally distributed.

4   **Transforming by data standardisation**:   A fourth method also seeks to control for residual values by **standardising** the data. This is achieved by calculating a Z-score for each data value:

$$Z = \frac{x - \mu}{\sigma}$$

where $\bar{x}$ is the mean, and σ is the standard deviation, of the data values. Figure 4.17(c) displays the standard scores for the example data.

5   **Transformation of the measurement scale**:   A fifth method, rather than seeking to retain the data in a ratio/interval format, transforms the data into **ordinal classes**. For example, we could convert the data in Figure 4.17(a) into classes of increasing (ranked) value (e.g., mirroring the histogram structure) through a simple process of categorisation, or into ranks or percentages.

Transforming data is not without its problems and you should be cautious in its use. For example, we could alter the data within Figure 4.4 by reassigning the values into classes of different sizes or by increasing or decreasing the number of classes. Not surprisingly, changing the classes data are assigned to, or using a non-linear transformation, can lead to an alternative conclusion once the data are analysed. Any interpretation must be of the transformed variable, which may not be straightforward. You should therefore transform data only with good reason and choose your method of transformation carefully. Under no circumstances should you 'massage' the data to give biased or untruthful messages. The prime safeguard

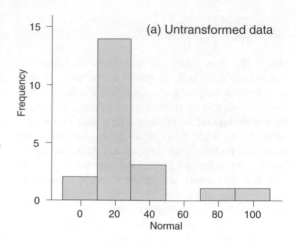

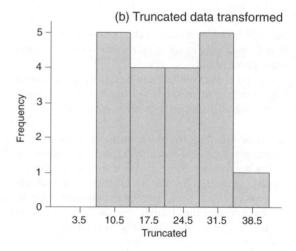

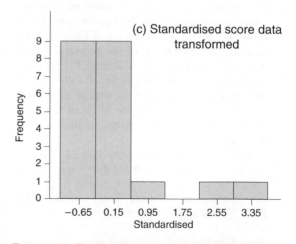

Figure 4.17   Transforming data range, standardisation and measurement scale.

against this is your honesty and integrity, although inadequate explanation as to why data were transformed will draw suspicion from readers of the final report (see Chapter 10).

4.8 Summary

After reading this chapter you should:

- be able to code quantitative data;
- be able to enter quantitative data into a computer for analysis;
- be able to check quantitative data for errors and mistakes;
- be able to use MINITAB;
- be able to perform an Initial Data Analysis;
- understand the basic concepts of probability;
- be able to transform your data.

In this chapter we have introduced the first steps in analysing quantitative data: preparation, exploration and description. Each is a vital component of quantitative analysis and should receive careful attention. Poor data preparation through hurried coding, rushed data entry or inadequate checking will produce data that contain errors and mistakes. Any further analysis of these data will retain these flaws and may well lead to invalid conclusions. Exploration and description of the data, whether statistical or graphical, provides some valuable background information that can be used to assess the viability of more sophisticated forms of analysis described in the next chapter. Again, failure to explore the data fully at this stage may lead to invalid analysis later. Whilst the discussion within this chapter has centred upon describing single variables, in the next chapter we turn to examining the associations and relationships between variables.

4.9 Questions for reflection

- *Outline the practical steps which you should adopt to check and safeguard your data.*

- *Can most data errors be spotted and rectified?*
- *Explain why computer programming your statistics is useful for analysing large quantities of data.*
- *How can you enter data into MINITAB?*
- *What is the purpose of an IDA?*
- *Compare and contrast the histogram and box-and-whisker plots as methods for the examination of the distribution of a data set.*
- *Which probability distribution might be suitable for modelling the predicted number of tornadoes in an area over the period of a year, if we know the average occurrence?*
- *Why transform data?*

Further reading

Bourque, L.B. and Clark, V.A. (1992) *Processing Data: The Survey Example*. Sage University Papers Series on Quantitative Applications in the Social Sciences 07-085. Sage, Newbury Park, CA.

Callender, J.T. and Jackson, R. (1995) *Exploring Probability and Statistics with Spreadsheets*. Prentice Hall, London.

Cooke, D., Craven, A.H. and Clarke, G.M. (1992) *Basic Statistical Computing*. Edward Arnold, London.

Cramer, D. (1997) *Basic Statistics for Social Research: Step by Step Calculations and Computer Techniques using MINITAB*. Routledge, London.

Erickson, B.H. and Nosanchuk, T.A. (1992) *Understanding Data*. University of Toronto Press, Toronto.

Hartwig, F. and Dearing, B.E. (1979) *Exploratory Data Analysis*. Sage University Papers Series on Quantitative Applications in the Social Sciences 07-016. Sage, Newbury Park, CA.

Press, W.H., Flannery, B.P., Teukolsky, S.A. and Vetterling, W.T. (1989) *Numerical Recipes – The Art of Scientific Computing*. Cambridge University Press, Cambridge.

Ryan, B.F. and Joiner, B.L. (1994) *MINITAB Handbook*. Duxbury Press, Belmont, CA.

CHAPTER 5

Analysing and interpreting quantitative data

This chapter covers

5.1   Introduction
5.2   Classifying tests
5.3   Tests of significance
5.4   Choosing the right test
5.5   The tests
5.6   Parametric tests
5.7   Non-parametric tests
5.8   What do the test results tell you?
5.9   Summary
5.10   Questions for reflection

5.1   Introduction

This chapter develops the themes discussed in Chapters 3 and 4 and is concerned with the formal statistical analysis of quantitative data, and the interpretation of results from such analysis. You should already be familiar with the process of setting up objectives or hypotheses, the construction of a sampling frame, the various types of data and the collection of data, as well as the pre-processing and the preliminary steps of data description which make up an Initial Data Analysis (IDA). Recall from Chapters 3 and 4 that these elements correspond to steps 1–4 of Chatfield's stages of a statistical investigation (see Box 5.1).

The remaining steps 5–7 are concerned with more formal statistical analysis procedures, the implications of a statistical investigation for placing the results in the wider context of established knowledge as well as in the more immediate context of the research project. It is step 5 which we shall examine in more detail in this chapter. Although elements of steps 6–7 are considered here, these elements are of more general application and are covered in more detail in Chapter 10.

It is not possible to cover in detail within the confines of one chapter all of the statistical methods of potential use to your research. We will restrict our

Box 5.1   Stages of a statistical investigation

1   Understand the problem and its objectives.
2   Collect the data.
3   Look at the quality of the data and pre-process the data.
4   Perform an initial examination of the data including description.
5   *Select and carry out appropriate statistical analysis, often suggested by the results of 4.*
6   *Compare the findings with any previous results.*
7   *Interpret and communicate the results.*

Source: Adapted from Chatfield 1995.

focus to some of the techniques concerned with **statistical inference** and **model construction** which have been widely used by geographers. For further information on the methods described here, and additional methods which are not covered (e.g., multivariate methods), you should try to make reference to one of the various texts available on statistical methods written for geographers and social scientists which are listed at the end of this chapter. We concentrate on some of the simpler statistical methods most likely to be of use for your research. We aim to illustrate the mechanics of each statistical method using a clear step-by-step description of the procedure, with worked examples using data from real research projects. Each worked example is explained using a handworked calculation, as well as a calculation using the MINITAB statistical package (see Chapter 4). The mathematical details are kept to a minimum throughout, and knowledge of nothing more complex than a simple summation (Σ) is assumed. Although some details of using MINITAB for windows are provided, further reference should be made either to the *MINITAB Handbook* (Ryan and Joiner, 1994) or to a book dealing with applications of MINITAB in social science (e.g., Cramer, 1997).

5.2 Classifying tests

After having performed an IDA on the data, you may suspect that a more formal statistical procedure may be appropriate. Recall from Chapter 4 that by this stage you should have a clear objective in mind, and one or more testable hypotheses about the data should have been established. The next step is to select an appropriate statistical method to apply. Many statistical methods are often known as **significance** (or **hypothesis**) **tests** as they examine the extent to which a hypothesis might be significant. In the main it is these tests that we shall focus upon in this chapter, although we will also detail some other methods.

Statistical tests can be subdivided into two groups which are known as **parametric** and **non-parametric**. Parametric tests make more assumptions about the data – usually about the underlying **normal distribution** (see Chapter 4) – but provide more powerful analysis because of the more detailed nature of the data. As a result, parametric analysis is more sensitive to the data and is more robust with less chance of error. In general, parametric tests are used to examine ratio or interval data. However, we often find that such data do not follow a normal distribution. In these circumstances, a mathematical transformation (e.g., logarithmic, square root, power function) of the data may be used to obtain an approximately normal distribution (see Chapter 4). It is worth noting that in the face of a non-linear transformation the most meaningful variable to interpret from an analysis is often the observed (untransformed) variable. As a general rule mathematical transformations should be avoided (Chatfield, 1995) and a non-parametric test used instead. Since distributional characteristics can be established only for interval and ratio data, non-parametric tests are used to examine nominal or ordinal data sets.

Data sets that are to be compared can be either **related** or **unrelated** in nature. Related data sets are those which represent the same sample population but are collected at different times or under different conditions. For example, related data sets might be comprised of the populations of the same towns over different times, or the performance of the same respondents to draw the same sketch map in interview and non-interview conditions. Unrelated data sets represent different populations. For example, unrelated data sets might consist of the amount of unemployment in large and small towns, or the performance of males and females on the same sketch map exercise.

5.3 Tests of significance

From some of the IDA methods detailed in the previous chapter, you may have detected what you think might be a significant difference between samples in your data. For example, you might have interviewed a random sample of 50 households from enumeration district (ED) A to compare against 50 households from ED B to determine whether income levels are different. Using some of the IDA methods described in Chapter 4 you have calculated a mean income of £18,434 for ED A and £20,210 for ED B. Although the estimate for ED B is larger, how do we determine if we have a *real* difference, or whether the difference has occurred purely *by chance*? The answer is to perform a **significance test.** There are four generic components common to each significance test which we describe below: **null/research hypothesis**; **calculated/critical values** of a statistic, **one/two-tailed tests**, and **significance level**. These steps follow the scheme of Kanji (1994) and Neave (1976).

Null/research hypothesis

Statistical tests start out with the assumption that there is *no real difference* between your data sets. In other words, the various treatments or changes in conditions which you might have varied to select your data sample appear not to have produced data which reflect such changes, or to have produced data in which any changes that exist are due purely to chance. Of course, from Chapter 4, such results may have been caused by the selection of too small a sample, but the possibility exists that the conditions you have varied in fact make *no real difference* to the data you have collected: i.e., there is an absence of a real effect.

This is termed the **null hypothesis** and is often denoted by H_0. Formally, the null hypothesis is usually expressed H_0: $\mu_1 = \mu_2$ where μ_n is the mean for each group (or often the median if a non-parametric test is being used), and the subscript n denotes the group. The alternative to this is called the **research hypothesis** (also referred to as the **alternative hypothesis**), denoted by H_1 and expressed by H_1: $\mu_1 \neq \mu_2$. It is important that you express both H_0 and H_1 in the context of your own research problem *before collecting your data and before starting your analysis*. In the enumeration district income example we might have formulated the following hypotheses:

H_0: $\mu_A = \mu_B$ ___There is no significant difference___ between the mean income of households in enumeration district A as compared with the mean income of households in enumeration district B: mean household income ___is not___ influenced by geographical location (i.e., location in a specific enumeration district).

H_1: $\mu_A \neq \mu_B$ ___There is a significant difference___ in the mean household income for households in enumeration district A as compared with enumeration district B: mean household income ___is___ influenced by geographical location (i.e., location in a specific enumeration district).

We would then collect a sample of data from each ED. In order to determine which hypothesis to accept, to determine whether or not your sampled data sets are consistent with the null hypothesis or the research hypothesis, we would perform a probability-based significance test. However, before such a test is carried out, we must determine how big any difference has to be, to be considered real beyond that expected due to chance.

Calculated/critical values

For each significance test, we can produce a probability distribution of a **test statistic**, termed a *sampling distribution under the null hypothesis*, calculated on the basis that the null hypothesis is true. A simple example might be the probability distribution curve of the Student's *t*-statistic in Figure 5.1. A large difference between data sets corresponds to a probability towards the tails of the distribution, i.e., differences occurring by chance are unlikely. We can determine whether any difference between the data sets is large enough not to have occurred by chance, by determining where the difference occurs in relation to the tails of the distribution. We can define a critical or **rejection region** as that part of the probability distribution

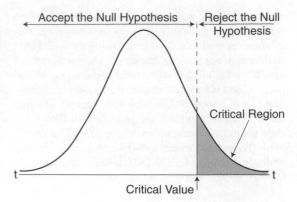

Figure 5.1   Rejection region for a probability distribution.

beyond a **critical value** of a test statistic at a certain probability, as displayed in Figure 5.1. We compare this critical value with the **test statistic** calculated from the data. If the calculated value of the test statistic falls within the rejection region, the differences in the data are *unlikely* to have occurred by chance. More specifically, we can say that data are inconsistent with the null hypothesis which can then be rejected, and the research hypothesis accepted. It is important to note that if our calculated value does not fall within the rejection region, *this does not prove the truth* of the null hypothesis, but merely fails to reject it. We have deliberately talked in general terms so far to introduce some of the concepts. To proceed, we will need to become more specific regarding our definition of the rejection region.

Significance level

The size of the rejection region chosen is determined by the **significance level**. This is the probability of obtaining a difference as great as or greater than the one observed, given that the null hypothesis is true. There are a variety of standard levels for such probability, those most commonly used being a 1 in 20 chance (5% or $p = 0.05$), and a 1 in 100 chance (1% or $p = 0.01$). By the selection of either a 5% or a 1% significance level, what we are saying is that we are willing to accept either a 5% or a 1% chance of making an error in rejecting the null hypothesis when it is in fact true. This is known as a **Type I error** (Table 5.1) and is often represented by the parameter α. Conversely, a **Type II error** represents the probability of not rejecting the null hypothesis when it is in fact false, and is often represented by the parameter β.

Table 5.1 Possible results from a significance test.

Decision	Null hypothesis is true	Null hypothesis is false
Null hypothesis not rejected	Correctly not rejected	Type II error
Reject null hypothesis	Type I error	Correctly rejected

Sources: Clark and Hosking 1986; Wright 1997.

In order to simplify the calculation of significance tests without computers, tables of critical values have been calculated at specific probabilities, usually for $\alpha = 0.05$ (5%) and 0.01 (1%). How do we choose between these significance levels? Strictly speaking this depends on whether you are more concerned to avoid making a Type I error or a Type II error. Choosing a 1% significance level *decreases* the possibility of making a Type I error but *increases* the probability of making a Type II error. Some statistical packages such as MINITAB directly report the significance of the calculated test statistic in terms of a probability value p. Where $p < 0.05$ this would indicate a significant result at a 0.05 (5%) level, and $p < 0.01$ would indicate a significant result at a 0.01 (1%) level, often termed 'highly significant'.

The results of significance tests are often reported in a standard form, for example: 'The difference was significant (F (2,16) = 16.59, $p < 0.05$)'. In this example, we are saying that the calculated value of F with 2 and 16 degrees of freedom is 16.59, which is significant at the 0.05 level. Do not worry if the meaning of 'F' and 'degrees of freedom' is unclear at this point: F is a calculated test statistic for the F distribution, and the degrees of freedom are determined by the sample size. We will explain these terms in the context of some of the tests in Sections 5.5–5.7 below. You might also find asterisks after the computed value of a significance test, for example $t = 3.12**$ (where t is from the Student's t-test discussed below). The asterisks represent the following probability ranges (Sokal and Rohlf, 1995):

$$* = 0.05 \geq p > 0.01 \qquad ** = 0.01 \geq p > 0.001$$
$$*** = p \leq 0.001$$

If we have a significant result, it is often useful to comment on the position of the calculated test statistic within the critical region: for example at the 0.05 significance level, a result of $p = 0.045$ would indicate *some* evidence that the null hypothesis should be rejected, whereas $p = 0.001$ indicates *considerable* evidence (Kanji, 1994).

One/two-tailed test

If we were to test our hypotheses about enumeration district income stated above, we would perform a **two-tailed test**. This is because we have to allow for the average income for enumeration district B to be either larger *or* smaller than that for enumeration district A. We could have chosen a slightly different research hypothesis, for example:

$H_1: \mu_A > \mu_B$. *The mean household income for households in enumeration district A is <u>significantly larger</u> as compared with enumeration district B: mean household income <u>is</u> influenced by geographical location (i.e., location in a specific enumeration district).*

This is termed a **one-tailed test** since we are only interested in a difference in one direction, in this case positive differences (larger). The research hypothesis H_1 is usually expressed as either $H_1: \mu_A > \mu_B$ or $H_1: \mu_A < \mu_B$. The choice of a two-tailed or one-tailed test will determine the distribution of the **rejection region** displayed in Figure 5.2 and, more importantly, the tabulated critical value that we use. Although the table of critical values listed in Appendix A (Table A.2) displays probabilities for both one- and two-tailed tests, often only the probabilities for a one-sided test are tabulated. In this case if we wish to use the conventional 0.05 (5%) significance level in conjunction with a two-tailed test, we must use the 0.025 (2.5%) value in the table. This means that it is easier to get significant results by adopting one-tailed tests. This does *not* mean that on the calculation of a given test statistic for a data sample you would be able to adjust your hypothesis to perform a one-tailed test, as your hypothesis should always be determined in advance of performing the test (Robson, 1994).

In the context of your research project, once you have rejected or failed to reject H_0, the next step would be to take the results and try to determine why you have the results you do. This is covered in more detail in Section 5.8.

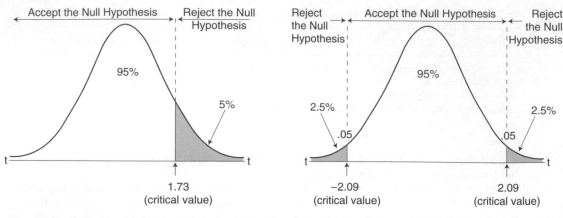

Figure 5.2 One- and two-tailed tests (source: Redrawn and adapted from Cohen and Holliday 1982).

5.4 Choosing the right test

If taken calmly and carefully, most students and researchers are able to perform the mechanics of the statistical tests considered below without difficulty. However, choosing the *right* test to use in a given analytical situation is more difficult. Consider the following hypothetical question and answer section taken from Chatfield (1995):

Q: What do you want to do with your data?
A: A t-test.

Q: No, I mean why have you carried out this study and collected these data?
A: Because I want to do a t-test (with a puzzled look).

Q: No. I mean what is your prior hypothesis? Why do you want to do a t-test?
A: My supervisor wants me to.

Q: No, I mean what is your objective in analysing these data?
A: To see if the results are significant . . .

The point here is that you do *not* undertake statistical tests in isolation, or test for significance in some arbitrary fashion. The main motivation for choosing a statistical test to apply to a set of data has to be driven by the *objectives* of your research project, and as indicated in Chapter 4, your project should have been designed and data sampled with a certain test or set of tests in mind. If not (for example, you may want to use secondary data that were collected by

someone else or for another purpose), then you might like to consider the following questions modified from Chatfield (1995) to suggest an appropriate method (or methods):

1   *What is the structure of your data? Univariate? Bivariate? Multivariate?*   We do not cover **multivariate methods** such as multiple regression, factor and principal components analysis in this chapter. You should refer to one of the books listed in the Further reading at the end of this chapter.

2   *Are the variables continuous or discrete? What is the data type – nominal, ordinal, interval or ratio?*   The data type will limit the tests which are available, usually into either a parametric or a non-parametric test (see Table 5.2). You can of course transform your data from one data type to another, as indicated in Chapter 4.

3   *Have you performed a similar type of analysis before?*   With a complex test, particularly if you have not used the statistical method before, it is advisable that you try out the method on another data set, ideally a pilot sample similar to your main data set. This can be very informative, and indeed might indicate whether a specific test is informative, or indeed worthwhile.

4   *Can you find a similar problem in a book or research paper?*   Often, not only will you not have used the statistical method before, but it might be impossible to try out the method on similar data to those you have collected. In other words, analysis of your own research data set will be your only experience! As an alternative, you might be able to find appropriate **worked examples** in

Table 5.2 A classification of tests.

Number of samples	Difference test	Association test
A. Tests for nominal data		
One, two, >two samples	Chi-squared	—
B. Tests for parametric data / interval or ratio data		
One sample	One-sample t	—
Two samples	Two-sample t, or two-sample t for paired samples	Pearson's product moment coefficient
>Two samples	ANOVA	Pearson's product moment coefficient (between pairs of samples)
C. Tests for non-parametric data / ordinal data		
One sample	Wilcoxon signed ranks, *or* Kolmogorov–Smirnov	—
Two samples	Mann–Whitney U, *or* Kolmogorov–Smirnov	Spearman's rank coefficient
>Two samples	Kruskal–Wallis	Spearman's rank coefficient (between pairs of samples)

books or research papers: some of these may be in a context not too far divorced from the data or research project in which you are interested. Even if you have access only to summary results, this will provide valuable material for comparison against results from your own analysis.

The choice of an appropriate statistical test can be determined by a consideration of the number of samples, the type and nature of data you have collected. With reference to Section 5.1 above and then Table 5.2, we can categorise the tests discussed below in terms of being parametric/non-parametric (or dealing with a certain data type), in terms of the number of data samples, and whether we are interested in establishing **differences** or examining **associations**.

How do you decide whether to use a non-parametric or a parametric test? Clearly if you are using nominal or ordinal data you are restricted to non-parametric tests. If you wish to use a parametric test, you should try to establish whether your data conform to the assumptions required for such tests. Although we list the main assumptions for each test in Sections 5.6–5.7 below, the common characteristics of both parametric and non-parametric tests are listed in Box 5.2. The evaluation of your data should consider the *normality* of the distribution, the presence/impact of *outliers*, the *homogeneity of variance* and the *sample size* (Pett, 1997). **Normality** can be assessed in three ways. First you could calculate the skewness (degree of symmetry)

Box 5.2 Common characteristics of parametric and non-parametric tests

Parametric
- **Independence of observations**, except where the data are paired.
- **Random sampling** of observations from a normally distributed population.
- **Interval scale measurement** (at least) for the dependent variable.
- **A minimum sample size** of about 30 per group is recommended.
- **Equal variances** of the population from which the data are drawn.
- **Hypotheses are usually made about the mean** (μ) of the population.

Non-parametric
- **Independence** of randomly selected observations except when paired.
- **Few assumptions** concerning the distribution of the population.
- **Ordinal or nominal** scale of measurement.
- **Ranks or frequencies** of data are the focus of tests.
- **Hypotheses are posed regarding ranks, medians or frequencies.**
- **Sample size requirements** are less stringent than for parametric tests.

Source: Adapted from Pett 1997.

and kurtosis (flat or peaked shape) of the distribution. Next, you could plot the shape of a distribution, using the histogram, stem-and-leaf and box-and-whisker plots from Chapter 4. Finally, there are several statistical tests (e.g., the Shapiro–Wilks test of normality) which could be used to try to fit the data to a normal distribution. **Outliers** can be identified using box-and-whisker plots. The **homogeneity of variance** can be assessed by comparing the variances between subgroups. These should be similar: a formal statistical test (e.g., the Levene test) can be applied if necessary. Although it is generally agreed that small samples are inappropriate for parametric analysis, it is not clear from the literature what size is *too small* (Pett, 1997). For further details, you should consult one of the basic statistical texts listed at the end of this chapter. A clear treatment can be found in Pett (1997), in particular in Chapter 3, 'Evaluating the characteristics of data'.

However, if you are still not sure, useful advice is to try *both* the parametric and non-parametric tests where possible. If you get similar results you would be more likely to believe them; if not, you may have to look more closely at the assumptions concerning the data (Chatfield, 1995). Of course, if your data appear to violate some of the underlying assumptions, you might wish to use a data transformation (see Section 4.7), and *then* use the appropriate parametric test. Again, a possible course of action might be to apply the test *before* and *after* the transformation, and see if the result changes.

5.5   The tests

In Sections 5.6 and 5.7 we describe the mechanics of some of the more common and useful statistical tests. We have included the calculation of both parametric and non-parametric correlation coefficients, and linear regression, which all include the calculation of tests of significance for the coefficients. Note that the coverage of tests in this book is *not* exhaustive: in particular we have not included several non-parametric tests such as the Friedman, McNemar or Cochran's Q. For further details you should consult Pett (1997) or Coshall (1989) for examples in a geographical context.

Methods are divided into parametric and their non-parametric equivalents. For each technique we have adopted a common format, similar to that used in Chapter 4:

Name of test	*e.g., unrelated t-test*
Experimental design	*Independent/dependent data sets*
Level of data	*Interval/ratio, nominal/categorical, ordinal*
Type of test	*Parametric or non-parametric*
Function	*e.g., test of difference*
What the test does	*A simple description of what the test is used for*
Assumptions	*The assumptions made in order to use the test*
Explanation and procedure	*Explanation of why and what you are doing, with step-by-step details of how to perform the test*
Example data	*A real geographical data set*
Handworked example	*Step-by-step using a calculator and the example data*
MINITAB example	*Step-by-step using MINITAB and the example data*
Interpretation	*A brief interpretation of the example data results*

If you know what type of test you wish to perform, we advise that you read through fully all of the above sections for the appropriate test, particularly the step-by-step description of the test and the explanation. Step-by-step details are also given for doing the calculations by hand and using the MINITAB statistical package. Even if you intend only to use MINITAB, it is often useful to have some experience of performing the calculations by hand – if only for the worked example, to understand how the test works. To understand the distinction between certain tests, you might have to read this section or indeed the chapter fully, but by this time we hope that you should understand the major elements of significance testing so that you are able to select, execute and interpret the results of an appropriate test.

We would like to encourage the use of the IDA techniques outlined in Chapter 4 as an essential preliminary to each test described below. This is mainly for two reasons. First, each test makes assumptions about the data being analysed. Parametric tests in particular make certain assumptions about the data distribution. Although some tests are robust and resistant to outliers in the data and non-normal distributions,

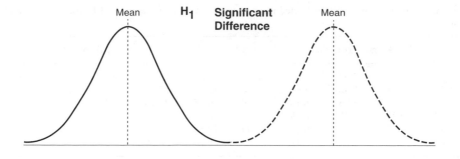

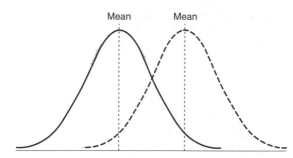

Figure 5.3 Differences in means and populations.

IDA techniques will help to establish whether or not the assumptions made about the data for each test appear to be warranted. This may well influence your choice of test, and we will discuss this further below. Second, the use of appropriate IDA techniques will give you a better understanding of, and feel for, your data, which can be helpful in determining whether you need a formal test at all, and perhaps more importantly will give you a good idea to anticipate what the results of any given test might be (Wright, 1997). Differences between groups may be so large as to be obvious without a test, and anticipating the results can help avoid errors in calculation.

5.6   Parametric tests

5.6.1   Unrelated *t*-test

Experimental design: Independent data sets.

Level of data: Interval/ratio.

Type of test: Parametric.

Function: Test of difference.

What the test does: Compares two unrelated data sets by inspecting the amount of difference between their means and taking into account the variability (spread) of each data set.

Assumptions:

1   Each data set is taken from a random sample.
2   The samples were taken independently of each other.
3   The data are drawn from a normally distributed population.

Explanation and procedure: With two groups of data, a two-sample *t*-test is used to determine if the difference in means is significant. The *t*-test used here is a comparison between two sample means allowing for the variability in the data. This is in the form:

$$t = \frac{\text{difference in means}}{\text{standard deviation of the difference in means}}$$

The larger the difference in the means, the more likely that a real, significant difference exists, and our samples come from different populations (Figure 5.3).

We do not have any information about the population variances (if we did, we could use a *Z*-test: see Section 4.6).

Table 5.3 Quantity of household waste collected in one week from separate samples of households using bags and wheeled bins.

Set A: bags (kg)						Set B: bins (kg)					
18.08	13.00	8.68	2.04	13.4	20.58	9.32	15.50	21.44	28.28	11.90	24.94
23.62	21.34	7.68	8.90	21.00	23.74	13.04	17.16	24.34	21.00	22.82	34.18
9.46	4.54	3.58	5.86	8.18	3.80	20.65	24.20	15.84	14.78	6.94	18.04
35.54	9.92	15.34	3.22	6.18	11.98	18.30	4.70	13.80	18.84	13.06	7.76
7.64	5.32	3.52	1.72	19.12	37.44	19.56	4.00	21.66	22.20	23.40	13.18
17.40	8.08	6.36	15.22	3.12		8.58	8.98				

Sources: MEL Research 1994.

Lay out the data such that you can calculate the squares of each case, or enter data into the MINITAB spreadsheet (**Step 1**). It is vital that you summarise and check the data for the presence of outliers or errors before starting to use the methods described in Chapter 4. Extreme values in the data could dramatically affect the results of a *t*-test, and could indicate that the data come from a non-normal distribution. If such values do not appear to exist, we can proceed with the test. Proceed with checking the assumptions, and the calculation of sample variances using sums, means and squares (**Steps 3–7**).

An important preliminary step is to determine whether the sample variances for each group are equal, for which we use an *F*-test (**Steps 8 and 9**). The variances are calculated for each group, and then the value of *F* is obtained:

$$F = \frac{\text{greater variance}}{\text{lesser variance}}$$

The next step is to compare the calculated value of *F* with a critical value of *F* at the 0.05 significance level with DF = $n - 1$ for each sample (**Step 10**). If the calculated *F* is greater than the critical value, the variances of each sample are significantly different. If the variances *are* significantly different, substitute the mean and variance and number of samples for each data set into the following formula to estimate *t*:

$$t = \frac{|\bar{x}_a - \bar{x}_b|}{\sqrt{\frac{s_a^2}{n_a} + \frac{s_b^2}{n_b}}}$$

where $\bar{x}_a$, $\bar{x}_b$ = mean, s_a^2, s_b^2 = variance, and n_a, n_b = number of samples of data sets A and B respectively. If the variances *are not* significantly different, a

pooled variance can be used. Substitute the mean and variance and number of samples for each data set into the following formula to calculate *t*:

$$t = \frac{|\bar{x}_a - \bar{x}_b|}{\sqrt{\left(\frac{s_a^2 \times (n_a - 1) + s_b^2 \times (n_b - 1)}{(n_a + n_b - 2)}\right) \times \left(\frac{1}{n_a} + \frac{1}{n_b}\right)}}$$

If the pooled variance has been used in Step 11 then calculate the **degrees of freedom** (DF) = $n_a + n_b - 2$ (**Step 12**). Otherwise, the DF can be calculated as follows:

$$DF = \frac{\left(\frac{s_a^2}{n_a} + \frac{s_b^2}{n_b}\right)^2}{\frac{(s_a^2/n_a)^2}{(n_a - 1)} + \frac{(s_b^2/n_b)^2}{(n_b - 1)}}$$

Lastly, check the critical value of *t* at the 0.05 significance level. If the calculated value of *t* is greater than the critical value, it is significant and we can reject the null hypothesis H_0 (**Step 13**).

To illustrate the *t*-test, we will make use of the data set in Table 5.3 on domestic household waste production. The calculations are given in Boxes 5.3 and 5.4.

Null hypothesis H_0: $\mu_1 = \mu_2$

There is *no difference* between the quantity of domestic waste produced from households using bags and the quantity produced from households using bins.

Research hypothesis H_1: $\mu_1 > \mu_2$

Households using bins produce a *larger* quantity of waste than households using bags.

Significance level: 0.05.

Interpretation: In the example above, the results of the *F*-test indicate that the sample variances are equal

Box 5.3   Handworked example: unrelated t-test

Set A (bags, kg)				Set B (bins, kg)			
x_a	x_a^2	x_a	x_a^2	x_b	x_b^2	x_b	x_b^2
18.08	326.89	2.04	4.16	9.32	86.86	21.66	469.16
23.62	557.90	8.9	79.21	13.04	170.04	28.28	799.76
9.46	89.49	5.86	34.34	20.65	426.42	21	441.00
35.54	1263.09	3.22	10.37	18.3	334.89	14.78	218.45
7.64	58.37	1.72	2.96	19.56	382.59	18.84	354.95
17.4	302.76	15.22	231.65	8.58	73.62	22.2	492.84
13	169.00	13.4	179.56	15.5	240.25	11.9	141.61
21.34	455.40	21	441.00	17.16	294.47	22.82	520.75
4.54	20.61	8.18	66.91	24.2	585.64	6.94	48.16
9.92	98.41	6.18	38.19	4.7	22.09	13.06	170.56
5.32	28.30	19.12	365.57	4	16.00	23.4	547.56
8.08	65.29	3.12	9.73	8.98	80.64	24.94	622.00
8.68	75.34	20.58	423.54	21.44	459.67	34.18	1168.27
7.68	58.98	23.74	563.59	24.34	592.44	18.04	325.44
3.58	12.82	3.8	14.44	15.84	250.91	7.76	60.22
15.34	235.32	11.98	143.52	13.8	190.44	13.18	173.71
3.52	12.39	37.44	1401.75				
6.36	40.45						

$\sum x_a = 424.6$ $\sum x_a^2 = 7881.3$ $\sum x_b = 542.39$ $\sum x_b^2 = 10761.41$

$\bar{x}_a = 12.13$ $n_a = 35$ $\bar{x}_b = 16.95$ $n_b = 32$

$(\sum x_a)^2 = 180285.16$ $(\sum x_b)^2 = 294186.91$

Step 1   Lay out the data in a table as above. Check the assumptions: normality can be assessed by constructing histograms.

Step 2   Prepare to calculate the variance of each sample for an F-test.

Step 3   Calculate the sum of data set x_a by adding together all the data values, and call the sum $\sum x_a$. Repeat for x_b. $\sum x_a = 424.6$ and $\sum x_b = 542.39$.

Step 4   Calculate the squares for each value in each set (i.e., $x_a \times x_a$) and insert into two new columns of data, x_a^2 and x_b^2 (columns 2, 4, 6 and 8 in the table above).

Step 5   Calculate the sum of the new x_a^2 column by adding together all the data values, and call the sum $\sum x_a^2$. Repeat for x_b^2. $\sum x_a^2 = 7881.3$ and $\sum x_b^2 = 10761.41$.

Step 6   Calculate the mean of x_a by dividing $\sum x_a$ by the number of entries in the data set (n_a): $\bar{x}_a = 424.6 \div 35 = 12.13$. Calculate the mean of x_b by dividing $\sum x_b$ by the number of entries in the data set: $\bar{x}_b = 542.39 \div 32 = 16.95$.

Step 7   Calculate the square of $\sum x_a$ (i.e., $\sum x_a \times \sum x_a$) and call the result $(\sum x_a)^2$. Repeat for $\sum x_b$. $(\sum x_a)^2 = 180285.16$ and $(\sum x_b)^2 = 294186.91$. Note that $(\sum x_a)^2$ is different from $\sum x_a^2$.

Step 8   Calculate the variance of each data set (see Section 4.5.4):

$$s_a^2 = \frac{7881.3 - \dfrac{180285.16}{35}}{35 - 1} = 80.30 \qquad s_b^2 = \frac{10761.41 - \dfrac{294186.91}{32}}{32 - 1} = 50.58$$

Step 9   Calculate $F = 80.30 \div 50.58 = 1.58$.

Step 10   From Table A.1 in Appendix A, for $DF_1 = 34$ and $DF_2 = 31$, the critical value of $F \approx 1.8$ at the 0.05 level. The calculated $F < F_{critical}$, and therefore the variances of each sample are not significantly different, and a pooled version of the t test can be used

Box 5.3   (cont'd)

Step 11   Substitute into the pooled formula and work through to calculate t:

$$t = \frac{|12.13 - 16.95|}{\sqrt{\left(\dfrac{8.96^2 \times (35 - 1) + 7.11^2 \times (32 - 1)}{(35 + 32 - 2)}\right) \times \left(\dfrac{1}{35} + \dfrac{1}{32}\right)}}$$

$$= \frac{4.82}{\sqrt{\left(\dfrac{(80.2816 \times 34) + (50.5521 \times 31)}{65}\right) \times \left(\dfrac{1}{35} + \dfrac{1}{32}\right)}}$$

$$= \frac{4.82}{\sqrt{\left(\dfrac{2729.5744 + 1567.1151}{65}\right) \times \left(\dfrac{1}{35} + \dfrac{1}{32}\right)}}$$

$$= \frac{4.82}{\sqrt{66.102915 \times \left(\dfrac{1}{35} + \dfrac{1}{32}\right)}} = \frac{4.82}{1.98856} = 2.42$$

Step 12   Calculate the degrees of freedom DF ($n_a + n_b - 2 = 65$).

Step 13   Check the critical value of t from Table A.2, Appendix A. At the 0.05 significance level for a one-tailed test with 65 DF, the critical value ≈ 1.67 which is less than our calculated value of 2.42; therefore we can reject the null hypothesis H_0.

Box 5.4   MINITAB example: unrelated *t*-test

Enter the bag data into C1 and the bin data into C2.

(a) Calculate sample variances and perform the *F*-test

Using pull-down menus:

➡ Calc
➡ Mathematical Expressions...
➡ Variable (new or modified)
Type in C3
➡ Expression:[ ]
Type in stdev(C1)*stdev(C1)

(Repeat for C2, replacing C1 by C2, and C3 by C4)

➡ Calc
➡ Mathematical Expressions...
➡ Variable (new or modified)
Type in C5
➡ Expression:[ ]
Type in C3/C4

Using command line:

MTB > Let C3 = stdev(C1)*stdev(C1).
MTB > Let C4 = stdev(C2)*stdev(C2).
MTB > Let C5 = C3/C4.

MINITAB results:

MINITAB will place the calculations for the variance of C1, C2 in C3 (= 80.30) and C4 (= 50.58), and the calculation of F in C5 = 1.587.

Box 5.4 (cont'd)

(b) Calculate the critical value

Using pull-down menus:

➡ Calc
➡ Probability Distributions➤
➡ F...
➡ Inverse Cumulative Probability
➡ Numerator degrees of freedom [ ]
Type in 34
➡ Denominator degrees of freedom [ ]
Type in 31
➡ Input constant
Type in 0.95
➡ OK

Using command line:

MTB > Invcdf 0.95;
SUBC > F 34 31.

MINITAB results:

Inverse Cumulative Distribution Function

F distribution with 34 d.f. in numerator and 31 d.f. in denominator

```
P(X <= x)            x
   0.9500         1.8055
```

(c) Since calculated $F < F_{critical}$, proceed with the *t*-test

Using pull-down menus:

➡ Stat
➡ Basic Statistics➤
➡ 2-Sample t...
➡ Samples in different columns
➡ First [ ]
Highlight C1
➡ Select
Highlight C2
➡ Select
➡ OK

Using command line:

MTB > Twosample C1 C2;
SUBC > Pooled.

MINITAB results:

```
MTB > twosample c1 c2;
SUBC > pooled.
```

Two Sample T-Test and Confidence Interval

```
Twosample T for C1 vs C2
        N      Mean     StDev   SE Mean
C1     35     12.13      8.96       1.5
C2     32     16.95      7.11       1.3

95% C.I. for mu C1 - mu C2: ( -8.8, -0.8)
T-Test mu C1 = mu C2 (vs not =): T = -2.42  P = 0.018  DF = 65

Both use Pooled StDev = 8.13
```

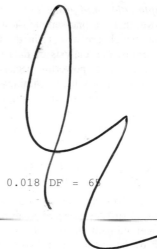

at the 0.05 significance level, and the calculation of t using the 'pooled' estimate produces a calculated value of t which exceeds the one-tailed critical value at the 0.05 significance level. The MINITAB results produce $p = 0.018$, which confirms the handworked result. We reject the null hypothesis. It would appear that at the 0.05 significance level, households using bins produce a larger amount of waste than households using bags. However, we cannot say at this point as to why bins produce more waste. This would need further investigation.

5.6.2   Related t-test

Experimental design: Related data sets.

Level of data: Interval/ratio.

Type of test: Parametric.

Function: Test of difference.

What the test does: Compares the actual size of differences between the matched/related scores in each condition or time. The test uses the mean difference for the group and also individual variation in the size of the differences for individual related pairs to compare data sets.

Assumptions:

1   Each data set is taken from a random sample.
2   The samples were taken independently of each other.
3   The data are drawn from a normally distributed population.
4   The variances of the two samples are not significantly different.

Explanation and procedure: In contrast to the test above, this is a one-sample t-test for use with related data, which are also termed 'matched pairs' data. Generally, the analysis of data in terms of matched pairs is more powerful because the design of a matched pairs scheme minimises unwanted 'background' variability which might be present in an independent randomised sampling scheme (Heath, 1995). This is reflected by a decrease in the standard error (denominator) of the t-test equation:

$$t = \frac{\text{difference in means}}{\text{standard deviation of the difference in means}}$$

The mechanics of the test are very similar to those of the standard t-test described above, except that we are concerned with the *differences* between each matched pair of data values from two groups.

To calculate the value of t by hand, lay out your data in two columns such that you can estimate the difference between each matched pair of values (**Step 1**). MINITAB will also require you to enter your data into the spreadsheet in this fashion, as you will have to obtain the differences similarly to the handworked method. Summarise and check the data for normality and outliers, and consider the use of a non-parametric technique or transformation of the variables. Next, calculate the mean of each group (**Step 2**). Then obtain the differences by subtracting each paired value from the group with the smaller mean from the corresponding paired value of the group with the larger mean. Insert this column into the data set. Square each of these differences to create another new column in the data set (**Step 3**). Proceed with the calculation of the sum of the differences (**Step 4**), and then the calculation of the standard deviation of those differences (**Steps 5 and 6**). Next calculate the value of t using the following formula (**Step 7**):

$$t = \frac{\sum_{i=1}^{n} d_i}{\sqrt{\left[\frac{n\sum_{i=1}^{n} d_i^2 - \left(\sum_{i=1}^{n} d_i\right)^2}{n-1}\right]}}$$

where d = the differences, and n = number of observations. Then determine the degrees of freedom (**Step 8**). Finally, obtain the tabulated critical value of t at the 0.05 level for comparison – Table A.2 in Appendix A (**Step 9**). If the calculated value of t is greater than the critical value, it is significant and we can reject the null hypothesis H_0 – we can be 95% confident that there is a significant difference between the two data sets.

To illustrate the related t-test, we have used data (Table 5.4) on occupied office space in the US, from Sui and Wheeler (1993): see Boxes 5.5 and 5.6.

Null hypothesis
$H_0: \mu_1 = \mu_2$ — There is *no significant difference* between the growth rates of occupied office space in large metropolitan areas in 1985 and 1990.

Research hypothesis
$H_1: \mu_1 \neq \mu_2$ — There is *a significant difference* between the growth rates of occupied office space in large metropolitan areas in 1985 and 1990.

Significance level: 0.05.

Table 5.4 Location and growth of primary occupied office space in selected large US metropolitan areas, 1985–1990.

	1985	1990		1985	1990
New York, NY	1.34	1.41	Rochester, NY	1.08	0.93
Hartford, CT	1.29	1.14	Cleveland, OH	1.05	0.97
Seattle, WA	1.27	1.08	Indianapolis, IN	1.03	1.06
Boston, MA	1.25	1.16	Orlando, FL	1.03	1.06
Houston, TX	1.21	1.18	Tampa, FL	1.02	1.01
Los Angeles, CA	1.17	1.14	Charlotte, NC	1.02	1.06
Atlanta, GA	1.14	1.10	Portland, OR	1.01	1.01
Richmond, VA	1.12	1.18	New Orleans, LA	0.99	1.10
St Louis, MO	1.10	0.98	Salt Lake City, UT	0.95	0.80
Miami, FL	1.09	1.11	Norfolk, VA	0.84	0.90
Kansas City, KS	1.08	1.01			

Source: Sui and Wheeler 1993

Box 5.5 Handworked example: related *t*-test

City	Location quotient x_a	Location quotient x_b	Difference d	Difference d^2
New York, NY	1.34	1.41	−0.07	0.0049
Hartford, CT	1.29	1.14	0.15	0.0225
Seattle, WA	1.27	1.08	0.19	0.0361
Boston, MA	1.25	1.16	0.09	0.0081
Houston, TX	1.21	1.18	0.03	0.0009
Los Angeles, CA	1.17	1.14	0.03	0.0009
Atlanta, GA	1.14	1.10	0.04	0.0016
Richmond, VA	1.12	1.18	−0.06	0.0036
St Louis, MO	1.10	0.98	0.12	0.0144
Miami, FL	1.09	1.11	−0.02	0.0004
Kansas City, KS	1.08	1.01	0.07	0.0049
Rochester, NY	1.08	0.93	0.15	0.0225
Cleveland, OH	1.05	0.97	0.08	0.0064
Indianapolis, IN	1.03	1.06	−0.03	0.0009
Orlando, FL	1.03	1.06	−0.03	0.0009
Tampa, FL	1.02	1.01	0.01	0.0001
Charlotte, NC	1.02	1.06	−0.04	0.0016
Portland, OR	1.01	1.01	0.00	0.0000
New Orleans, LA	0.99	1.10	−0.11	0.0121
Salt Lake City, UT	0.95	0.80	0.15	0.0225
Norfolk, VA	0.84	0.90	−0.06	0.0036

$\bar{x}_a = 1.09906$ $\bar{x}_b = 1.06619$ $\Sigma d = 0.69$ $\Sigma d^2 = 0.1689$ $(\Sigma d)^2 = 0.4761$

Step 1 Lay out the data and check the assumptions.

Step 2 Calculate the mean for each data set by summing the data values and dividing by the number of data entries (e.g., $23.08 \div 21 = 1.09905$). $\bar{x}_a = 1.09905$ and $\bar{x}_b = 1.06619$.

Step 3 Calculate the difference between the data sets by subtracting the 1990 data (b) from the 1985 data (a) (which has the larger mean) to produce column 3, in the table above. Square each difference to produce column 4.

Step 4 Calculate the sum of all the differences, Σd in column 3 = 0.69.

Box 5.5   (cont'd)

Step 5   Calculate the sum of the differences squared, $\sum d^2$ in column 4 = 0.1689.

Step 6   Calculate the square of the sum of differences, $(\sum d)^2$ = 0.4761.

Step 7   Input these values into the formula and work through to calculate t:

$$t = \cfrac{0.69}{\sqrt{\left[\cfrac{21 \times 0.1689 - 0.4761}{21 - 1}\right]}} = \cfrac{0.69}{\sqrt{\left[\cfrac{3.5469 - 0.4761}{20}\right]}}$$

$$= \frac{0.69}{\sqrt{0.1535}} = \frac{0.69}{0.391791} = 1.76114$$

Step 8   Calculate the degrees of freedom DF = 21 − 1 = 20.

Step 9   Check the critical value of t from Table A.2, Appendix A. At the 0.05 significance level for a two-tailed test with 20 DF the critical value is 2.086, which is greater than our calculated value of 1.76. Therefore we fail to reject the null hypothesis H_0.

Box 5.6   MINITAB example: related t-test

Enter the 1985 location quotient into C1 and the 1990 location quotient into C2. Calculate the difference between the two data sets by subtracting the data set with the lower mean from the other (e.g., Type: Let C3 = C1 − C2).

Using pull-down menus:

➥ Stat
➥ Basic Statistics➤
➥ 1-Sample t…
➥ Samples in different columns
➥ First [ ]
Highlight C3
➥ Select
➥ Test Mean
➥ OK

Using command line:

Type: Ttest 0 C3.

MINITAB results:

```
MTB > ttest 0 C3

Test of MU = 0.0000 vs MU N.E. 0.0000

        N      Mean      StDev    SE Mean      T    P value
C3     21    0.0329    0.0855     0.0187    1.76      0.094
```

Interpretation: In the example above, the calculated value of t is less than the critical value of t at the 0.05 significance level. The MINITAB results produce $p = 0.094$ which confirms the handworked result. We therefore fail to reject the null hypothesis. At the 0.05 significance level, there is no significant difference between the growth rates of occupied office space in large metropolitan areas in the US in 1985 and 1990 (note that there is a difference at the 0.1 level).

5.6.3   Analysis of variance (one-way)

Experimental design: Independent data sets.

Level of data: Interval/ratio.

Type of test: Parametric.

Function: Test of difference.

What the test does: The test examines the differences between three or more groups by looking at each group mean. The null hypothesis H_0 states that there is no difference between group means and any differences have been generated by chance.

Assumptions:

1   Sample data must be of interval/ratio type.
2   Sample data must be drawn from a normal distribution.
3   Sample data must be drawn from independent random samples.
4   The populations from which the samples are drawn must have similar variances.

Explanation and procedure: If you have data which are classified into groups and you want to find out if the means of each are different, you perform an analysis of variance (ANOVA). Why not just do lots of *t*-tests? If you use 0.05 as the significance level, the probability of getting a significant result using a *t*-test when there is really no difference between the means is 5%. But if you make multiple comparisons, say between 10 pairs of means, then the probability of getting a significant result purely by chance is (Grant, personal communication):

$$1 - 0.95^{10} = 0.4$$

If we pool all data – ignoring any classification into groups for the moment – we can calculate an overall total variance. This can be partitioned into two components: the variance *between* groups and the variance *within* groups (or residual variance). Here the model we are working from is:

Total variance = between-group variance + within-group variance

The between-group variance is in the form of the squared deviations between each group mean and the global mean, known as the **between-group sum of squares** (SSB), which, divided by the degrees of freedom, gives the **between-group mean square** (MSB). Similarly, the within-group variance is obtained from the sum of squared deviations between each group observation and the group mean, known as the **within-group sum of squares** (SSW), which, divided by the degrees of freedom, gives the **within-group mean square** (MSW). The ratio between the MSB and the MSW is known as the ***F*-test**. This is approximately 1 when the null hypothesis H_0 is true, since we are saying that we expect MSB – MSW ≈ σ^2. When there are large differences between group means, the MSB will be larger and the *F*-ratio will be larger than one.

The differences in means and distributions under both a null hypothesis and a research hypothesis for three groups are shown in Figure 5.4.

The analysis of variance makes several assumptions about the data. The variance within each group is similar; the data are normally distributed and come from an independent random sample. You should organise your data into a table (**Step 1**) and perform an IDA on the groups to check these assumptions prior to launching into an ANOVA. Individual group standard deviations can be compared but it is often sufficient to compare the mean, median and range (**Step 2**). Note that with small groups the range may be used to measure spread rather than the standard deviation. It may be useful to examine the results graphically using a box-and-whisker plot for each group. This may also reveal group differences to be obvious, which may indicate that an ANOVA is not necessary (**Step 3**). Small variations between groups are not too important. Larger variations between groups can be dealt with by a data transformation as explained in Chapter 4 (**Step 4**), or by the adoption of a non-parametric technique such as the Kruskal–Wallis test rather than ANOVA.

The within-group sum of squares (SSW) (**Step 5**) is defined as:

$$SSW = \sum_{j=1}^{n} (x_{1j} - \bar{x}_1)^2$$

where x_{1j} is each observation in group 1, and $\bar{x}_1$ is the mean for group 1. The SSWs of all groups are then added together. The between-group sum of squares (SSB) (**Step 6**) is defined as:

$$SSB = \sum_{i=1}^{k} n_i \times (\bar{x}_i - \bar{x})^2$$

where $\bar{x}_i$ is the mean of each group of n_i values. We can divide the SSB and the SSW by the between-group degrees of freedom $(k - 1)$ and the within-group DF $(n - k)$ (**Step 7**), where n denotes the total number of observations to obtain the mean square (MS) (**Step 8**) for the group means:

$$MSB = SSB/(k - 1)$$

and within each group:

$$MSW = SSW/(n - k)$$

The *F*-statistic is calculated (**Step 9**) from:

$$F = MSB/MSW$$

The results of an ANOVA are usually presented in the following form (**Step 10**):

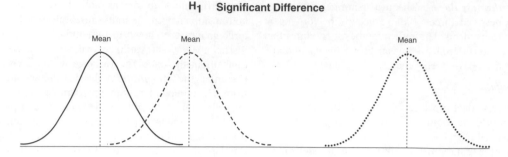

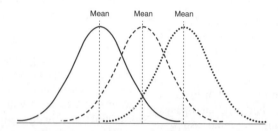

Figure 5.4   Differences in means and populations for the analysis of variance.

Source of variation	DF	SS	MS	F
Between groups	$k - 1$	SSB	SSB/$(k - 1)$	MSB/MSW
Within groups	$n - k$	SSW	SSW/$(n - k)$	
Total	$n - 1$	SSB + SSW		

Note that the *within-* and *between-groups* of variation are additive and will sum to the total sum of squares, as will the degrees of freedom. This provides a useful check on the calculations. If $F < 1$, there is less variation among the means than you would have expected by chance, and you have obtained a non-significant result. For $F > 1$, using $DF_1 = (k - 1)$ and $DF_2 = (n - k)$, look up the critical value of F from a table of percentage points of F distribution at 0.05 level (see Table A.1, Appendix A) (**Step 11**) and determine whether or not the H_0 can be rejected (**Step 12**).

If the ANOVA produces a significant result, how do we tell which groups are different from each other? There are a variety of tests which can be used, termed **multiple comparison tests** (see Sokal and Rohlf, 1995, for a survey), the choice of which depends initially on whether you are making **planned**

comparisons (*a priori*) which are planned before we perform the sampling and ANOVA. In this case, if the number of comparisons is orthogonal it is simply a matter of performing a modified *t*-test for each comparison, but if the number of comparisons is non-orthogonal, the overall *error rate* must be proportionally reduced: see Sokal and Rohlf (1995) and Wright (1997) for details. If, however, you are not sure what comparisons to make in advance of performing the test, you need to adopt an **unplanned comparisons** (*a posteriori*) procedure. Again, there are a variety of methods which can be used here, one of the more popular being known as 'Tukey's honestly significant difference method' – the **Tukey test** for short – which is the simplest method to use, ideally in the presence of equal sample sizes. The idea is to obtain a critical difference based on sample characteristics and a critical look-up value. For this a parameter T can be defined as follows:

$$T = Q \sqrt{\frac{MSW}{n_i}}$$

where Q is a special statistic for significance testing called the **studentised range**. The parameter Q is obtained at the 0.05 level using the number of groups k and $DF = (n - k)$ (see Table A.3, Appendix A).

Table 5.5 Distance estimates from blind people over three trials

Blind individuals	Trial 1	2	3
A	206.7	219.2	219.5
B	176.3	193.4	184.4
C	176.8	204.4	187.5
D	196.8	200.1	193.3
E	200.4	190.2	203.3
F	190.2	198.7	223.0
G	190.3	215.2	201.0
H	197.6	207.9	214.7

Substitute Q into the equation above to calculate T. If the difference between any pair of means (e.g., $\bar{x}_1 - \bar{x}_2$) exceeds or equals T then those means are significantly different.

As an example, Table 5.5 details the total distance estimates by blind individuals A–H between a number of locations along a route after walking the route on three separate (trial) occasions. The data were generated using a ratio-scaling technique, where respondents are given a ruler which represents the total length of the route and are asked to estimate the length between pairs of locations as a proportion of the ruler's length. Boxes 5.7 and 5.8 show the calculations.

Null hypothesis $H_0: \mu_1 = \mu_2$ — There is *no significant difference* between the distance estimates of blind people on different trials.

Research hypothesis $H_1: \mu_1 \neq \mu_2$ — There is *a significant difference* between the distance estimates of blind people on different trials.

Significance level: 0.05.

Interpretation: In our example above, the F value did not exceed the critical value at the 0.05 significance level, and the MINITAB results produce $p = 0.108$. This means that the sample variances of the data within each condition (trial) did not differ significantly. Note that, although the value of F is not significant, we have produced the results for the Tukey test to give an example of the MINITAB output. Since all of the pair comparison confidence intervals contain a zero, we deduce that none of the pairs is significantly different – not a surprise, given the test

results. Given that the respondents were learning a new route, and that the trials represented the first three times this route was walked, we might expect there to be a significant difference between trials as respondents' knowledge of the relative distances between places increased. The fact that we have not found such a difference can be interpreted in two ways: (1) either it takes longer than three trials before there are significant increases in the accuracy of distance estimation, or (2) the respondents were very accurate in their responses, right from the first trial. The only way to determine this is through further testing, relating all (remember we have used only the totals) the cognitive estimates of the blind respondents to the real distances. Such analysis has been performed by Jacobson *et al.* (1998) using linear regression (see below), and revealed that, contrary to what might be thought, blind respondents were very accurate in their distance estimations, right from the first trial.

5.6.4 Pearson's product moment correlation coefficient

Experimental design: Independent data sets.

Level of data: Interval/ratio.

Type of test: Parametric.

Function: Test of association.

What the test does: Determines if two variables are interdependent, i.e., the degree to which they vary together.

Assumptions:

1 Sample data must be of interval/ratio type.
2 Sample data must be drawn from a normal distribution.

Explanation and procedure: Pearson's product moment correlation coefficient provides a measure of the linear association between two variables and varies between −1 (perfect negative correlation) and +1 (perfect positive correlation). After laying out the data and checking the assumptions (**Step 1**), an important second step is to produce a scatter plot of one variable against the other to assess whether the calculation of a correlation is appropriate (**Step 2**). In simple terms Pearson's correlation coefficient is a ratio of the variance shared between the two variables to the overall variance of the two variables (Cramer, 1997):

Box 5.7   Handworked example: ANOVA

x_{1j}	$(x_{1j}-\bar{x}_1)$	$(x_{1j}-\bar{x}_1)^2$	x_{2j}	$(x_{2j}-\bar{x}_2)$	$(x_{2j}-\bar{x}_2)^2$	x_{3j}	$(x_{3j}-\bar{x}_3)$	$(x_{3j}-\bar{x}_3)^2$
206.7	14.81	219.41	219.2	15.56	242.19	219.5	16.16	261.23
176.3	−15.59	242.97	193.4	−10.24	104.81	184.4	−18.94	358.63
176.8	−15.09	227.63	204.4	0.76	0.58	187.5	−15.84	250.83
196.8	4.91	24.13	200.1	−3.54	12.51	193.3	−10.04	100.75
200.4	8.51	72.46	190.2	−13.44	180.57	203.3	−0.04	0.00
190.2	−1.69	2.85	198.7	−4.94	24.38	223	19.66	386.61
190.3	−1.59	2.52	215.2	11.56	133.69	201	−2.34	5.46
197.6	5.71	32.63	207.9	4.26	18.17	214.7	11.36	129.11

$\sum x_{1j} = 1535.1$

$\sum (x_{1j}-\bar{x}_1)^2 = 824.61$

$\bar{x}_1 = 191.8875$

$\sum x_{2j} = 1629.1$

$\sum (x_{2j}-\bar{x}_2)^2 = 716.90$

$\bar{x}_2 = 203.6375$

$\sum x_{3j} = 1626.7$

$\sum (x_{3j}-\bar{x}_3)^2 = 1492.62$

$\bar{x}_3 = 203.3375$

Step 1   Organise data into a table similar to that above.

Step 2   Calculate summary statistics (mean, median and range) for each group.

Step 3   Construct a box-and-whisker plot for each group (see Section 4.5).

Step 4   Transform the data if necessary (see Section 4.7).

Step 5   For each group k, calculate the difference between each observation and the group mean, and place in a new column. Square to obtain $(x_{kj}-\bar{x}_k)^2$ and sum for each group and then all groups to obtain SSW = 3034.13.

Step 6   Obtain the overall grand mean (199.62), and then substitute it into the equation below to obtain SSB:

$$SSB = \sum_{i=1}^{k} n_i \times (\bar{x}_i - \bar{x})^2$$

$$= 8 \times (191.8875 - 199.62)^2 + 8 \times (203.6375 - 199.62)^2 + 8 \times (203.3375 - 199.62)^2$$

$$= 718.01$$

Step 7   Calculate the between-group $DF_1 = (3 - 1) = 2$ and the within-group DF_2 as $(24 - 3) = 21$.

Step 8   Calculate the mean squares for group means MSB = 718.01 ÷ 2 = 359.01 and within each group MSW = 3034.13 ÷ 21 = 144.48.

Step 9   Calculate the F-statistic as the ratio MSB/MSW = 359.01 ÷ 144.48 = 2.48.

Step 10   Present results as follows:

Source of variation	DF	SS	MS	F
Between groups	2	718.01	359.01	2.48
Residual	21	3034.13	144.48	
Total	23	3752.14		

Step 11   Using $DF_1 = 2$ and $DF_2 = 21$, look up the critical value of F from Table A.1, Appendix A, which at the 0.05 significance level with 2 and 21 DF is 3.47.

Step 12   The calculated value of F does not exceed the critical value; therefore we fail to reject the null hypothesis H_0.

Box 5.8 MINITAB example: ANOVA

In MINITAB a one-way analysis of variance can be performed using the command 'aovoneway', which requires each group to be in a separate column, or 'oneway' which has all data listed in the first column and the group membership in the second column. We will make use of the latter alternative. Note that MINITAB performs the Tukey multiple comparisons test by printing out a confidence interval for the population value of the difference between each pair of means. If this interval includes zero, the pair of means is not significant.

Place each length estimate into C1, placing a number representing the group membership (termed a subscript) in C2. For example:

C1	C2
206.7	1
176.3	1
...	
197.6	1
219.2	2
etc.	

Using pull-down menus:

➥ Stat
➥ ANOVA
➥ Oneway...
➥ First [ ]
Highlight C1
➥ Select
Highlight C2
➥ Select
➥ OK

Using command line:

MTB > Oneway C1 C2;
SUBC > Tukey 0.05.

MINITAB results:

One-Way Analysis of Variance

```
      Analysis of Variance on 1
Source   DF     SS    MS      F      p
2         2    718   359   2.48  0.108
Error    21   3034   144
Total    23   3752
```

```
                          Individual 95% CI'S For Mean
                              Based on Pooled StDev
 Level   N    Mean    StDev  --+---------+---------+---------+----
   1     8  191.89   10.85   (----------*----------)
   2     8  203.64   10.12                 (-----------*-----------)
   3     8  203.34   14.60                 (---------*----------)
                            --+---------+---------+---------+----
Pooled StDev = 12.02        184.0    192.0    200.0    208.0
```

Tukey's pairwise comparisons

```
   Family error rate = 0.0500
Individual error rate = 0.0200

Critical value = 3.56

Intervals for (column level mean) - (row level mean)

             1         2

   2     -26.88
           3.38

   3     -26.58    -14.83
           3.68     15.43
```

Table 5.6 Child accidents and socio-economic characteristics for selected enumeration districts in Norwich.

C1	C2	C3	C4	C5	C6
1	4.066	4.9485	0.1788	0.7459	5.146
2	3.321	4.2020	0.0915	0.3586	0.029
3	5.020	6.3516	0.1440	0.6313	3.381
4	2.634	4.2098	0.0419	0.0930	−3.184
5	4.316	5.7413	0.1150	0.6145	2.784
6	3.691	5.6869	0.0937	0.3393	0.937
7	3.212	4.8775	0.1569	0.6583	4.312
8	4.770	9.2448	0.2151	0.7132	4.443
9	3.461	9.8701	0.1125	0.5275	2.522
10	3.251	4.7002	0.0586	0.1088	−1.522
11	2.451	4.2972	0.0565	0.1178	−2.208
12	4.367	8.5385	0.1906	0.7663	5.102
13	4.748	9.1357	0.1534	0.7204	3.626
14	3.147	5.9883	0.0377	0.0596	−2.440
15	3.314	5.5404	0.0687	0.2270	−1.012
16	3.563	5.9448	0.0942	0.3389	0.940
17	2.912	4.1245	0.0372	0.0728	−3.088
18	4.102	6.8702	0.1803	0.6899	5.077
19	4.686	5.8212	0.0959	0.5747	2.515
20	2.833	4.6989	0.0571	0.1537	−1.698
21	4.323	6.2101	0.1456	0.4533	2.436

C1 Area identifier (these are homogeneous social areas made up from census EDs)
C2 Child accident rate per 1000 days at risk (i.e., per 1000 days living in area)
C3 Percent of population aged 0–4 years
C4 Proportion of economically active males unemployed
C5 Proportion of households in rented accommodation
C6 Townsend material deprivation score (a composite of standard scores for unemployment, overcrowded households, households with no car and rented accommodation)

Source: Reading *et al.*, 1997. Data supplied by kind permission of Robin Haynes and Andrew Lovett.

$$r = \frac{\text{covariance of variables X and Y}}{\sqrt{(\text{variance of variable X}) \times (\text{variance of variable Y})}}$$

The covariance and variance are based on the difference between the scores and the mean of each variable, and the calculations to estimate covariance/variance can be broken down into a number of simple steps. First (**Step 3**), create two new columns of data consisting of the square of each x score (i.e., x_i^2 in one) and the square of each y score (i.e., y_i^2 in the other). Next (**Step 4**), create a fifth column of data by calculating the product of each x and y score (i.e., $x_i \times y_i$). Calculate the sum Σx_i of all the observed x scores and the sum Σy_i of all the observed y scores (**Step 5**). Obtain the squares of these sums, i.e., $(\Sigma x_i)^2$ and $(\Sigma y_i)^2$ (**Step 6**), and sum the column of x_i^2 to give Σx_i^2 and similarly with y_i^2 to give Σy_i^2 (**Step 7**). Finally, sum the remaining column of $x_i \times y_i$ products to give $\Sigma(x_i \times y_i)$ (**Step 8**). The correlation coefficient r can then be calculated (**Step 9**):

$$r = \frac{\sum_{i=1}^{n}(x_i \times y_i) - \frac{1}{n}\left(\sum_{i=1}^{n} x_i \times \sum_{i=1}^{n} y_i\right)}{\sqrt{\left(\sum_{i=1}^{n} x_i^2 - \frac{1}{n}\left(\sum_{i=1}^{n} x_i\right)^2\right) \times \left(\sum_{i=1}^{n} y_i^2 - \frac{1}{n}\left(\sum_{i=1}^{n} y_i\right)^2\right)}}$$

If the variance shared between the two variables is high, then this ratio will approach either +1 or −1, i.e., a perfect positive or negative correlation. The significance of the ratio can be calculated by calculating t and looking up a value of t in Table A.2, Appendix A (**Step 10**), where:

$$t = r \times \sqrt{\frac{n-2}{1-r^2}}$$

The choice of a one- or two-tailed test will be dependent on the directionality (if any) of the research hypothesis.

As an example, we will use two variables selected from Table 5.6 (columns C2 and C6), compiled by researchers at the School of Environmental Sciences, University of East Anglia. We will make further use of the data in this table for demonstrating Spearman's correlation coefficient and linear regression later in this chapter. Boxes 5.9 and 5.10 show the calculations.

Null hypothesis
$H_0: \mu_1 = \mu_2$

There is *no significant association* between the child accident rate and the Townsend material deprivation score.

Research hypothesis
$H_1: \mu_1 \neq \mu_2$

There is *a positive association* between the child accident rate and the Townsend material deprivation score.

Significance level: 0.05.

Interpretation: In the example, we have used data from Table 5.6 which is a data set collected to examine accidents to pre-school children (aged 0–4 years) and the influence of social variables in 21 areas in Norwich. The results above, correlating childhood accidents with the Townsend material deprivation score (see Townsend *et al.*, 1988), produce an $r = 0.807$, and indicate a strong positive association between the two variables, which we might expect since childhood accidents are most common in households living in poor circumstances. Further, this positive association is significant at the 0.05 significance level.

5.6.5 Regression analysis

Experimental design: Independent data sets.

Level of data: Interval/ratio.

Type of test: Parametric.

Function: Examining trends.

What the test does: Bivariate regression analysis is concerned with the prediction of the values of one variable, knowing the other, the estimation of causal relationships between two variables, and the description of functional relationships where only one of the variables is known exactly. Strictly speaking, it is *not* a test but rather seeks to examine trends within data. The expressions of the trend can, however, be tested for significance.

Assumptions:

1 The independent variable x is not a set of sampled values, or if it is, then the values in the independent variable must at least have been measured with a negligible amount of error.

2 For any given value of x, all the y values are normally distributed. If there are only single y values for each, they are collectively normally distributed.

3 The expected value for variable y for any given x is described by a linear function $y = a + bx$.

4 The variance of the dependent variable is constant for all values of the independent variable and independent of the magnitude of x or y.

5 The values of the residuals have a normal distribution, with an expected value of zero.

6 The values of the residuals are independent of each other, i.e., there is an absence of correlation.

Explanation and procedure: In the previous section, we examined how the strength of a bivariate relationship could be measured in terms of a **correlation coefficient**. However, we often need to know exactly how one variable *varies with respect to another*. This usually means that we have to fit a mathematical equation – often called a **line of best fit** – to the paired data values. Unless the two variables are perfectly correlated (i.e., fall on a straight line), there will be a degree of scatter between them and subsequently many different linear and non-linear mathematical equations can be fitted to the data. We will only consider linear equations within this chapter for situations where there are individual data points comprising each pair. You should make reference to a more specialised text such as Bates and Watts (1988) for details on fitting non-linear functions.

One group of techniques fits a linear function, described as a **least-squares linear regression**, on the basis of minimising the squared deviations between the data points and the function. Deviations can be minimised in either an x-direction, a y-direction or even perpendicular to the line of best fit. For a least-squares regression of the y variable on x, we would minimise the deviations in the y direction as seen in Figure 5.5.

The variables that we assign to x and y in this case depend on our understanding of the relationship between them. The variable we assign to x varies in either a known fashion (**Type I regression**) or an unknown independent fashion (**Type II regression**), but is somehow related to the variation observed in y. In other words, values of y are **dependent** on the values of x. Following convention, y refers to the **dependent variable** (also known as the *response* variable) and x refers to the **independent variable** (also known as the *predictor* variable). In the worked example above, we return to the data in Table 5.6 on child accidents in Norwich and focus on the Child

---

Box 5.9   Handworked example: Pearson's correlation coefficient

x	y	x^2	y^2	$x \times y$
5.146	1.403	26.481	1.967	7.218
0.029	1.200	0.001	1.441	0.035
3.381	1.613	11.431	2.603	5.455
−3.184	0.969	10.138	0.938	−3.084
2.784	1.462	7.751	2.138	4.071
0.937	1.306	0.878	1.705	1.224
4.312	1.167	18.593	1.362	5.032
4.443	1.562	19.740	2.441	6.942
2.522	1.242	6.360	1.541	3.131
−1.522	1.179	2.316	1.390	−1.794
−2.208	0.896	4.875	0.804	−1.979
5.102	1.474	26.030	2.173	7.521
3.626	1.558	13.148	2.427	5.648
−2.440	1.146	5.954	1.314	−2.797
−1.012	1.198	1.024	1.436	−1.213
0.940	1.271	0.884	1.614	1.194
−3.088	1.069	9.536	1.142	−3.301
5.077	1.411	25.776	1.992	7.166
2.515	1.545	6.325	2.386	3.885
−1.698	1.041	2.883	1.084	−1.768
2.436	1.464	5.934	2.143	3.566

$\sum x_i = 28.098$
$\sum y_i = 27.176$
$(\sum x_i)^2 = 789.498$
$(\sum y_i)^2 = 738.535$
$\sum x_i^2 = 206.059$
$\sum y_i^2 = 36.042$
$\sum (x_i \times y_i) = 46.151$

Step 1   Lay out the data in a table as above and check the assumptions. In this case, we have transformed the accident rate (C2 in Table 5.6 above) by taking natural logs to produce the data in column y_i above. This makes the distribution more normal, though the correlation could be performed with the raw (untransformed) data.

Step 2   Plot the data in the form of a scatterplot of one variable against the other to assess whether the calculation of a correlation is appropriate.

Step 3   Take the square of each x score (i.e., x_i^2) and insert in column 3. Do the same for each y score to produce y_i^2 and insert in column 4.

Step 4   Insert the product of each x and y score (i.e., $x_i \times y_i$) in column 5.

Step 5   Calculate the sum $\sum x_i$ of all the observed x scores and the sum $\sum y_i$ of all the observed y scores: $\sum x_i = 28.098$, $\sum y_i = 27.176$.

Step 6   Square $\sum x_i$ and $\sum y_i$ to produce $(\sum x_i)^2$ and $(\sum y_i)^2$: $(\sum x_i)^2 = 789.498$ and $(\sum y_i)^2 = 738.535$.

Step 7   Sum the column of x_i^2 to give $\sum x_i^2$; and similarly with y_i^2 to give $\sum y_i^2$: $\sum x_i^2 = 206.059$, $\sum y_i^2 = 36.042$.

Step 8   Sum the remaining column of $x_i \times y_i$ products (i.e., column 5) to give $\sum (x_i \times y_i) = 46.151$.

Step 9   Substitute in and calculate the value of r from the following formula:

$$r = \frac{46.151 - \dfrac{28.098 \times 27.176}{21}}{\sqrt{\left(206.059 - \dfrac{789.498}{21}\right) \times \left(36.042 - \dfrac{738.535}{21}\right)}}$$

$$= \frac{46.151 - 36.3615}{\sqrt{(206.059 - 37.595) \times (36.042 - 35.168)}} = \frac{9.7895}{\sqrt{168.464 \times 0.874}} = 0.8068$$

Step 10   Calculate t using the following formula:

$$t = 0.8068 \times \sqrt{\frac{21 - 2}{1 - 0.8068}} = 0.8068 \times \sqrt{\frac{19}{0.1932}} = 0.8068 \times \sqrt{98.344} = 8.001$$

and look up the one-tailed value of t in Table A.2, Appendix A, at the 0.05 significance level using $n - 2 = 19$ degrees of freedom. The critical value of $t = 1.729$, so the calculated value of t exceeds the critical value. We can therefore reject the null hypothesis H_0.

Box 5.10 MINITAB example: Pearson's correlation coefficient

From Table 5.6, place C2, 'Child Accident Rate', in C1; and C6, 'Townsend Material Deprivation Score', in C2. Transform C1 by taking natural logs: place the result in C3. The 'plot' command (see Box 5.12) can be used to produce a scatterplot of one variable plotted against the other.

Using pull-down menus:

➥ Stat
➥ Basic Statistics➤
➥ Correlation…
Highlight C2
➥ Select
Highlight C3
➥ Select
➥ OK

Using command line:

Type: Correlation C2 C3.

MINITAB results:

```
MTB > Correlation C2 C3.
```

Correlations (Pearson)

```
Correlation of C2 and C3 = 0.807
```

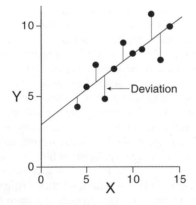

Figure 5.5 Least-squares regression.

Accident and Percentage Unemployed variables. In this case, the 'Percentage Unemployed' is the independent *x* variable, and 'Child Accidents' is the dependent *y* variable, which we think may somehow be linked to variations in the unemployment rate (which we are using as a surrogate for deprivation). In fact, the same could be said of any of the other variables in Table 5.6 other than Child Accidents: these variables were all selected as potential *causal factors* explaining the Child Accident variable. Estab-

lishing causal relationships constitutes one important use of linear regression in human geography.

The least-squares form of regression used here is appropriate only under specific circumstances: usually that the independent variable *x* is *free of error*, or if error exists it is known to be negligible. This is the **Type I regression** mentioned above, which we have described in this section. Where there are errors in *both* variables, a **Type II regression** is more appropriate. In this situation, other techniques for fitting functions to paired data, commonly known as structural analysis, are more appropriate, but are not discussed here.

We can model the data shown by the points in Figure 5.5 in terms of the relationship

$$y = a + bx + e$$

where $e = y - \hat{y}$ represents the deviations which we minimise. In contrast, the form of the equation fitted in Figure 5.5 is

$$y = a + bx$$

We have discussed the meaning of the terms *x* and *y*, but the final two terms in the above equation also need to be explained. Both are constants and together they are known as **regression coefficients**. The *a* coefficient is the intercept, which is the value of *y* obtained when x is zero (i.e., when the line of best fit *intercepts* the y-axis). The b coefficient is the slope (gradient) of the line, which tells us how much change we will observe in the y variable with a specific change in the x variable.

To perform a linear regression we need to estimate the a and b coefficients in the equation above. For the b coefficient:

$$b = \frac{\text{covariance}(xy)}{\text{variance}(x)} = \frac{(\Sigma xy - n\overline{xy})/(n-1)}{(\Sigma x^2 - n\bar{x}^2)/(n-1)}$$

Note that the equations for the variance and covariance are equivalent to the definitions in Sections 4.5.4 and 5.6.4. We make use of an abbreviated form here. In order to be a best-fit line, the regression line must pass through the means of both the x and y variables, and once we have obtained b, the a coefficient can be obtained from

$$a = \bar{y} - b\bar{x}$$

The model of variance in our regression is often expressed in a form similar to that used with the analysis of variance discussed earlier in this chapter:

Sum of squares = Total sum of + Sum of squares explained by the   squares (TSS)   unexplained by regression (RSS)   the regression (ESS)

We can assess the goodness-of-fit of the linear regression by looking at the ratio of the total sum of squared deviations (TSS) to the sum of squared deviations explained by the regression (RSS), with the resulting parameter R^2 known as the **coefficient of determination**. When the regression is a good description of the relationship, the two terms above will be similar in size and the value of R^2 will be high. However, when the regression is a poor description of the relationship, the RSS will be small, as will the value of R^2.

Once we have fitted a regression model, it is important to assess the significance of the estimated coefficients. It is possible to calculate the **standard error** around regression coefficients with associated confidence limits, and to perform **tests of significance** upon regression parameters. There are a variety of methods which can be used, and hypotheses which can be tested. For example, to test whether the regression coefficient b comes from a population where the population slope $\beta = 0$, we can also make use of a Student's t-test. To obtain the calculated value of t we make use of a form of the formula which we encountered with the Pearson correlation coefficient:

$$t = r \times \sqrt{\frac{n-2}{1-r^2}} = \frac{b}{\text{SE}(b)}$$

which can then be compared with the critical value of t at $n-2$ degrees of freedom. If the calculated value of t exceeds this, then β is significantly different from 0. In a similar fashion the coefficient a can be tested to see if the population intercept α is significantly different from 0.

The calculation of standard errors and confidence limits should never be the final step in a regression analysis. The examination of the residuals in a least-squares linear regression can supply much useful information concerning the appropriateness of the linear model and insights into the structure of the data. Residuals are defined as

$$\text{residual} = \text{data} - \text{fit} = y - \hat{y}$$

They are often converted into **leveraged coefficients**, and then calculated in a standardised form. One of the main aims of the analysis of residuals is to identify influential observations. Chatfield (1995) defines an **influential observation** as one whose removal would lead to substantial changes in the fitted regression model. **Cook's distance** D is a measure of the influence of individual observations: values of D which exceed 1 indicate that the observation is influential. In addition, residuals can be plotted against both the calculated fits, and the x and y variables. If the linear regression provides a good description of the data, such a plot should reveal no particular pattern, but it could reveal (1) a few much larger residuals which could be **outliers**; (2) a curved regression on fitted values; (3) change in the variability of residuals; and (4) a skewed or non-normal distribution of residuals.

The data displayed in Figure 5.6 have been taken from Anscombe (1973). Although they display four very different patterns, they produce the same R^2 and the same regression parameters. The results presented here, although extreme, provide examples of the different effects which may make the final calculated regression parameters misleading. This demonstrates the necessity of not only carefully plotting both the x and y variables in the form of a scatterplot prior to the regression analysis, but also plotting the residuals after the analysis.

To calculate linear regression, first check the assumptions, and construct a scatterplot to determine if a linear model is appropriate (**Step 1**). Regression is often used inappropriately (see above for explanation). For each variable, sum and divide by the sample size n to obtain the means $\bar{x}$ and $\bar{y}$. Obtain x^2, y^2 and $x \times y$, and the respective sums of each: (Σx^2),

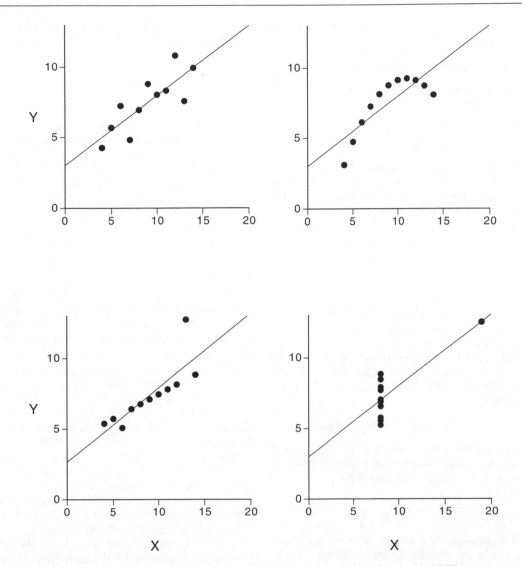

Figure 5.6 Patterns of residuals from a linear regression (redrawn from Anscombe 1973).

(Σy^2) and (Σxy) (**Step 2**). Remember that x is the independent variable and y the dependent variable. To obtain the regression slope coefficient b, substitute into the following equation (**Step 3**):

$$b = \frac{\Sigma xy - n\overline{xy}}{\Sigma x^2 - n\bar{x}^2}$$

Next calculate the intercept coefficient a by substituting into the following equation (**Step 4**):

$$a = \bar{y} - b\bar{x}$$

The sum of squares *explained* by the regression (RSS) is obtained (**Step 5**) from:

$$\Sigma(\hat{y} - \bar{y})^2 = \frac{(\Sigma xy - n\overline{xy})^2}{\Sigma x^2 - n\bar{x}^2}$$

and the sum of squares *unexplained* by the regression (ESS) is obtained (**Step 6**) from:

$$\Sigma(y - \hat{y})^2 = \Sigma y^2 - n\bar{y}^2 - \frac{(\Sigma xy - n\overline{xy})^2}{\Sigma x^2 - n\bar{x}^2}$$

The total sum of squares (TSS) is most simply obtained (**Step 7**) from:

$$\Sigma(y - \bar{y})^2 = \text{RSS} + \text{ESS}$$

Finally, the coefficient of determination (R^2) is obtained from:

$$\frac{\text{RSS}}{\text{TSS}} = \frac{\Sigma(\hat{y} - \bar{y})^2}{\Sigma(y - \bar{y})^2} = R^2$$

Note: The following steps concern the estimation of the standard error and confidence limits around some of the regression coefficients calculated above and should be used only as required. Many of the quantities used have already been calculated, and can simply be substituted into the appropriate equation.

The standard error (SE) around the slope coefficient b is calculated from the equation below (**Step 8**):

$$\text{SE}(b) = \sqrt{\frac{(\Sigma(y - \hat{y})^2)/(n - 2)}{\Sigma(x - \bar{x})^2}}$$

The 95% confidence limits around the slope can be obtained using the Student's t-distribution with the appropriate value of t obtained using $n - 2$ degrees of freedom:

$$b - [t_{0.05} \times \text{SE}(b)], \quad b + [t_{0.05} \times \text{SE}(b)]$$

We can then test the regression slope coefficient b, making use of the Student's t-distribution (**Step 9**) where the value of t is calculated from:

$$t = \frac{b}{\text{SE}(b)}$$

The standard error (SE) around the intercept a is calculated from the equation below (**Step 10**):

$$\text{SE}(a) = \sqrt{\frac{\Sigma(y - \hat{y})^2}{n - 2} \times \left(\frac{1}{n} + \frac{\bar{x}^2}{\Sigma(x - \bar{x})^2}\right)}$$

Again, the 95% confidence limits can be obtained using the Student's t-distribution with the appropriate value of t obtained using $n - 2$ degrees of freedom:

$$a - [t_{0.05} \times \text{SE}(a)], \quad a + [t_{0.05} \times \text{SE}(a)]$$

In a similar fashion to the b coefficient above, we then can test the intercept regression coefficient a:

$$t = \frac{a}{\text{SE}(a)}$$

Similarly we can calculate the standard error (SE) around the mean value of the dependent variable (**Step 11**):

$$\text{SE}(\bar{y}) = \sqrt{\frac{(\Sigma(\hat{y} - \bar{y})^2)/(n - 2)}{n}}$$

with the 95% confidence limits obtained, again making use of the Student's t-distribution (as above):

$$\bar{y} - [t_{0.05} \times \text{SE}(\bar{y})], \quad \bar{y} + [t_{0.05} \times \text{SE}(\bar{y})]$$

As an example (Boxes 5.11 and 5.12), we will make use of data from Table 5.6, except this time we will examine the relationship between the child accident rate and the proportion of economically active males who are unemployed. Note that we will include the calculation and plotting of residuals from this regression only in the MINITAB example (Box 5.12).

Interpretation: The equation for the linear model is $y = 0.974 + 2.90x$, with an R^2 of 57.6%. Note that the results of the significance tests for both the intercept and slope coefficients are significant at the 0.05 significance level. The advantage of linear regression over a measure of correlation is that we can predict a measure of y (in this case, the child accident rate) from a measure of x (in this case, unemployed economically active males). Note, however, that although the pattern of residuals shows no clear trend, the percentage of variation which is explained by the linear regression is only 57.6%, which still leaves some 42.4% unexplained. Although there is a significant linear relationship, we would not have considerable confidence in any predictions of y based on x. In fact, a more useful model would be constructed if we regressed the child accident rate against *several* of the variables from Table 5.6 in a **multiple regression**, although we will not consider this in this book. You should consult one of the more advanced texts listed at the end of this chapter for further details.

Box 5.11   Handworked example: linear regression

x	y	x^2	y^2	xy
0.178	1.402	0.031	1.967	0.250
0.091	1.200	0.008	1.440	0.109
0.144	1.613	0.021	2.603	0.232
0.041	0.968	0.001	0.937	0.040
0.115	1.462	0.013	2.138	0.168
0.093	1.305	0.008	1.705	0.122
0.156	1.166	0.024	1.361	0.183
0.215	1.562	0.046	2.440	0.336
0.112	1.241	0.012	1.541	0.139
0.058	1.178	0.003	1.389	0.069
0.056	0.896	0.003	0.803	0.050
0.190	1.474	0.036	2.172	0.280
0.153	1.557	0.023	2.426	0.238
0.037	1.146	0.001	1.314	0.043
0.068	1.198	0.004	1.435	0.082
0.094	1.270	0.008	1.614	0.119
0.037	1.068	0.001	1.142	0.039
0.180	1.411	0.032	1.992	0.254
0.095	1.544	0.009	2.385	0.148
0.057	1.041	0.003	1.084	0.059
0.145	1.463	0.021	2.143	0.213

$\bar{x} = 0.110$
$\bar{y} = 1.294$
$\sum x^2 = 0.317$
$\sum y^2 = 36.042$
$\sum xy = 3.182$

Step 1   Check the assumptions. Place the variables in a table as above, with the independent variable (unemployed males) assigned to the x column, and the dependent variable (child accidents) assigned to the y column. Note: only the log transform of child accidents is included in the table above. Then, plot one variable against the other to see if a linear model is appropriate. See the explanation section above for more details. (Throughout these calculations, we have rounded the figures down to three decimal places for display purposes, though full precision should be used in the calculations.)

Step 2   Sample size $n = 21$, mean $\bar{x} = 0.110$ and mean $\bar{y} = 1.294$. Place x^2, y^2 and $x \times y$ into new columns 3, 4 and 5 in the table above. $\sum x^2 = 0.317$, $\sum y^2 = 36.042$ and finally the sum $\sum x \times y = 3.182$.

Step 3   To obtain the regression slope coefficient b, substitute into the equation to give:

$$b = \frac{3.182 - 21(0.110 \times 1.294)}{0.317 - 21 \times (0.110)^2} = \frac{3.182 - 3.009}{0.317 - 0.257} = \frac{0.173}{0.059} = 2.895$$

Step 4   To calculate the regression intercept coefficient a, substitute into the equation to give:

$$a = 1.294 - 2.895 \times (0.110) = 0.973$$

Step 5   The sum of squares *explained* by the regression (RSS) can be obtained:

$$\text{RSS} = \frac{(3.182 - 21 \times (0.110 \times 1.294))^2}{0.317 - 21 \times (0.110)^2} = \frac{(0.173)^2}{0.059} = 0.502$$

Step 6   The sum of squares *unexplained* by the regression (ESS) is then obtained:

$$\text{ESS} = 36.042 - 21 \times (1.294)^2 - \frac{(0.173)^2}{0.059} = (36.042 - 35.169 - 0.502) = 0.369$$

Step 7   The total sum of squares (TSS) is obtained by substituting in:

$$\text{TSS} = 0.502 + 0.369 = 0.871$$

Then, to obtain the coefficient of determination (R^2), substitute into the equation to give:

$$R^2 = \frac{0.502}{0.871} = 0.576$$

Box 5.11   (cont'd)

Step 8   The standard error (SE) around the slope coefficient b becomes:

$$SE(b) = \sqrt{\frac{0.369/19}{0.0599}} = \sqrt{0.3242} = 0.5694$$

From Table A.2 in Appendix A, the two-tailed Student's t-value at a 0.05 significance level, using $n - 2 = 19$ degrees of freedom = 2.093; therefore the 95% confidence interval is:

$$2.895 - [2.093 \times 0.5694], \quad 2.895 + [2.093 \times 0.5694]$$

i.e., from 1.70 to 4.09.

Step 9   To test the regression slope coefficient b, substitute into the equation to give:

$$t = \frac{2.895}{0.5694} = 5.08$$

In this case, we compare this value with the two-tailed critical value at a 0.05 significance level with $n - 2$ degrees of freedom from above (2.093). The calculated value exceeds the critical value, therefore b is significantly different from zero.

Step 10   The standard error (SE) around the intercept a can then be obtained:

$$SE(a) = \sqrt{\frac{0.369}{19} \times \left(\frac{1}{21} + \frac{0.0121}{0.0599}\right)} = \sqrt{0.01942 \times (0.04762 + 0.202)}$$

$$= \sqrt{0.01942 \times 0.2496} = \sqrt{0.004847} = 0.069$$

For the 95% confidence interval substitute in $SE(a)$ and the value of t from above (2.093):

$$0.973 - [2.093 \times 0.069], \quad 0.973 + [2.093 \times 0.069]$$

i.e., from 0.83 to 1.12.

Step 11   To test the regression slope coefficient a, substitute into the equation to give:

$$t = \frac{0.973}{0.069} = 14.1$$

Again, we compare this value with the two-tailed critical value of t at $n - 2$ degrees of freedom from above (2.093). The calculated value considerably exceeds the critical value, therefore a is significantly different from zero.

Step 12   The standard error (SE) around the mean value of the dependent variable is obtained from:

$$SE(\bar{y}) = \sqrt{\frac{0.502/19}{21}} = \sqrt{\frac{0.026}{21}} = 0.035$$

with the 95% confidence interval:

$$1.294 - [2.093 \times 0.035], \quad 1.294 + [2.093 \times 0.035]$$

i.e., from 1.22 to 1.37.

Box 5.12   MINITAB example: linear regression

In addition to the regression equation and R^2, MINITAB provides as standard the SE ('StDev') and t tests for significance ('t-ratio') of the regression coefficients, as well as ANOVA statistics which could also be used for testing the coefficients. Any 'unusual' observations are also identified. In addition, we have requested residuals and fits to be stored in columns C4 and C5. We will only graph fits against residuals, although you should also consider graphing residuals against the original variables or producing a histogram of the residuals for further analysis.

Place the data for the proportion of economically active males unemployed in C1 and the accident rate per 1000 days at risk in C2. Transform C2 using natural log, placing the results in C3.

Using pull-down menus:	*Using command line:*

(1) Scatterplot

➡ Graph   MTB > Plot C3 C1.
➡ Plot...
➡ Graph variables: [y]
Highlight C3
➡ Select
➡ Graph variables: [x]
Highlight C1
➡ Select
➡ OK

(2) Regression

➡ Stat   MTB > Regress C3 1 C1;
➡ Regression ➤   SUBC > Residuals C4;
➡ Regression...   SUBC > Fits C5.
Highlight C3
➡ Select
Highlight C1
➡ Select
➡ OK

(3) Residuals

➡ Graph   MTB > Plot C3 C1.
➡ Plot...
➡ Graph variables: [y]
Highlight C3
➡ Select
➡ Graph variables: [x]
Highlight C1
➡ Select
➡ OK

Box 5.12 (cont'd)

MINITAB results:

(1) Scatterplot

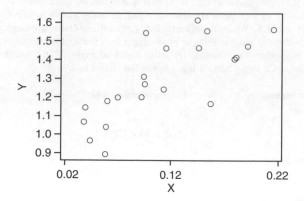

Figure 5.7 MINTAB output: scatterplot.

(2) Regression

```
Regression
The regression equation is
C3 = 0.974 + 2.90 C1

Predictor      Coef      Stdev    t-ratio        p
Constant    0.97351    0.07004      13.90    0.000
C1           2.8956     0.5697       5.08    0.000

s = 0.1395    R-sq = 57.6%    R-sq(adj) = 55.4%
```

```
Analysis of Variance

SOURCE        DF         SS         MS        F        p
Regression     1    0.50286    0.50286    25.83    0.000
Error         19    0.36985    0.01947
Total         20    0.87270

Unusual Observations
Obs.      C1         C3       Fit    Stdev.Fit    Residual    St.Resid
 19    0.096     1.5446    1.2512       0.0316      0.2934       2.16R

R denotes an obs. with a large st. resid.
```

(3) Residuals

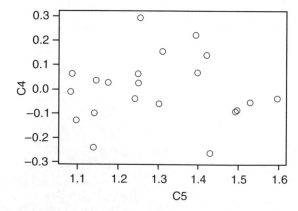

Figure 5.8 MINTAB output: residuals.

5.7 Non-parametric tests

5.7.1 Chi-square (χ^2)

Experimental design: Frequency or count data.

Level of data: Nominal/categorical.

Type of test: Non-parametric.

Function: Comparing proportions.

What the test does: This test helps to establish whether there is a significant difference in the distribution of data between two or more variables or whether any differences are due to chance. A comparison is made of observed frequencies with the expected frequencies if chance is the only factor operating.

Assumptions:

1 Should be used only with **count** data; inappropriate for **magnitudes** or **percentages**.
2 There must be few categories for which the expected frequency is small (i.e., less than 5).

Explanation and procedure: The first step (**Step 1**) is to check whether the data satisfy the test assumptions. Then assemble the data into a **contingency table** consisting of the categories of one variable for each row i against the categories of the other variable for each column j, with the **observed frequency** O in each cell. Next (**Step 2**), sum the total of observed frequencies O_{ij} for each row i and column j. Obtain the grand total, checking that this is the same for summing all rows as for summing all columns. Once Steps 1 and 3 are completed, you may suspect that your data follow a trend: there may appear to be more counts in one category than another. But how large do the differences have to be in order to be significant and not to have arisen by chance? The **chi-square** goodness-of-fit test allows you to test whether any observed differences in your data table are in fact significant. More specifically, it tests the null hypothesis H_0 that 'rows and columns are independent', in other words that the probability of an observation O_{ij} falling in any particular column j does not depend on the row i the observation is in (and vice versa) (Chatfield, 1995).

The test is based on estimating the **expected frequency** E_{ij} of observations in each cell. This is obtained by distributing the data according to the proportions each row and column make with the overall total (**Step 3**). E_{ij} is calculated for each cell by multiplying the row total by the column total and dividing by the grand total, i.e.:

$$E_{ij} = \frac{\sum_{i=1}^{r} O_{ij} \times \sum_{j=1}^{c} O_{ij}}{\sum_{j=1}^{c}\sum_{i=1}^{r} O_{ij}}$$

Now, examine the distribution of expected frequencies E_{ij}. If more than 20% of the expected values are less than 5, merge these categories with adjacent, related categories in the table and then recalculate the expected frequencies (**Step 4**). This formula calculates the difference between the observed and expected frequencies, squaring the result and then dividing by the expected frequency. Calculate the total **degrees of freedom** (DF) available, which for a 1×1 table is (number of columns − 1) or for a 2×2 (or greater) table is (number of rows − 1) × (number of columns − 1) (**Step 5**). If DF > 1, calculate the chi-square statistic using the standard formula as follows. For each cell take the absolute difference between the observed and expected cell values $O_{ij} - E_{ij}$ (**Step 6**). Take the square of the result and sum together for all cell values:

$$\chi^2 = \sum_{i=1}^{r}\sum_{j=1}^{c} \frac{(O_{ij} - E_{ij})^2}{E_{ij}}$$

A further correction is made if you are using a small 2×2 table: this is termed **Yates' continuity correction** (**Step 7**). This is necessary because the chi-square statistic is derived from the continuous probability distribution, which is being used as an approximation to the discrete probability distribution of observed frequencies. The calculation of χ^2 is the same as in Step 6 except for the subtraction of 0.5 from $O_{ij} - E_{ij}$:

$$\chi^2 = \sum_{i=1}^{r}\sum_{j=1}^{c} \frac{(|O_{ij} - E_{ij}| - 0.5)^2}{E_{ij}}$$

If the sample size is reasonably large this correction has little effect, and the standard formula should be used (see **Step 6**). Once applied to each cell in the table, the results are then summed to produce a calculated value of chi-square (χ^2) which can then be compared against the critical value of χ^2 (see Table A.4, Appendix A) (**Step 8**). The chi-square test is a one-tailed test since differences will always lead to a large value of χ^2 (Chatfield, 1995). If the calculated value of the chi-square statistic is greater than the critical value, the null hypothesis H_0 can be rejected and it can be concluded that there is a significant association between the variables (**Step 9**). Everitt (1992) provides an exhaustive summary of the analysis of contingency tables, including alternatives to pooling

Table 5.7   Household head income (weekly) by dwelling type for those aged under 65 in Aberdeen.

	Multi-storey and flatted	Tenement	Terraced and semi-detached	Total
Under £50	18	20	12	50
£50–£74	13	18	20	51
£75–£99	23	19	45	87
£100+	21	30	58	109

Source: Williams *et al.*, 1986.

Box 5.13   Handworked example: chi-square

	Multi-storey and flatted	Tenement	Terraced and semi-detached	Total
Under £50	18	20	12	50
£50–£74	13	18	20	51
£75–£99	23	19	45	87
£100+	21	30	58	109

Step 1   Set out your data (observed frequencies) in a table as above.

Step 2   Obtain the row and column totals and check that the overall total (297) is the same for summing row totals (50 + 51 + 87 + 109 = 297) as for summing column totals (75 + 87 + 135 = 297).

Step 3   Set out a new table of expected frequencies, checking that rows and columns still sum to the totals above:

	Multi-storey and flatted	Tenement	Terraced and semi-detached	Total
Under £50	12.626	14.646	22.727	50
£50–£74	12.879	14.939	23.182	51
£75–£99	21.970	25.485	39.545	87
£100+	27.525	31.930	49.545	109
Total	75	87	135	297

For example, for the first row:

$$\frac{75 \times 50}{297} = 12.626 \qquad \frac{87 \times 50}{297} = 14.646 \qquad \frac{135 \times 50}{297} = 22.727$$

Step 4   There are no expected observations less than 5, therefore it is not necessary to merge adjacent categories and recalculate.

Step 5   DF = (4 − 1) × (3 − 1) = 6.

Step 6   Calculate (observed − expected)^2 ÷ expected, for each cell, and sum to obtain χ^2. For example, for the first cell: $(18 − 12.626)^2 ÷ 12.626 = 2.287$.

$$
\begin{array}{ccccccc}
2.287 & + & 1.957 & + & 5.063 & + & \\
0.001 & + & 0.627 & + & 0.437 & + & \\
0.0483 & + & 1.650 & + & 0.752 & + & \\
1.547 & + & 0.117 & + & 1.443 & = & 15.929
\end{array}
$$

Step 7   The critical value of χ^2 at the 0.05 significance level with 6 degrees of freedom is 12.59 (see Table A.4, Appendix A).

Step 8   The calculated value of $\chi^2 = 15.929$ exceeds the critical value and the null hypothesis H_0 can be rejected.

Box 5.14   MINITAB example: chi-square

Enter the income data into C1 (Multi-storey), C2 (Tenement) and C3 (Terraced).

Using pull-down menus: *Using command line:*

➥ Stat   MTB > Chisquare c1-c3
➥ Tables➤
➥ Chisquare Test…
Highlight C1
➥ Select
Highlight C2
➥ Select
Highlight C3
➥ Select
➥ OK

MINITAB results:

```
MTB > chisquare c1-c3.
```

Chi-Square Test

```
Expected counts are printed below observed counts

          C1       C2       C3    Total
    1     18       20       12       50
       12.63    14.65    22.73

    2     13       18       20       51
       12.88    14.94    23.18

    3     23       19       45       87
       21.97    25.48    39.55

    4     21       30       58      109
       27.53    31.93    49.55

Total     75       87      135      297

ChiSq = 2.287 + 1.957 + 5.063 +
        0.001 + 0.627 + 0.437 +
        0.048 + 1.650 + 0.752 +
        1.547 + 0.117 + 1.443 = 15.929
df = 6,  p = 0.015
```

---

categories where the expected values are less than 5. Also, for a small 2×2 table, a simplified version of the χ^2 formula may be used (see Everitt, 1992, Section 2.3).

To illustrate this test, we will use a data set (Table 5.7) which compares dwelling type against income for a sample of those aged under 65 in Aberdeen: see Boxes 5.13 and 5.14.

**Null hypothesis
H_0:**   Dwelling type is independent of tenant income (i.e., there is *no significant difference* in tenant income with dwelling type).

**Research hypothesis
H_1:**   Dwelling type depends on tenant income (i.e., there is *a significant difference* in tenant income with a change in dwelling type).

Significance level: 0.05.

Interpretation: In the handworked example above, the calculated value of chi-square exceeds the critical value and is therefore significant at the 0.05 significance level. Similarly, using MINITAB we obtain $p = 0.015$, which is less than 0.05 and therefore significant. We

Table 5.8 Quantity of household waste (kg) collected in one week from separate samples of households using bags and wheeled bins.

	0–4.99	5–9.99	10–14.99	15–19.99	20–24.99	25–29.99	30+
Set A: bags	2	10	9	5	5	2	2
Set B: bins	0	3	5	8	9	5	2

Source: MELResearch 1994.

can reject the null hypothesis H_0 and be 95% certain that dwelling type depends on tenant income.

5.7.2 Kolmogorov–Smirnov

Experimental design: Frequency data in ordered classes (one or two samples).

Level of data: Ordinal.

Type of test: Non-parametric.

Function: Comparing distributions.

What the test does: Compares the data sample with some expected population distribution (one sample) or compares the distributions of two data samples.

Assumptions:

1 Samples are unrelated (for the two-sample case).
2 Data must be in ordered classes.

Explanation and procedure: The Kolmogorov–Smirnov test can be used with both a single sample and two samples. In the single-sample case, it is used to determine whether the **goodness of fit** of a sample is similar to a particular distribution, i.e., whether the observed frequencies of the sample coincide with what we would expect given a particular probability distribution. In the two-sample case, we are assessing whether the populations from which the samples have been drawn differ. Coshall (1989: 17) notes that the Kolmogorov–Smirnov test makes more complete use of the available data than the chi-square test, as it does not require the 'lumping together' of categories, and because it takes advantage of the ordinal nature of the data. That said, it is rarely used in human geography applications.

The Kolmogorov–Smirnov test makes use of the **cumulative frequency** (or probability) distribution. The first step in the analysis is to assemble the data for each sample in a frequency form, which should be placed in an ordered table (**Step 1**) and then in cumulative frequency form (**Step 2**). The cumulative frequency proportions are obtained by dividing the cumulative frequencies by the total number of observations in each sample (**Step 3**). The absolute differences between the cumulative frequency proportions

for each column are obtained (**Step 4**) with the D statistic defined as the maximum value of the differences (**Step 5**). The critical value of D (at the 0.05 significance level) is obtained from a table of critical values (Table A.5, Appendix A) or, if either sample size n is greater than 25, from the equation (**Step 6**):

$$D = 1.36 \sqrt{\frac{n_1 + n_2}{n_1 \times n_2}}$$

If the calculated value of D exceeds the critical value, then the null hypothesis is rejected (**Step 7**).

As an example data set, we have made use of the waste data from Table 5.3 above expressed in terms of class frequencies (see Table 5.8): see Boxes 5.15 and 5.16.

Null hypothesis H_0:	There is *no significant difference* between the quantity of domestic waste produced from households using bags and that from households using bins: the samples come from similar distributions.
Research hypothesis H_1:	There is *a significant difference* between the quantity of domestic waste produced from households using bags and that from households using bins: the samples come from different distributions.

Significance level: 0.05.

Interpretation: From the data listed above, we find that at the 0.05 significance level, the two distributions do differ significantly, which matches the result obtained for the unrelated *t*-test above (Section 5.6.1). Pett (1997) notes that this test has the advantage over the *t*-test in that it compares the entire distribution, not just measures of central tendency.

5.7.3 Mann–Whitney *U* test

Experimental design: Independent data sets.

Level of data: At least ordinal.

Type of test: Non-parametric.

Function: Comparison.

Box 5.15 Handworked example: Kolmogorov–Smirnov

	0–4.99	5–9.99	10–14.99	15–19.99	20–24.99	25–29.99	30+
Set A: Bags	2	10	9	5	5	2	2
Set B: Bins	0	3	5	8	9	5	2

Step 1 Assemble data into an ordered table (as above) with classes in columns, variables in rows and the frequency in each cell.

Set A: Bags	2	12	21	26	31	33	35	Check $n = 35$
Set B: Bins	0	3	8	16	25	30	32	Check $n = 32$

Step 2 Sum the cells in each row and enter (as above) to obtain the cumulative frequency distribution, where the final sum in each row should equal the sample size.

Step 3 Calculate the cumulative frequency proportion C1 and C2 of the two samples by dividing the cumulative frequency in each cell by the row total:

$$\text{C1: } \frac{2}{35} = 0.057 \quad \frac{12}{35} = 0.343 \quad \frac{21}{35} = 0.6 \quad \frac{26}{35} = 0.743 \quad \frac{31}{35} = 0.886 \quad \frac{33}{35} = 0.943 \quad \frac{35}{35} = 1.0$$

$$\text{C2: } \frac{0}{32} = 0.0 \quad \frac{3}{32} = 0.094 \quad \frac{8}{32} = 0.25 \quad \frac{16}{32} = 0.5 \quad \frac{25}{32} = 0.781 \quad \frac{30}{32} = 0.937 \quad \frac{32}{32} = 1.0$$

Step 4 Calculate the absolute differences between the cumulative frequency proportions for each column to obtain:

$$0.057 \quad 0.249 \quad 0.35 \quad 0.243 \quad 0.105 \quad 0.006 \quad 0.0$$

Step 5 Calculate the D statistic as the maximum value from Step 4, $D = 0.35$.

Step 6 Since both sample numbers are greater than 25, calculate the critical value of D (0.05 significance level) as:

$$D = 1.36 \times \sqrt{\frac{n_1 + n_2}{n_1 \times n_2}} = 1.36 \times \sqrt{\frac{32 + 35}{32 \times 35}} = 1.36 \times \sqrt{\frac{67}{1120}} = 1.36 \times 0.245 = 0.333$$

If n_1 and $n_2 < 25$ see Table A.5, Appendix A, for the critical value.

Step 7 Compare the critical and calculated values of D at a 0.05 significance. In this case the calculated value, 0.35, does exceed the critical value and we can reject the null hypothesis H_0.

Box 5.16 MINITAB example: Kolmogorov–Smirnov

MINITAB does not perform a Kolmogorov–Smirnov test as of version 10.1. However, you *can* get the software to do most of the hard work for you, and if either sample size exceeds 25, you can also calculate the critical value of D. Place Set A in C1 and Set B in C2.

Using pull-down menus:

➥ Calc
➥ Mathematical expressions...
➥ Variable (new or modified): [ ]
Type in C3
➥ Expression [ ]
Type in parsum(C1)/sum(C1)
etc.

Using command line:

MTB > Let C3 = parsum(C1)/sum(C1).
MTB > Let C4 = parsum(C2)/sum(C2).
MTB > Let C5 = max(abs(C3-C4)).
MTB > Let C6 = 1.36*(sqrt((sum(C1)+ sum(C2))/(sum(C1)*sum(C2)))).

Repeat for C4, C5 and C6, substituting the expressions listed for the command line instructions above.

MINITAB results:

The calculated D statistic will be in C5 (0.35) and the critical value of D will be in C6 (0.3326).

What the test does: Determines whether two independent samples are from the same population. The **scores** in each data set are compared by ranking the individual scores and determining whether the ranks are evenly divided.

Assumptions:

1   Each data set is taken from a random sample.
2   The samples were taken independently of each other.
3   The data sets have approximately the same variances (shape).
4   The data are obviously non-normally distributed.

Explanation and procedure: The Mann–Whitney U test is the non-parametric counterpart of the t-test for unrelated (independent) data. The test is used to determine whether ordinal data collected in two different samples differ significantly. For example, we might want to determine whether towns in two different regions have differing mortality rates or whether men and women assign the same familiarity ratings to an area. The test calculates whether there is a significant difference in the distribution (based on the median) of data by comparing the **ranks** of each data set.

The first step in the calculation is to check whether the data fulfils the test assumptions (**Step 1**). Next, rank all data as if one group giving the lowest score are ranked as 1. At this stage maintain group identity (**Step 2**). Now, find the sums R_a and R_b of the ranks $R(x)$ in data sets A and B. For example, for set A (**Step 3**):

$$R_a = \sum_{j=1}^{n_a} R(x_j)$$

where n_a is the number of values in data set A. If these sums differ substantially, then there is reason to suspect that the two samples come from different populations (Pett, 1997). Then calculate U_s for the smaller sample using the following formula (**Step 4**):

$$U_s = \left(n_1 \times n_2 + \frac{n_1 \times (n_1 + 1)}{2} - R_1 \right)$$

where R_1 is the sum of ranks for the smaller sample, n_1 is the size of the smaller sample, and n_2 is the size of the larger.

Calculate U_l for the larger sample:

$$U_l = n_1 \times n_2 - U_s$$

Select the smaller of U_s and U_l and call it U (**Step 5**). Check the critical value of U in Table A.8, Appendix

A (**Step 6**). If U is *less than* the critical value it is significant and we can reject the null hypothesis – there is a significant difference between the two data sets. If one or both samples are greater than 20, convert the U value into a Z-score using the formula:

$$Z = \frac{U - \mu_U}{\sigma_U} = \frac{U - \dfrac{n_1 \times n_2}{2}}{\sqrt{\dfrac{n_1 \times n_2 \times (n_1 + n_2) + 1}{12}}}$$

When ties occur between the ranked observations of two samples, the Z-score can be adjusted as follows:

$$Z = \frac{U - \dfrac{(n_1 \times n_2)}{2}}{\sqrt{\dfrac{n_1 \times n_2}{n \times (n - 1)} \times \left(\dfrac{n^3 - n}{12} - \sum T_j \right)}}$$

where $T_j = (t_j^3 - t_j)/12$, t_j is the number of observations tied at a particular rank j, and N is the total number of observations in both samples. The significance of Z is assessed by checking the Z-score in the critical table (Table A.8, Appendix A). The test is illustrated by a data set (Table 5.9) comparing the performance of football teams in the UK before and after a merger in 1921: see Boxes 5.17 and 5.18.

Null hypothesis H_0:   Teams from the old Southern and Northern Leagues have *not differed significantly* in their success since 1921.

Research hypothesis H_1:   There is *a significant difference* in the degree of success enjoyed by teams from the old Southern and Northern Leagues since 1921.

Significance level: 0.05.

Interpretation: These results (see Boxes 5.17 and 5.18) reveal that the two data sets are significantly different, and we can reject the null hypothesis at the 0.05 significance level. The MINITAB results reveal that if these data are two independent random samples, the median of data set A (Southern) is between -41.0 and -16.99 units higher/lower than the median of data set B (Northern) and that the chance of observing two samples as separated as these, when in fact the populations from which they are drawn do not differ, is only 0.0004. We can therefore be confident that the teams from the old Southern and Northern Leagues have differed in their success since 1921. Note that we cannot say why, only that we are 95% confident that there is a difference.

Table 5.9 Highest league position ever obtained by original members of the Southern and Division 3 (North) after merger in 1921.

Southern	Position			Division 3	Position		
Brentford	5	Newport County	48	Accrington	48	Lincoln	30
Brighton	16	Northampton	21	Ashington	60	Nelson	43
Bristol Rovers	28	Norwich	5	Barrow	52	Rochdale	47
Cardiff	2	Plymouth	26	Chesterfield	26	Southport	52
Crystal Palace	13	Portsmouth	1	Crewe	56	Stalybridge	58
Exeter	52	QPR	2	Darlington	37	Tranmere	44
Gillingham	48	Reading	35	Durham	66	Walsall	47
Luton	7	Southampton	2	Halifax	47	Wigan B.	51
Merthyr Tydfil	57	Southend	51	Hartlepool	67	Wrexham	37
Millwall	23	Swansea	6				

Source: Waylen and Snook, 1990.

Box 5.17 Handworked example: Mann–Whitney *U* test

	Southern			Northern	
	x(position)	$R(x)$ (rank)		x(position)	$R(x)$ (rank)
Brentford	5	5.5	Accrington	48	26.0
Brighton	16	10.0	Ashington	60	36.0
Bristol R.	28	15.0	Barrow	52	31.0
Cardiff	2	3.0	Chesterfield	26	13.5
Crystal Palace	13	9.0	Crewe	56	33.0
Exeter	52	31.0	Darlington	37	18.5
Gillingham	48	26.0	Durham	66	37.0
Luton	7	8.0	Halifax	47	23.0
Merthyr Tydfil	57	34.0	Hartlepool	67	38.0
Millwall	23	12.0	Lincoln	30	16.0
Newport C.	48	26.0	Nelson	43	20.0
Northampton	21	11.0	Rochdale	47	23.0
Norwich	5	5.5	Southport	52	31.0
Plymouth	26	13.5	Stalybridge	58	35.0
Portsmouth	1	1.0	Tranmere	44	21.0
QPR	2	3.0	Walsall	47	23.0
Reading	35	17.0	Wigan B.	51	28.5
Southampton	2	3.0	Wrexham	37	18.5
Southend	51	28.5			
Swansea	6	7.0			

$n_a = 20$ $R_a = 269$ $n_b = 18$ $R_b = 472$

Step 1 Check the assumptions. The data are ordinal in nature (rank/relative position) and have approximately even distributions.

Step 2 Rank both columns of data as if they were one data set, to create two new columns (columns 2 and 4).

Step 3 Sum the ranks in each of the new rank columns and call the results R_a and R_b. $R_a = 269$ and $R_b = 472$.

Step 4 In this case set B is the smaller sample ($n_b = 18$). Assign n_b to n_1 and R_b to R_1 and substitute in to calculate U_s the U_1:

$$U_s = 18 \times 20 + \frac{18 \times (18 + 1)}{2} - 472$$

$$= 360 + \frac{342}{2} - 472 = 59$$

$$U_1 = 20 \times 18 - 59 = 301$$

Step 5 Select the smaller of U_s and U_1 and call it U. $U = 59$.

Step 6 Check U in critical values Table A.8, Appendix A, for $n_a = 20$ and $n_b = 18$. At the 0.05 significance level, the critical value is 123. The calculated value of U (59) is *less than* 123; therefore we can reject the null hypothesis H_0.

Box 5.18   MINITAB example: Mann–Whitney *U* test

Enter the test data into the Data Screen with the Southern League data entered into C1 and the Division 3 (North) entered into C2. Return to the MINITAB environment and using either the pull-down menus or command line calculate the Mann–Whitney statistic.

Using pull-down menus:

➥ Stat
➥ Non-parametric ➤
➥ Mann-Whitney…
Highlight C1
➥ Select
Highlight C2
➥ Select
➥ OK

Using command line:

MTB > Mann-Whitney C1 C2.

MINITAB results:

```
MTB > mann-whitney c1 c2.

Mann-Whitney Confidence Interval and Test

C1          N = 20        Median =   18.50
C2          N = 18        Median =   47.50
Point estimate for ETA1-ETA2 is -29.00
95.2 Percent C.I. for ETA1-ETA2 is (-41.00, -15.00)
W = 269.0
Test of ETA1 = ETA2 vs. ETA1 > ETA2 is significant at 0.0004
The test is significant at 0.0004 (adjusted for ties)
```

5.7.4   Wilcoxon signed ranks test

Experimental design: Related data sets.

Level of data: At least ordinal.

Type of test: Non-parametric.

Function: Comparison.

What the test does: The test examines the differences between data from the same phenomenon collected in two different conditions or times by examining the ranks of the difference in values over the two conditions.

Assumptions:

1   The data are paired across conditions or time.
2   The data are symmetrical but need not be normal or any other shape.

Explanation and procedure: The Wilcoxon signed ranks test is the non-parametric counterpart of the *t*-test for related data or paired *t*-test. The test is used to determine whether ordinal data collected from the same phenomenon differ between conditions or times. For example, we may want to know whether a town's mortality rate changes significantly between dates or

whether the conditions under which a questionnaire or interview is conducted influence the findings of a study significantly. The test calculates whether there is a significant difference by examining whether the ranks of individual phenomena differ between conditions or times.

To calculate a Wilcoxon signed ranks test, first check the test assumptions (**Step 1**). If the test assumptions are met then calculate the difference between the two sets of data by subtracting one data set from the other (**Step 2**). Next, calculate the rank of the differences, ignoring the sign (−/+) of the difference. Any pairs with zero values should be ignored and omitted from the ranking and further analysis. When two or more differences have the same rank, take the average of the ranks they would have received had they differed and assign this average to each (**Step 3**). Sum the ranks for each sign, with the smaller of the two assigned to *T* (**Step 4**). Check the critical value of *T* in Table A.6, Appendix A (**Step 5**). Remember that *n* is the number of samples in the data set minus the number of entries where there was no difference between the two data sets. If *T* is *less than* the critical value it is significant at that level and we can reject the null hypothesis H_0, i.e., there is a significant difference between the two data sets. If the number of samples n exceeds the table limits, the T value has to be transformed to a Z-score using the formula (Cohen and Holliday, 1982):

$$Z = \frac{T - \frac{n(n+1)}{4}}{\sqrt{\frac{n(n+1)(2n+1)}{24}}}$$

When extensive ties occur between ranks (e.g., the majority of the data are assigned to the same five or six tied ranks) then a correlation factor is introduced:

$$Z = \frac{T - \frac{n(n+1)}{4}}{\sqrt{\frac{n(n+1)(2n+1)}{24} - \frac{\sum u^3 - \sum u}{48}}}$$

where u is the total number of tied ranks. The significance of Z is assessed by checking the Z-score in critical Table A.7, Appendix A.

To illustrate this test, we will examine the scores obtained by individuals participating in a sketch map exercise (Table 5.10): see Boxes 5.19 and 5.20.

Null hypothesis
H_0:

There is *no significant difference* between median sketch map scores in each of the two conditions.

Research hypothesis
H_1:

There is *a significant difference* between median sketch map scores in each of the two conditions.

Significance level: 0.05.

Interpretation: From our calculations we can conclude that at the 0.05 significance level there is a difference between the performance of respondents in interview and non-interview conditions, and that this difference did not occur by chance. This is confirmed by the MINITAB test, with $p = 0.041$. Note that this result tells us only that there is a significant difference but tells us nothing about why the difference has occurred.

5.7.5 Kruskal–Wallis

Experimental design: Three or more samples.

Level of data: At least ordinal.

Type of test: Non-parametric.

Function: Comparison.

Table 5.10 Difference in content score for respondent in interview/non-interview condition when completing a sketch map.

Respondent	Interview	Non-interview
1	11	12
2	12	9
3	15	10
4	7	9
5	12	8
6	9	9
7	12	7
8	13	10
9	11	10
10	8	10
11	9	6
12	10	8

Assumptions:

1 Populations from which the samples are drawn have similar distributions.
2 The samples are drawn at random.
3 The samples are independent of each other.

Explanation and procedure: The Kruskal–Wallis test is the non-parametric equivalent to the one-way ANOVA test discussed in Section 5.6.3. This test requires that we rank all data from the largest to the smallest ignoring group subdivision, and average those ranks which are tied. Similar to the Mann–Whitney U test discussed above, if the populations from which we have sampled are not significantly different, we would expect that the rank sums for each group would be approximately the same.

The first step in the calculation is to check the assumptions and lay out the data in a table (**Step 1**). Next, rank all the observations taken together as one group where the lowest rank = 1 (**Step 2**). For tied ranks, allocate the average rank to each tie. Now create new columns in the data table and enter the ranks (**Step 3**). Sum the ranks separately for each group k to give R_k, using the formula (**Step 4**):

$$R_k = \sum_{j=1}^{n_k} R(x_j)$$

where n_k is the number of observations in group k. Next, substitute into the following formula to calculate H (**Step 5**):

$$H = \frac{12}{n \times (n+1)} \times \sum_{j=1}^{k} \frac{(R_k)^2}{n_j} - [3 \times (n+1)]$$

Box 5.19   Handworked example: Wilcoxon signed ranks test

Respondent	Interview	Non-interview	d	$R(d)$
1	11	12	−1	1.5
2	12	9	3	7.0
3	15	10	5	10.5
4	7	9	−2	4.0
5	12	8	4	9.0
6	9	9	0	*
7	12	7	5	10.5
8	13	10	3	7.0
9	11	10	1	1.5
10	8	10	−2	4.0
11	9	6	3	7.0
12	10	8	2	4.0

Step 1 Check the assumptions. The data are ordinal in nature (rank/relative position) and have approximately even distributions.

Step 2 Subtract the non-interview score from the interview score to create a new column (d) of data showing the differences. One pair indicates no difference (*) which we omit.

Step 3 Rank these new differences (d) to create a new column of ranked data $R(d)$, ignoring the +/− sign. Note that 1 occurs twice and has an average rank of 1.5 ((1 + 2) ÷ 2). Similar averaging is done for 2 ((3 + 4 + 5) ÷ 3 = 4.0), 3 ((6 + 7 + 8) ÷ 3 = 7.0) and 5 ((10 + 11) ÷ 2 = 10.5).

Step 4 Sum the ranks of the differences for the positive rankings = 56.5 and for the negative rankings = 9.5. The smaller of these values is T.

Step 5 Compare T to the critical value in Appendix A, Table A.6, for a 0.05 significance level with n = sample size (12) − number of pairs with a zero difference (1) = 11. The critical value at 0.05 is 10. The calculated value of T is *less than* 10 and therefore we can reject the null hypotheses H_0.

Box 5.20   MINITAB example: Wilcoxon signed ranks test

Enter the interview data into C1 and the non-interview data into C2. Then calculate the differences between the two and store in C3: C3 = C1 − C2. We want the two-tailed statistic, so we select 'not equal to' in the menu sequence below. You would select 'less than' or 'greater than' if you were performing a one-tailed test. The equivalent selection at the command line is 'Alternative 0' for a two-tailed test, with the replacement of the '0' with a −1 or +1 for a one-tailed test. Note that MINITAB reports the *larger* sum of ranks as the Wilcoxon statistic. Some statistics packages and tables are based on this sum rather than the smaller sum.

Using pull-down menus:
➡ Stat
➡ Nonparametrics ➤
➡ 1-Sample Wilcoxon…
Variables
Highlight C3
➡ Select
Select Test median
➡ Alternative [not equal to]
➡ OK

Using command line:
MTB > WTest 0.0 C3;
SUBC > Alternative 0.

MINITAB results:

```
Wilcoxon Signed Ranks Test

TEST OF MEDIAN = 0.000000 VERSUS MEDIAN N.E. 0.000000

            N FOR     WILCOXON              ESTIMATED
      N     TEST      STATISTIC   P-VALUE   MEDIAN
C3    12    11        56.5        0.041     1.750
```

Table 5.11 Numbers of retail outlets in 21 English towns.

Group number (population size)			
1 (11001–15000)	2 (15001–19000)	3 (19001–23000)	4 (23001–27000)
156	168	256	304
99	198	193	256
115	186		323
160	198		195
195	146		175
164	223		
206			
170			

Source: Coshall 1988.

where n = total number of all ranks. If there are tied scores, the formula in **Step 5** must be adjusted. Otherwise, go straight to **Step 9**. For each tie, the number of observations included in the tie t is used to obtain the parameter T. For each group of ties j, the parameter $T_j = (t_j^3 - t_j)$ is calculated from $(t_j - 1) \times t_j \times (t_j + 1)$ (**Step 6**). Sum together all the T_j, and substitute into the following equation to calculate parameter D (**Step 7**):

$$D = 1 - \frac{\sum_{j=1}^{m} T_j}{(n-1) \times n \times (n+1)}$$

where m = the number of groups of ties.

Now calculate the corrected value of H by dividing by D (**Step 8**). Compare this with a critical value of χ^2 (see Table A.4, Appendix A) and if H exceeds this critical value then reject the null hypothesis – there is a significant difference between the two data sets (**Step 9**). Similar to an analysis of variance, this does not tell you *which* groups are significantly different. To do this requires a similar *post hoc* approach to multiple comparisons as used in ANOVA, with a variety of options available. The Mann–Whitney U test can be used to establish whether pairs of groups are significantly different (Pett, 1997: 219). The reader is directed to Coshall (1988) or Pett (1997) for further details.

The data set we have chosen to illustrate the Kruskal–Wallis test is a set of data (Table 5.11) collected by Coshall (1988) to examine the trends between numbers of retail outlets and population: see Boxes 5.21 and 5.22.

Null hypothesis H_0: There are *no significant differences* in the numbers of retail outlets in the four samples of towns of varying population sizes.

Research hypothesis H_1: There is *a significant difference* in the numbers of retail outlets in the samples of varying population sizes.

Significance level: 0.05.

Interpretation: In our handworked example, the null hypothesis is rejected, and we can conclude that at the 0.05 significance level there is a significant difference in the numbers of retail outlets in the towns when classified according to their population sizes. Note how marginal the results are: MINITAB provides $p = 0.049$. Similar to the analysis of variance we cannot tell which pairs are different, only that a difference exists.

5.7.6 Spearman's rank order correlation coefficient

Experimental design: Independent data sets.

Level of data: Ordinal.

Type of test: Non-parametric.

Function: Comparison.

What the technique does: Describes the degree of association between two sets of ranked data.

Assumptions:

1 Data are at the ordinal level of measurement.
2 The sample is randomly selected.

Explanation and procedure: Spearman's rank correlation coefficient (also known as *Spearman's rho*)

Box 5.21 Handworked example: Kruskal–Wallis

Groups and ranks							
1	$R(x)$	2	$R(x)$	3	$R(x)$	4	$R(x)$
156	4	168	7	256	18.5	304	20
99	1	198	14.5	193	11	256	18.5
115	2	186	10			323	21
160	5	198	14.5			195	12.5
195	12.5	146	3			175	9
164	6	223	17				
206	16						
170	8						
R_k	54.5		66.0		29.5		81.0

Step 1 Check the assumptions and lay out the data into a table (as above).

Step 2 Rank all the observations taking the data together as one group to produce $R(x)$.

Step 3 Insert the ranks into the table.

Step 4 Obtain the sum of the ranks for each group k to produce R_k as at the bottom of the table above.

Step 5 Square each R_k and divide by n_k to obtain the term $(R_k)^2 \div n_k$, e.g., for Group 1:

$$\frac{(54.5)^2}{8} = 371.281$$

Then sum together to obtain 2844.606 and substitute into the equation below:

$$H = \frac{12}{21 \times (21+1)} \times \sum_{j=1}^{k} \frac{(R_k)^2}{n_j} - [3 \times (21+1)]$$

$$= \frac{12}{21 \times (21+1)} \times 2844.606 - [3 \times (21+1)]$$

$$= 0.025974 \times 2844.606 - 66 = 7.886$$

Step 6 In this example there are three groups of tied scores. For each tie calculate T, e.g., for the tie on 195, $T = (2-1) \times 2 \times (2+1) = 6$.

Step 7 $\sum T = 6 + 6 + 6 = 18$. Now calculate D:

$$D = 1 - \frac{18}{(21-1) \times 21 \times (21+1)} = 1 - \frac{18}{9240} = 0.998$$

Step 8 The corrected value of $H = 7.886 \div 0.998 = 7.901$.

Step 9 The critical value of χ^2 at the 0.05 significance level with $k - 1 = 3$ degrees of freedom is 7.81. The calculated value exceeds the critical value, therefore we reject the null hypothesis H_0 (just!).

provides a measure of the linear association between two ranked variables and varies between –1 (perfect negative correlation) and +1 (perfect positive correlation). It is the non-parametric equivalent of Pearson's correlation coefficient discussed in Section 5.6.4. The basis of the statistic is the amount of disagreement between two sets of rankings. In simple terms, the value of the coefficient is a ratio of the actual amount of disagreement with the maximum possible amount of disagreement (e.g., when one rank order is the reverse of the other) subtracted from 1:

$$r = 1 - \frac{2 \times \text{actual amount of disagreement}}{\text{maximum possible disagreement}}$$

and is scaled so that r ranges from +1 to –1. The actual amount of disagreement is expressed in terms of differences between ranks, and the maximum possible disagreement is expressed directly in terms of sample size.

Box 5.22 MINITAB example: Kruskal–Wallis

Place the data in C1. Since the results are significant, the Mann–Whitney U test discussed earlier in this chapter can be used to establish which groups are different.

Using pull-down menus:

➥ Stat
➥ Non-parametric ➤
➥ Kruskal-Wallis...
Highlight C1
➥ Select
Highlight C2
➥ Select
➥ OK

Using command line:

MTB > Kruskal-Wallis C1 C2.

MINITAB results:

```
MTB > Kruskal-Wallis C1 C2.

  LEVEL       NOBS     MEDIAN    AVE. RANK    Z VALUE
    1           8       162.0         6.8      -2.43
    2           6       192.0        11.0       0.00
    3           2       224.5        14.8       0.90
    4           5       256.0        16.2       2.15
  OVERALL      21                    11.0

  H = 7.89 d.f. = 3 p = 0.049
  H = 7.90 d.f. = 3 p = 0.049 (adjusted for ties)

  * NOTE * One or more small samples
```

The first step (**Step 1**) in calculating Spearman's rank correlation coefficient is to plot the data in the form of a scatterplot of one variable against the other to assess whether the calculation of a correlation is appropriate. Next, create two new columns of data by ranking each of the variables separately, if not already in ranked form, giving a rank of 1 to the highest score (**Step 2**). Obtain the difference d between each pair of ranks and the difference squared d^2, then sum the results to obtain $\sum d^2$ (**Step 3**). Now, calculate the denominator $(n + 1) \times n \times (n - 1)$, where n is the number of values ranked (**Step 4**). If there are no tied ranks for each variable, calculate the correlation coefficient r using the following formula (**Step 5**):

$$r = 1 - \frac{6 \times \sum_{i=1}^{n} d_j^2}{(n-1) \times n \times (n+1)}$$

However, if there are tied ranks on one or both of the variables then the formula becomes (Pett, 1997):

$$r = \frac{(n^3 - n) - 6 \sum_{i=1}^{n} d_j^2 - (T_x + T_y)/2}{\sqrt{(n^3 - n)^2 - [(T_x + T_y) \times (n^3 - n)] + (T_x \times T_y)}}$$

where the number of ties on each variable is obtained (e.g., for T_x) from:

$$T_x = \sum_{i=1}^{g} (t_i^3 - t_i)$$

with g = the number of groupings of tied ranks, and t = the number of ties in each group.

The calculated value of r will be $-1 \le r \le 1$ (**Step 6**). To test the significance of this result that the two variables are uncorrelated, we make use of the Student's t-distribution (**Step 7**):

$$t = r \sqrt{\frac{n-2}{1-r^2}}$$

Simply look up the value in Table A.2, Appendix A, at the appropriate significance level, using a one- or two-tailed test depending on the research hypothesis.

In the example detailed here, we make use of data from Table 5.6, concerned with child accidents and socio-economic characteristics for selected enumeration districts in Norwich. The example correlates child accident rate (x) with the proportion of households in rented accommodation (y): see Boxes 5.23 and 5.24.

Null hypothesis
$H_0: \mu_1 = \mu_2$

There is *no significant association* between the child accident rate and the proportion of houses in rented accommodation.

Research hypothesis
$H_1: \mu_1 \neq \mu_2$

There is *a positive association* between the child accident rate and the proportion of houses in rented accommodation.

Significance level: 0.05.

Interpretation: The calculated correlation coefficient between these variables is 0.777. A *t*-test is used to determine the significance, and in the example above, at the 0.05 significance level, we can reject the null hypothesis that the two variables are uncorrelated: there is a significant positive association between the child accident rate and the proportion of houses in rented accommodation.

5.8   What do the test results tell you?

Many statisticians and researchers who use statistical methods are often uncomfortable about the widespread use of inferential statistical tests. This is often because of the specification of inappropriate hypotheses, the enforcement of rigid probability value p cutoffs, and the misinterpretation of the results of the statistical test. Chatfield (1995) notes that:

1   A *significant effect* is not the same as an *interesting effect*.

2   A *non-significant effect* is not necessarily the same as *no difference*.

With respect to (1) we can produce significant effects in several ways: by selecting a research hypothesis which is obviously true, or a null hypothesis which is silly. Returning to the enumeration district example at the start of Section 5.3, would we *really* expect average income to be the same for two areas? Perhaps, but if the example used was for the city of Washington DC, and we selected enumeration district A on the east side of the city in an area with a predominantly low-income population, and enumeration district B in the west of the city in an area with a predominantly high-income population, we might

Box 5.23   Handworked example: Spearman's rank order correlation coefficient

x	y	$R(x)$	$R(y)$	d	d^2	
4.066	0.7459	13	20	−7	49	
3.321	0.3586	9	10	−1	1	
5.020	0.6313	21	15	6	36	$\sum d^2 = 344$
2.634	0.0930	2	3	−1	1	
4.316	0.6145	15	14	1	1	
3.691	0.3393	12	9	3	9	
3.212	0.6583	6	16	−10	100	
4.770	0.7132	20	18	2	4	
3.461	0.5275	10	12	−2	4	
3.251	0.1088	7	4	3	9	
2.451	0.1178	1	5	−4	16	
4.367	0.7663	17	21	−4	16	
4.748	0.7204	19	19	0	0	
3.147	0.0596	5	1	4	16	
3.314	0.2270	8	7	1	1	
3.563	0.3389	11	8	3	9	
2.912	0.0728	4	2	2	4	
4.102	0.6899	14	17	−3	9	
4.686	0.5747	18	13	5	25	
2.833	0.1537	3	6	−3	9	
4.323	0.4533	16	11	5	25	

Box 5.23   (cont'd)

Step 1   Produce a scatterplot of one variable against the other as shown in Figure 5.9.

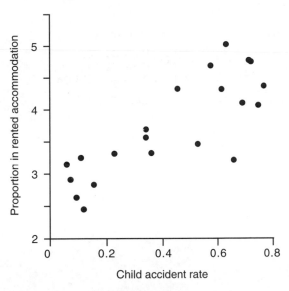

Figure 5.9   Scatterplot of child accidents against proportion in rented accommodation.

Step 2   Put the data into two separate columns in a table, and rank each variable separately to produce $R(x)$ and $R(y)$, then place in columns 3 and 4 as in the table above.

Step 3   Place the difference d between each pair of ranks in column 5, e.g., for the first pair of ranks $d = 13 - 20$ $= -7$, and then square each d to obtain d^2 and place in column 6 and sum. Here, $\Sigma d^2 = 344$.

Step 4 $n = 21$, and $(n - 1) \times n \times (n + 1) = 9240$.

Step 5   There are no tied ranks, and therefore we can calculate r as:

$$r = 1 - \frac{6 \times 344}{9240} = 1 - \frac{2064}{9240} = 1 - 0.2234 = 0.776$$

Step 6   To test the significance of r, we perform a t-test:

$$t = 0.776 \times \sqrt{\frac{21 - 2}{1 - 0.776^2}} = 0.776 \times \sqrt{\frac{19}{0.397824}}$$

$$= 0.776 \times \sqrt{47.759} = 0.776 \times 6.910 = 5.363$$

Step 7   In this case with DF $= 21 - 2 = 19$, at the 0.05 significance level, the one-tailed critical value $= 1.729$ (see Table A.2, Appendix A) which is *less than* our calculated value, and therefore we reject the null hypothesis H_0.

Box 5.24   MINITAB example: Spearman's rank order correlation coefficient

Place the child accident rate data in C1 and the proportion renting in C2. You should use the 'plot' command (explained in Box 5.12) to determine whether the two variables appear to be related. To perform a Spearman's rank correlation in MINITAB, you will first have to rank the data using the command 'rank', and then perform a product-moment correlation on the ranks.

Using pull-down menus:

➥ Manip
➥ Rank...
Highlight C1
Rank Data in [ ]
➥ Select
Highlight C3
Store ranks in [ ]
➥ Select
➥ Rank...
Highlight C2
Rank Data in [ ]
➥ Select
Highlight C4
Store ranks in [ ]
➥ Select
➥ Stat
➥ Basic Statistics➤
➥ Correlation...
Highlight C3
➥ Select
Highlight C4
➥ Select
➥ OK

Using command line:

MTB > Rank C1 C3.
MTB > Rank C2 C4.
MTB > Correlation C3 C4.

MINITAB results:

```
MTB > Rank C1 C3.
MTB > Rank C2 C4.
MTB > Correlation C3 C4.
Correlation of C3 and C4 = 0.777
```

expect a significant difference in the average household income. Any sizeable difference ought to be clear from the IDA performed on the data, prior to using the statistical methods above. Furthermore, since statistical tests were often developed with small samples in mind, we often find that it is easy to produce significant results with large sample sizes even if the size of the difference is small. With respect to (2), a non-significant effect can be produced by using too small a sample.

Advice from statistical textbooks suggests that it is also unwise to stick rigidly to probability value p cutoffs. Freedman *et al.* (1978) suggest that if we calculate $p = 0.052$, this in fact means *much the same*

thing as $p = 0.049$, even if it is strictly beyond the 0.05 significance level cutoff, in contrast to a p value of 0.9. What is more useful is to determine whether a similar result can be replicated from another independent set of data, i.e., that the results are repeatable (Chatfield, 1995).

It is important to reiterate that you should not interpret a significant result as indicating that your research hypothesis is true or has been proven; we simply do not have the evidence to reject it at that significance level (Robson, 1994). Neither does the significance test answer the question 'What caused the difference?' – all the test answers is 'Is the difference real?' (Freedman *et al.*, 1978). This is an important point: as Freedman *et al.* point out, a test of significance does *not* check the design of a study. For a well-designed study, establishing a real difference may prove the investigator's point, but in a poorly designed study, establishing a real difference tells us nothing. The importance of good project design, with the planned use of certain statistical methods rather than their adoption as an afterthought, is essential.

- *You have a data set comprised of five independent groups of interval data. What tests would be appropriate to test for (a) a difference between all the groups and (b) a difference between pairs of groups?*
- *If we can use a linear regression to describe the relationship between two variables, and the calculated slope coefficient is 0.5, how much change will we observe in the Y variable with every unit change in the X variable?*

5.9 Summary

After reading this chapter you should:

- understand the processes of hypothesis significance testing;
- be able to define and distinguish parametric and non-parametric tests;
- be able to choose an appropriate test of difference or association given your data;
- be able to perform the mechanics of statistical tests upon quantitative data;
- be able to interpret the results from statistical tests.

In this chapter, we have introduced statistical analysis procedures capable of determining the validity of a hypothesis. By the use of the techniques of Initial Data Analysis (IDA) detailed in the previous chapter, it is important that you test the assumptions required for your chosen test. Conducting a test using data that violate the test assumptions may lead to an invalid analysis and incorrect results. Similarly, you will also need to take care with both the calculation of the statistic (especially if done by hand) to prevent the production of an erroneous result, and also the interpretation of the result such that you do not draw a significant conclusion where none exists. In the next chapter, we turn our attention to the mapping and analysis of quantitative spatial data.

5.10 Questions for reflection

- *Under what conditions would you use a parametric significance test?*
- *Can we ever prove the truth of a statistical null hypothesis? If so, how?*
- *Distinguish between one- and two-tailed significance tests.*

Further reading

General statistics texts

Ehrenberg, A.S.C. (1978) *Data Reduction: Analysing and Interpreting Statistical Data*. Wiley, Chichester.

Freedman, D., Pisani, R. and Purves, R. (1978) *Statistics*. W.W. Norton, London.

Robson, C. (1994) *Experiment, Design and Statistics in Psychology*. Penguin Books, London.

Rowntree, D. (1981) *Statistics without Tears: A Primer for Non-mathematicians*. Penguin Books, Harmondsworth.

Wright, D.B. (1997) *Understanding Statistics: An Introduction for the Social Sciences*. Sage, London.

Statistical methods in geography

Clark, W.A.V. and Hosking, P.L. (1986) *Statistical Methods for Geographers*. John Wiley, New York.

Ebdon, D. (1985) *Statistics in Geography*. Blackwell, Oxford.

Griffith, D.A. and Amrhein, C.G. (1997) *Statistical Analysis for Geographers*. Prentice Hall, Englewood Cliffs, NJ.

Griffith, D.A. and Amrhein, C.G. (1997) *Multivariate Statistical Analysis for Geographers*. Prentice Hall, Englewood Cliffs, NJ.

Shaw, G. and Wheeler, D. (1994) *Statistical Techniques in Geographical Analysis*. David Fulton Publishers, London.

Silk, J. (1979) *Statistical Concepts in Geography*. Allen and Unwin, London.

Walford, N. (1995) *Geographical Data Analysis*. John Wiley, Chichester.

Wrigley, N. (1985) *Categorical Data Analysis for Geographers and Environmental Scientists*. Longman, London.

CHAPTER 6

Spatial analysis

This chapter covers

6.1 Introduction
6.2 Maps
6.3 Geographical Information Systems
6.4 Current issues in the use of GIS for socio-economic applications
6.5 Sources of digital spatial data
6.6 Planning and implementing analysis using GIS
6.7 Spatial statistics
6.8 Summary
6.9 Questions for reflection

6.1 Introduction

Previous chapters have covered the quantitative analysis of data using non-spatial techniques. However, your research may need to consider spatial information, and perform spatial analysis alone or in association with the non-spatial statistical analysis techniques described in Chapter 5. This may involve the mapping of spatial data, spatial analysis and/or spatial statistical analysis. In the modern geography department, these tasks will usually require the use of a computer-based system for data selection, analysis and/or graphical display. Although there are a variety of mapping/spatial analysis systems available from the very simple to the quite complex, you are likely to encounter such tools in the form of a Geographical Information System (GIS). The aim of this chapter is to consider mapping and the various applications of GIS, typical spatial operations and spatial data availability for research in human geography, as well as more common spatial statistics. Although the widespread commercial interest in GIS technology has led to the availability of a wide variety of GIS systems, we shall try to avoid specific system terminology, and consider the generic operations common to most systems. Spatial statistics are still poorly catered for in many GIS systems, so we will make use of the GASP package written for the PC to illustrate some of the more common techniques.

6.2 Maps

The traditional role of the map has been to represent spatial information – usually in relation to the surface of the earth. Maps are powerful graphical tools that classify, represent and communicate spatial relations. Maps represent a concentrated database of information on the location, shape and size of key features of the landscape and the connections between them (Hodgkiss, 1981), and are a method of visualising a world that is too large and too complex to be seen directly (MacEachren, 1995). Maps aid navigation, facilitate understanding and mark out territories. Well-designed maps are effective sources of communication because they exploit the mind's ability to see relationships in physical structures, providing a clear understanding of a complex environment, reducing search time and revealing spatial relations that may otherwise not be noticed. This 'power of maps' (Wood, 1993) has long been recognised. Maps have been used by civilisations for thousands of years to depict geographical distributions across the earth. For example, Thrower (1996) reports that the oldest known plan of an inhabited site is the Bedolina map from northern Italy of *ca.* 2000–1500 BC. Historic maps can reflect the prevailing level of knowledge or dogma of the time of their construction, but the information they present is in a hardcopy and passive form (i.e., on paper). With the arrival of computer technology, the map has become less passive, more interactive and more virtual (i.e., on screen). Today, map-making is no longer the prevail of a small group of skilled cartographers, but can be undertaken by anyone with access to the data and appropriate technology. This does not, however, necessarily mean that anyone can construct good maps!

6.2.1 Mapping geographic space

Throughout history the nature of map-making has changed, becoming more sophisticated with time (see Figure 6.1). The traditional Cartesian maps we use today (e.g., Rand McNally or Ordnance Survey maps) have their origins in the work of Ptolemy, librarian at Alexandria in the second century AD. Combining geography with mathematics, he documented instructions for making map projections, devised a system of latitude and longitude, and constructed a series of maps of the known world. These ideas, while influential in Is!amic science, failed to be adopted in Europe until the thirteenth century when refugees fleeing from the Byzantine capture of Constantinople reached Italy in 1410 (Thrower, 1996). Ptolemy's ideas radically changed the face of European cartography, underlying the mapping revolution of the Renaissance

period and igniting the shift from symbolic representations towards scientific representations. Prior to the Renaissance period, land maps were primarily cartographic representations of culture rather than accurate geographical representations of spatial relations, and sea charts, although relatively spatially accurate, were charted from experience rather than formal surveying (Livingstone, 1992).

At first, (world) maps were patchy and incomplete due to poor knowledge and weak surveying and mapping techniques, but with time, particularly after the Renaissance, map making became more sophisticated as cartographers became more skilled and versed in geometry and the problems of surveying and representing a spheroid on a single sheet of paper. For example, by 1509 Martin Waldseemuller had produced a map of the world including North and South America, and a number of map projections had been developed culminating in the now commonly used Mercator projection in 1569 (Thrower, 1996). Although maps have become more accurate since this time, the role of maps as symbolic agents has continued, as it does today. For further thoughts on this subject, you should consult the text by Dorling and Fairbairn (1997) which considers the role of the map in a variety of human geographic contexts.

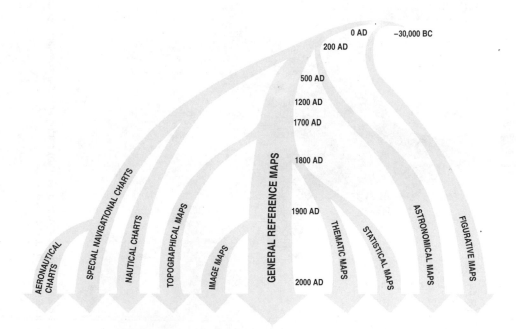

Figure 6.1 The changing nature of cartography (source: Robinson *et al.* 1995).

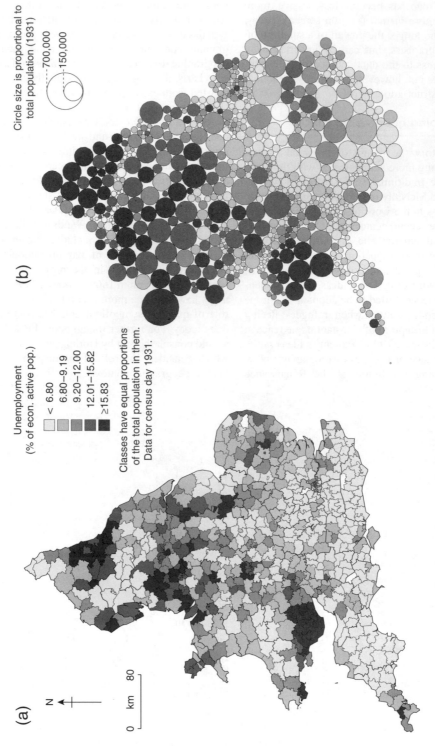

Figure 6.2 (a) Choropleth map, and (b) cartogram representation, of unemployment in England and Wales in 1931 using 1898 Rural Districts as a base (source: Gregory *et al.* in press).

In today's modern cartography there are a number of standardised methods for mapping the geographic world. These extend beyond standard Cartesian maps to include **choropleths** (Figure 6.2(a)) and **cartograms** (see Figure 6.2(b) and Dorling, 1996, for further examples), as well as conceptual and topological maps that seek to provide relative rather than absolute spatial accuracy. As Wood (1993) notes, there is no one accepted typology of current maps. Indeed, he provides a number of examples: *The Map Catalogue* (Makower, 1986) provides a threefold taxonomy of land, sky and water; the *US National Report to ICA* (Loy, 1987) classes maps into the three classes of government mapping, business mapping and university cartography; and Southworth and Southworth (1982) provide an eightfold taxonomy of land form; built form; networks and routes; quantity, density and distribution; relation and comparison; time, change and movement; behaviour and personal imagery; and simulation and interaction. In other words, contemporary mapping is *reactive* to its purpose and seeks to communicate most effectively a message to the intended target audience.

The historical dimension of map making is all very interesting, but unless your research project is directly concerned with representational issues in historical maps, you will probably find it of limited use. Contemporary maps and map construction are more relevant. In recent years, a new cartographic revolution has been taking place. While cartographic conventions have remained relatively stable, how geographic maps are displayed and used has been changing with the advent of the digital age. **Geographic Information System** (GIS) technology can provide maps that can be manipulated, transformed and selectively interrogated by their users. Whereas with paper maps, the map was the database, within a GIS any one map is just one particular spatial representation of the database (Burrough and McDonnell, 1998). GIS allows the construction of maps at different scales from the local to the regional and beyond, which combine different data variables and use different classification schemes, all from the same database and all in seconds; feats not readily achievable using paper maps which would need to be redrawn at great cost, in terms of both money and time. Furthermore, maps, once static, qualitative representations that needed to be drawn by cartographic 'experts', have become dynamic, virtual, quantitative models (although not necessarily interactive) that can be easily constructed by people with little cartographic training within a GIS. Maps within GIS can be easily queried in ways that are difficult using a paper map. For example, *'where are all locations of feature A?'* would have involved a large manual search of one or several paper maps, whereas in a GIS all locations could be identified and mapped without difficulty. The consequence of such developments is that you can use maps and related spatial analysis tools directly and dynamically as part of your research project. We will detail some of the typical functionality of a GIS in Section 6.3 below.

It is important to note that the impact of technology is not limited to GIS. Extending beyond the revolution in the mapping medium are new methods of mapping **visualisation** (see Hearnshaw and Unwin, 1994). New methods of visualising data can allow a perspective on your data not previously possible. For example, it is now possible to create three-dimensional terrain models that can be viewed from different angles, flown through and draped with various relevant, variable overlays (Figure 6.3). Increased computing power and sophisticated algorithms are being used to create displays that would have been difficult and time-consuming to perform by hand. The digital-based, cartographic revolution of the late 1980s and early 1990s has also seen geographic space come online. Maps of all geographic scales are accessible over the Internet and new types of maps have started to appear. The map plays a central role in the visual medium of the World Wide Web. Interactive, hypermedia maps are now quite common, with some used as a means to manipulate external databases to retrieve information (see Martin *et al.*, 1998).

In the course of your research, you may be concerned only with the construction of maps such as that shown in Figure 6.2 to help locate your study, or to explain some of your findings. Although we will consider some general elements of map design in Chapter 10, we will not consider in detail elements of map construction in this book. There are a variety of useful texts available, such as *Elements of Cartography* by Robinson *et al.* (1995), and the classic text *Maps and Diagrams* by Monkhouse and Wilkinson (1971), which may be of use in this context.

6.2.2 Map distortion

The data displayed on maps can be distorted both intentionally and unintentionally. If you are using maps as a data source for your research project, particularly if you are using digitised maps as an input to a GIS, it is important that you are aware of the various types of **map distortions**; some of the more important are summarised in Box 6.1.

Figure 6.3   Virtual reality model of Westminster, London (source: Batty in Longley *et al.* 1998).

Box 6.1   Sources of distortion in maps

Scale	Representative fraction which determines size of features, and degree of generalisation shown on a map. The smallest object which can be shown on a map is about 0.2–0.5 mm across.
Projection	Mathematical transformation to project the Earth's curved surface onto a flat sheet. *Conformal* (orthomorphic) projections preserve small shapes and directions but distort areas. In contrast, *equal area* projections preserve areas but distort shape.
Cartographic generalisation	Makes visible features on a map, involving the selection and symbolisation of cartographic objects.

If you intend to work with maps for your research project, it is imperative that you utilise a map of large enough scale. **Scale** is a measure of the ratio between the size of a feature in the real world and its representation on a map. Usually this is expressed in terms of a representative fraction or ratio: 1:1250, or 1 unit on the map is equivalent to 1250 units in the real world. This is a large-scale map, whereas in comparison 1:2,500,000 is a small-scale map. Often, these terms are used synonymously with 'extent', particularly by non-geographers, which in fact reverses the meaning! You should always exercise caution when you encounter such terms in the course of your reading. Scale is important: variations in scale can be introduced by the choice of a map projection, and the choice of a scale controls the degree of generalisation, and hence the degree of information present in a map.

In order to represent the 3D curved surface of the earth on a 2D flat map, we require some sort of a **projection**. Projections are simply mathematical transformations which project points from the globe on to a flat map. Projections can be classified in terms of the geometric form of the flattenable (developable) surface used: *conic, cylindrical, azimuthal,* and a *miscellaneous* category based on no particular geometric form. The relative orientation (aspect) of the flattenable surface with respect to the earth's polar axis can also vary. For example, for the normal aspect of a *cylindrical projection*, the surface is parallel to the polar axis; with the transverse aspect it is at 90° (Figure 6.4).

The point of contact between the flattenable surface and the curved surface is known as the **standard line** or the standard point (Robinson *et al.*, 1995). This is the only part of the flattenable surface at

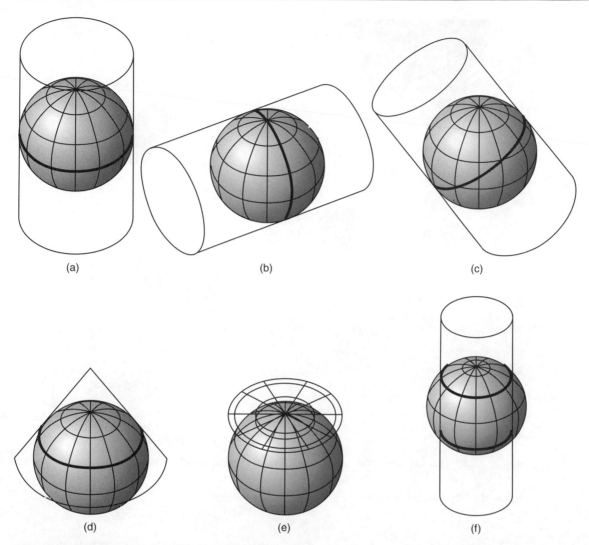

Figure 6.4 Main types of projection: (a) normal aspect cylindrical; (b) transverse aspect cylindrical; (c) oblique aspect cylindrical; (d) conical; (e) azimuthal, with bearing on earth projected to bearing on plane; (f) normal aspect secant cylindrical. Standard lines are emphasised in black (source: Dorling and Fairbairn 1997).

the true scale. Away from this point the presence of distortion will result in changes in scale. For a *transverse cylindrical projection*, the standard line corresponds with a meridian (line of longitude); for a *standard cylindrical projection* this line corresponds with a parallel (line of latitude) (see Figure 6.4). The further away from this line, the greater the distortion.

Distortions are manifest as changes in areas, angles, shapes, distances and directions. In any given projection, only one or two of these can be preserved at the expense of the others. **Conformal** (also termed *orthomorphic*) projections preserve small shapes and directions, but distort areas. An example of such a projection is the *cylindrical Mercator projection* (Figure 6.5). In this projection, direction is maintained by equal stretching in a north-south direction as in an east–west direction. For the normal aspect of this projection, the standard line is at the equator, with the maximum distortion in area towards the poles. In contrast, **equal-area** (also termed *equivalent*) projections preserve areas but distort shapes, and an example of such a projection is the *Mollweide*

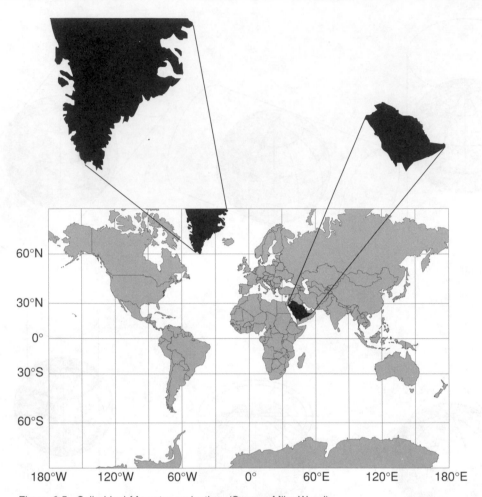

Figure 6.5 Cylindrical Mercator projection. (Source: Mike Wood)

projection (Figure 6.5(b)). Here, areas are maintained, but as the parallels reduce in size towards the poles, meridians intersect at increasingly acute angles, distorting shapes towards the edges of the map. The final class are the **azimuthal** projections. An example of such a projection is the *azimuthal equidistant*, often used for the polar regions since direction and distance are true along the meridians (however, this projection is neither conformal nor equal-area).

For small areas, particularly those near the standard line, the degree of distortion introduced by the projection type is minimal. However, away from this line, the distortion introduced can be significant, particularly for large areas. Compare the representation of Greenland and Saudi Arabia in the conformal and equal-area projections displayed in Figures 6.5 and

6.6. The area of both is approximately the same (2,175,600 km^2 and 2,149,690 km^2 respectively), yet Greenland in Figure 6.5 appears considerably larger, and squashed in Figure 6.6.

For projections based on geometric flattenable surfaces, the pattern of this distortion is systematic away from the standard line; however, it is less so for those projections based on no particular geometric form. The Mollweide projection displayed in Figure 6.6 possesses minimal distortion in the mid-latitudes and is often used to display those regions. Indeed, the amount of deformation at any point of a projection can be visualised by observing the distortions in form across the projection of an object which appears as a small circle when located at the standard line. This is known as *Tissot's Indicatrix*.

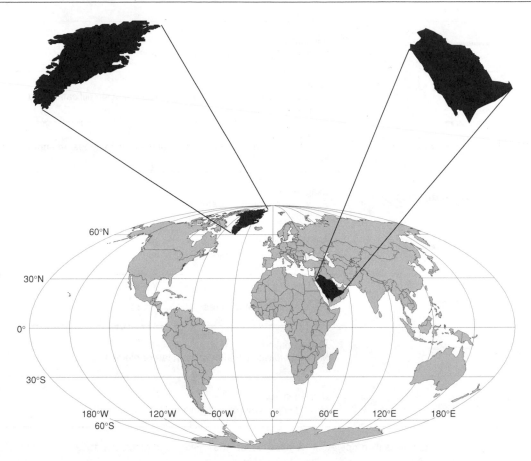

Figure 6.6 Mollweide projection, with Greenland and Saudi Arabia enlarged for comparison. (Source: Mike Wood)

For those areas of the world which have a greater north–south extent than east–west (e.g., the British Isles), almost all maps are based on the transverse aspect of Mercator's projection, where the standard line is in a north–south direction. Topographic maps in the US and many other countries are also based on a version of the transverse Mercator, called the Universal Transverse Mercator (UTM), which consists of north–south strips of 6° longitude width. Locations can be expressed either in terms of absolute geographical coordinates, or in terms of the local coordinate system. For example, the city of Belfast has an absolute geographical position of longitude 6°57′ west, latitude 55°45′ north, and a grid reference of 334374 using the Irish grid (based on a transverse Mercator projection).

It is important to be aware that the features and symbols on the map can be distorted, often intentionally, by the cartographer. This can be due to the choice of map scale: changing scale from a large to a small scale will necessitate the generalisation of features on the map if they are to remain clear. Conversely, changing scale from small to large will require the introduction of new detail. Shea and McMaster (1989) considered that a scale change could result in the *congestion, coalescence, conflict, complication, inconsistency* and *imperceptibility* of map features. The response to such conditions is termed **cartographic generalisation**. There are several classifications of the operations involved in cartographic generalisation (see Monmonier, 1991; Robinson *et al.*, 1995; Shea and McMaster, 1989). Jones (1997) recognises two types of generalisation: **semantic generalisation** and **geometric generalisation**. Semantic generalisation is concerned with the process of selecting information to include in the map. Here, elements can be selected from hierarchical structures based on classification (e.g., city, town, village) as well as aggregation (e.g.,

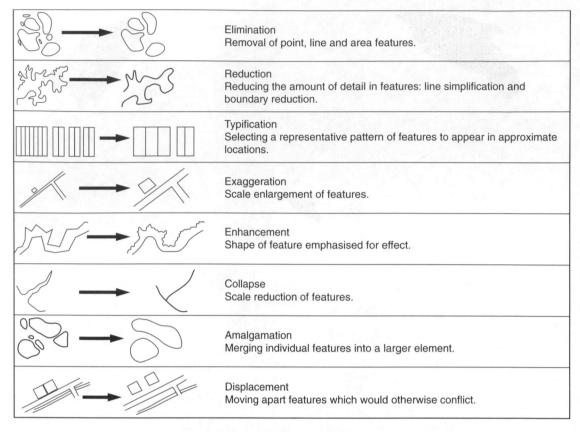

Figure 6.7 Geometric generalisation operators (sources: Jones 1997; Shea and McMaster 1989).

city district, subdivision, block). Geometric generalisation is motivated primarily by symbolisation, the need to make visible features on the map. Common geometric operators are listed in Figure 6.7.

Some mapping organisations have been known to introduce deliberate errors such as non-existent roads, to aid the process of detecting infringements of copyright. Remember, maps compiled at different dates, by different cartographers, represent as much the subjective symbolisation chosen by the cartographer as the actual pattern of reality.

6.3 Geographical Information Systems

6.3.1 What is a GIS?

Geographical Information Systems (GIS) have been developed over the last 30 years as tools for the analysis of spatial data. The various contributory factors which have led to the development of what we now

call a GIS are quite varied. These include the emergence of computer technology to assist with map-based analysis, the associated benefits of automating map production, and the development of a layer-based and computerised representation of various disparate data including population censuses and environmental variables (Longley *et al.*, 1999). There are consequently a variety of definitions of GIS, for example:

- *A system for capturing, storing, checking, manipulating, analysing and displaying data which are spatially referenced to the earth* (Report of the Chorley Committee–DoE, 1987).
- *An organized collection of computer hardware, software and geographic data designed to efficiently capture, store, update, manipulate and display all forms of geographically referenced information* (Dangermond, 1992).
- *An institutional entity, reflecting an organisational structure that integrates technology with a database, expertise and continuing financial support over time* (Carter, 1989).

Data (spatial, attribute)

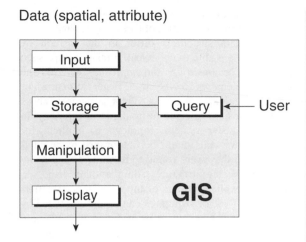

Figure 6.8 GIS defined in terms of component operations on data.

The first two definitions are typical of a more mechanistic view of GIS, emphasising the technological elements and the data transformations which can take place. It is possible, however, to see GIS in a broader context incorporating the use of technology for organisational decision support, as in the third definition. This is no less valid, but we will focus on the mechanistic definition, summarised in Figure 6.8, since this is more useful in the context of this book as it provides a summary of the role GIS can play in your research project.

Not surprisingly, GIS has close relationships to a variety of other computer-based technologies, particularly those that deal with similar data. For example, there are many common elements between GIS technology, digital remote sensing, database management and computer science. Although the distinctions between these technologies are becoming increasingly blurred, the distinguishing characteristic of GIS is that we are primarily concerned with **spatial data** (also known as geographic or geospatial data).

6.3.2   What can a GIS be used for?

It is not an understatement to say that the existing and potential applications of GIS are legion. This is particularly so as the functionality of GIS finds its way into more general mainstream desktop computing applications. GIS is used to answer the sort of questions posed in Box 6.2. These specific spatial queries can be viewed in the context of more general questions that we can pose (Rhind, 1990):

- **Location:** *What is at . . . ?*
- **Condition:** *Where is it . . . ?*
- **Trend:** *What has changed . . . ?*
- **Routing:** *What is the best way . . . ?*
- **Pattern:** *What is the pattern . . . ?*
- **Modelling:** *What if . . . ?*

Let us examine the components of GIS as displayed in Figure 6.8. The data used in a GIS usually comprise two elements: a **spatial** (or location) element and an **attribute** (or non-spatial) element. An example of a spatial element might be a point referenced as a six-figure grid reference, or given a name such as 'Chicago'. The attribute element is the variable we might wish to assign to the spatial location, for example the population count, or percentage unemployed. In contrast to many physical variables, the representation of socio-economic attributes in a GIS is not straightforward (Martin, 1999) – we will discuss problems of representation in Section 6.5 below. Spatial data have historically been input to GIS manually direct from paper maps, and, although an increasing amount of data is now available in a digital form (see Section 6.5 below), maps still provide an important source of data, particularly for student research. The spatial information from such a paper map is usually **digitised**; that is, converted from an analogue to a digital form using a digitising table, which consists of an electromagnetic table onto which the map is placed, the features being 'traced' around using a cursor or pen. If you intend to use such paper maps for analysis in a GIS, all your maps will need to be registered to a common projection/coordinate system. GIS systems are usually able to transform data from one type of projection to another, as long as you know the projection of your original data. For example, a commonly used GIS such as ARC/INFO supports some 46 different map projections. We will discuss some of the more practical considerations of incorporating GIS into your research in Section 6.6 below.

What we can do with the GIS largely depends on the conceptual model we have of reality, and the approximation of this in the form of both a data model and the physical data structure in the GIS. A **conceptual model** describes our mental view of reality: do we break down the human landscape into *discrete entities*, or do we consider the landscape in a more *continuous* and integrated manner? A **data model** is concerned with the logical organisation of data, and represents the implementation of a conceptual model. In the context of GIS, data models of reality are in the form of either an empty space littered with a series of discrete **objects** in classes such as points, lines and areas, or as a continuously varying **field** of values. Objects can be defined as the representation of physical things in the digital world, and a two-dimensional field can be defined as a single-valued function of location in a two-dimensional space (Mark, 1999). Since many socio-economic phenomena are discrete, for example houses, schools, roads, towns, individual people, etc., their representation in terms of objects would seem to be more logical. However, it is possible and often more appropriate to view certain socio-economic data such as population in a continuous form as a field or surface. Figure 6.9 displays examples of both objects and a continuous elevation field derived from digital photogrammetry of a 1:10,000 scale aerial photograph.

In contrast to a data model, a **data structure** is concerned with the practicalities of data organisation within a computer. Historically, data structures in a GIS have come in two main forms: **raster** structures and **vector** structures (Figure 6.10). The raster structure is based on a grid comprised of **cells** (also known as pixels) which are most commonly square, of uniform size. Features cannot be resolved if they are smaller than the cell size. Each grid is often termed a **tessellation**, and completely fills space with cells required for all areas of geographic space, even where there are no features present. Each cell contains or references a single attribute value, and each attribute occupies a separate layer (or coverage). In contrast, spatial phenomena in the vector system are represented by coordinate-based spatial objects in the form of **points**, **lines** and **areas**. Here, attribute information must be managed separately in tables, and linked to each spatial object using a database.

These data structures differ in the fashion by which they are able to code spatial relationships such as *sharing*, *connectivity* and *adjacency* between the different elements (cells in a raster structure, or points, lines and areas in a vector structure). These relationships are referred to by the term **topology**. In the basic raster system, topology is implicit: for example, two cells could be determined to be adjacent only if they were found to be in adjoining rows/columns. Vector structures, on the other hand, do allow the explicit storage of topology. Indeed, different types of vector structures can be differentiated by the amount of topological information they contain; for example, a polygon-based structure does not encode any topological information (Figure 6.11(a)). In contrast, a segment-based structure is based on **directed arc**, **node** and **polygon** information; e.g., US Census Bureau DIME format contains information on left- and right- polygons and from- and to- nodes (Figure 6.11(b)). The advantage of topology is more efficient storage of information. Shared objects are stored once only, and there is an increase in the speed of performing operations based on the relatedness of information. Topology also aids the detection of errors in spatial data, for example dangling nodes in a polygon coverage (Section 6.4.3 below).

A certain amount of time in GIS courses is often put aside to consider the relative merits of raster and vector structures, the transformation of data from one structure to the other, and the various more efficient methods of packing and referencing raster data (e.g., quadtrees, block and run-length encoding). The adage 'raster is faster, but vector is corrector' has often been used to summarise the merits of each. Lower-resolution rasters offer fast processing but with poor geometric fidelity and often large file sizes. Higher-resolution vectors offer more efficient yet also more complex data storage. Trends in computer hardware prices with reducing costs of memory and increasing computer power make the arguments for and against the use of raster and vector systems in terms of efficiency in storage and analysis increasingly redundant. In addition, these days, commercial off-the-shelf GIS systems (e.g., ARC/INFO) are able to accommodate data in both a vector and a raster form. From the perspective of your research, it is perhaps more appropriate to consider which model of reality – object-based or field-based – is appropriate for the phenomena in your study. It is easier to accommodate a field-based model within a raster

(a)

(b)

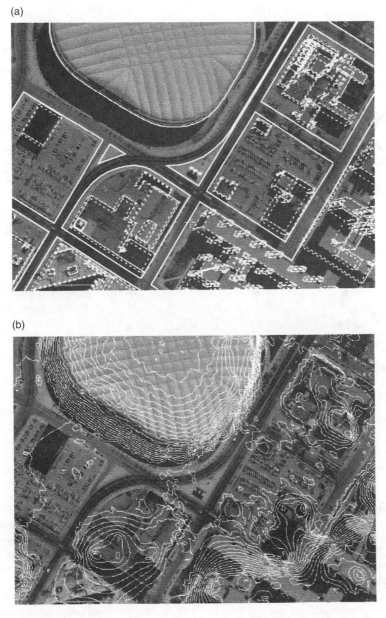

Figure 6.9 Extract from a 1:10,000 scale photograph displaying (a) aspects of the built environment represented as objects, (b) the built environment represented as a continuous elevation surface (images courtesy of LH Systems Ltd).

structure, and an object-based model within a vector structure, although both models can be represented in either structure. For example, we can take an object-based model and convert from a raster (Figure 6.12(a)) to a vector data structure (Figure 6.12(b)). Similarly, we can take a field-based model, and convert from a raster (Figure 6.12(c)) to a vector data structure in the form of contoured surface (Figure 6.12(d)).

Irrespective of the data structure used, both spatial data and non-spatial attribute data need to be stored

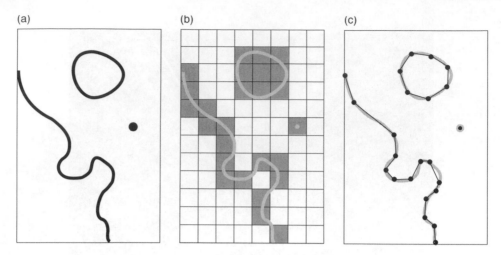

Figure 6.10   (a) Reality, modelled in (b) raster, and (c) vector form.

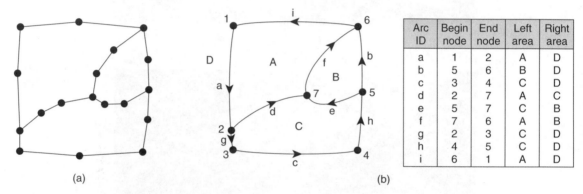

Arc ID	Begin node	End node	Left area	Right area
a	1	2	A	D
b	5	6	B	D
c	3	4	C	D
d	2	7	A	C
e	5	7	C	B
f	7	6	A	B
g	2	3	C	D
h	4	5	C	D
i	6	1	A	D

Figure 6.11   Vector representation in (a) spaghetti, and (b) node, directed arc and area form.

and accessed efficiently within the GIS. This is achieved using a **database management system** (DBMS). There are a variety of models of organising information in a DBMS, including **hierarchical**, **network**, **relational** and **object-oriented** structures, with the relational structure the most common. For the purposes of your research project, the details of specific DBMSs are not really relevant, and we will provide no further elaboration here. If you do require further details, the text by Worboys (1995) should provide a useful starting point.

6.3.3   Spatial analysis

Addressing the queries listed in Box 6.2 is at the core of what GIS is all about, and in order to do so, you need to perform **spatial analysis** on your data. Such spatial analysis tasks may constitute the simple map-

ping and visualisation of location and patterns, or the cartographic modelling of single variables. Alternatively, they may involve more complex analysis such as network analysis, the modelling of multiple criteria/variables in the form of a spatial decision support system (e.g., multi-criteria decision making or multiobjective decision making), and the incorporation of models of spatial dependency and geometric and attribute fuzziness. We will outline these analysis tasks below. What type of spatial analysis tools you use will of course depend on the specific context of your research project, and the questions you wish to answer. We will first outline some of the more generic spatial analysis functions common to most GIS packages, and then consider some more specific data transformation, integration and modelling tasks which can be undertaken using a GIS. It is impossible to do justice in a single chapter to the variety of

(a)

(b)

(c)

(d)

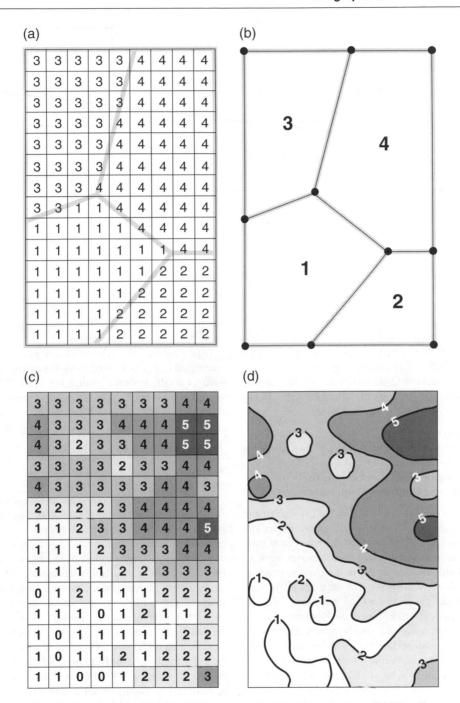

Figure 6.12 Object-based model in (a) raster, and (b) vector structure; field-based model in (c) raster, and (d) contoured vector structure.

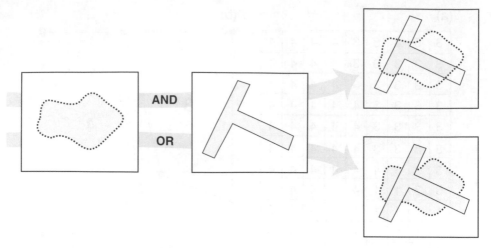

Figure 6.13 Boolean AND and OR overlays on two spatial data layers.

functions and the diversity of applications for which GIS has been used. Therefore, for further information, reference should be made to some of the books listed at the end of this chapter.

There are various categorisations of the spatial analysis tools available in the standard GIS. Berry (1987) recognised four main categories:

1 **Reclassifying** maps
2 **Overlaying** maps, consisting of logical, arithmetic/statistical operations
3 **Measuring distances** and **connectivity**
4 **Context (neighbourhood) operations**

The ability to **reclass** and **overlay** data is one of the mainstays of much GIS analysis. Reclassing data is a useful tool, often used to categorise a numerical variable, or to reduce the number of categories prior to the combination of data layers in the overlay process. Arithmetic overlay allows us to add, subtract, multiply or divide the elements of one data layer by another. Logical overlay, based on Boolean algebra, allows us to select various combinations of elements of spatially coincident data sets. This is similar to the set combinations of Venn–Euler diagrams which were discussed in Chapter 4. The two most common forms of logical overlay are the logical AND (union) and the logical OR (intersection). Refer back to Figure 4.7 to remind yourself which elements of two sets are included in these operations. These overlay tools give us the ability to identity those areas where one criterion is met (OR), and those areas where multiple criteria are met (AND) (Figure 6.13). Individual layers

can also be weighted differently to emphasise the importance of different factors.

This process of reclassification and overlay for several variables is termed **sieve mapping**, and in the form of **multi-criteria evaluation** (MCE) is central to applications such as site selection used in the planning process. Overlay is generally simpler to implement in a raster GIS, since it operates on pairs of spatially coincident cells. In a vector GIS, however, overlay is more complex, as all data layers must be intersected, new polygons formed, and all locations classified according to their aggregate data characteristics.

Measuring distances and connectivity involve the use of **connectivity functions** which accumulate values over an area being traversed. These functions can be further subdivided into *contiguity measures*, *proximity measures*, *spread functions* and *network functions*. (Aronoff, 1989) A common measure of contiguity is the size of an area, which may be a limiting factor in a GIS analysis (for example, in the selection of a candidate site for the building of a new facility). Proximity is often defined in terms of the straight line distance between objects. The reclassification of such distances allows the construction of **buffers**, one of the most useful operations using a GIS. Points, lines and areas can be buffered, as seen in Figure 6.14(a)–(c). These are examples of external buffers: areas can also be buffered internally to produce *setbacks* (Figure 6.14(d)). It is possible to combine the results of separate buffer operations and overlay the results. Figure 6.15 displays the resultant zone *inside* the buffer around a road, but *outside* a buffer around surface hydrology features.

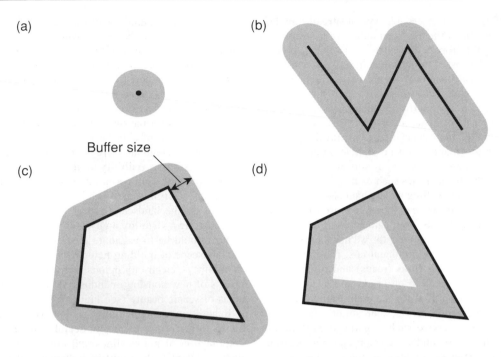

Figure 6.14   Buffer generated around (a) a point, (b) a line, and (c) an area; (d) internal buffer or 'setback'.

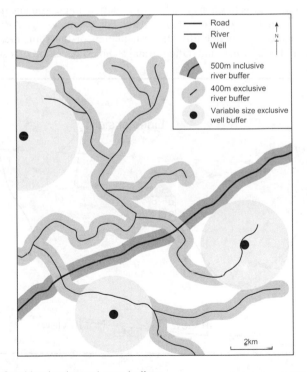

Figure 6.15   Combined point and area buffers.

Buffers are in fact a simple type of **spread function**. These functions literally accumulate values outwards across space from a central point or object, proportional to the cost of movement across space, which can be varied with location to represent different phenomena. (Aronoff, 1989) In this fashion it is possible to calculate a 'cost surface' which may represent the costs in terms of distance, or other variables such as travel time or the financial costs of travelling across space. In this manner, it is possible to model the ease of transport across space to any sort of service centre, e.g., health care facilities such as hospitals. If we are constrained to travelling along channels of flow, such as a road system, we can then make use of **network analysis** using a GIS. Two main types of analysis are possible using networks. Firstly, we may be concerned with finding either a minimum cost route through a network comprising **links** (roads) and **nodes** (junctions or turns) between an origin and a destination, such as might be used for the optimisation of routes for the emergency services, or simply the optimal provision of a service such as garbage collection or letter delivery. Secondly, we can perform a **location–allocation** analysis to allocate the elements of the network to the nearest service centre, examples of which might be a school or a fire station.

Context operations can be defined as functions which create a new layer of information in the GIS, based on an existing layer and the context (or neighbourhood) within which each feature in this layer is found. Often the method for the analysis of such data makes use of a circular or rectangular search window within which some statistical measure of the values is obtained, e.g., *sum*, *maximum*, *minimum* and *stand-ard deviation*. (Aronoff, 1989) Where the data values are in a continuous form such as an elevation or population surface, it is possible to transform data into slope gradient and slope aspect information. Where we have elevation data, in the form of either a Digital Elevation Model (DEM) or a Triangulated Irregular Network (TIN), we can also perform a **visibility analysis**. This makes use of lines of sight from certain points on a topographic surface to determine the visibility to or from certain features on the landscape (see Figure 6.16). Examples of using a visibility analysis in a socio-economic context could be to calculate the visual intrusion on the landscape of building new artificial features such as chimneys, electricity pylons and wind-mills, or the effects of new housing or industrial developments in areas of scenic beauty (see Figure 6.17).

In addition to the methods listed above, we can use a GIS to perform a variety of searches based on the combination of points, lines and areas. Perhaps the most common is the **point-in-polygon**, which determines whether or not a point is within a particular polygon. Within most GIS systems, we can also perform some simple statistical analysis of the attribute data. For example, the ARC/INFO GIS allows the calculation of the sum, mean, minimum, maximum and standard deviation of a selection of attribute data.

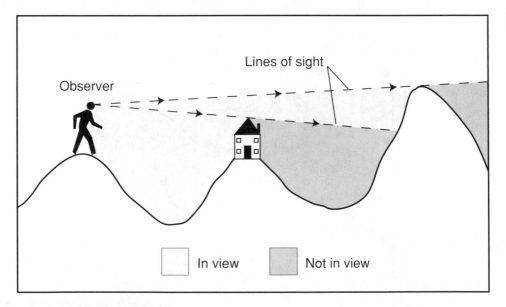

Figure 6.16 Viewshed definition.

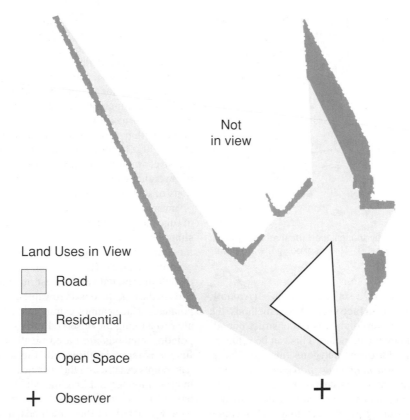

Not
in view

Land Uses in View

Road

Residential

Open Space

╋ Observer

Figure 6.17 The viewshed in front of a property in an urban area of Glasgow
(image courtesy of Iain Lake). Reproduced by permission of the Ordnance
Survey. © Crown Copyright. Licence No NC/02/8481.

More complex statistical analysis is available, but usually as an 'add-on' to the GIS system, or as some sort of an interface with an external statistical package.

6.3.4 Data transformation and integration

One of the strengths of GIS is the ability to transform and integrate spatial data, from one object class to another, which we often need to carry out when our input data are sampled insufficiently across space. Where data are in the form of sampled points, this often involves some sort of **interpolation.** Interpolation can be defined as the process of defining objects at unmeasured locations from nearby objects at measured locations. The basis of all interpolation methods is what has become known as Tobler's (1970) first law of geography: 'everything is related to everything else, but near things are more related than distant things'.

There are many forms of interpolation, which include the interpolation of points/lines to create sur-

faces, and the interpolation of areas from one scale/level of aggregation to another. Some of the more common techniques are listed in Box 6.3.

The quality of any given interpolation method depends on both the accuracy, number and distribution of known points used in the calculation, and how well the mathematical function correctly models the phenomenon; therefore, there is often more than one solution

---

Box 6.3 Interpolation methods for point and area data

1 **Point-to-surface methods:**	Boundary definition methods
	Global techniques
	Local techniques
2 **Area-to-area methods:**	Overlay method
	Pycnophylactic interpolation
	Zone centroid techniques

Source: Burrough (1986), Martin (1996)

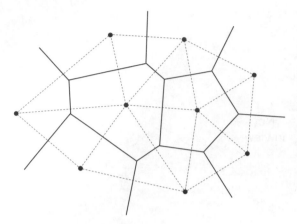

Figure 6.18 Delaunay triangulation (dashed lines) and the creation of proximal or 'Thiessen' polygons around a set of points (solid lines).

$$\hat{z}_0 = \sum_{i=1}^{N} w_i$$

where w_i are the weights which are a function of distance. In this case, the simplest weight is a switch, where a weight of 1 is given to all those points within a certain distance of the point to be estimated, and a weight of 0 to all others, i.e., the average of all the points within a window of a certain radius. (Goodchild, 1991) In order to decrease the influence of progressively more distant points, it is often more useful to use weights which are continuously decreasing functions of distance d^{-k}, for example inverse distance d^{-1} or inverse distance squared d^{-2}. The main advantage of distance-weighted averages, in addition to their simplicity, is that they are adaptable: the weighting function can be changed to suit different requirements or circumstances. However, the main disadvantage is that all interpreted values must lie between the largest and smallest of the observed values, such that the interpolated surface cannot extrapolate a trend outside the area containing the observed values. Other local interpolation methods are based on the use of piecewise functions termed **splines** fitted to a small number of data points exactly, and the statistical method of optimal interpolation termed **kriging**, which identifies weights based on the spatial autocorrelation function of the data. For details of these methods, and further details on some of the methods discussed above, you should consult Lam (1983) and the relevant chapters in Burrough (1986) and McDonnell (1998). We will return to discuss the subject of autocorrelation below in Section 6.7.

to the interpolation of a set of data values. (Aronoff, 1989) For **point-to-surface** interpolation methods, the simplest form of interpolation is to construct boundaries between known data points. This can be achieved by constructing **Thiessen polygons** (also known as *Voronoi* or *natural-neighbour polygons*). For this method, lines adjoining adjacent points are constructed, and then perpendicular bisectors are created mid-way between neighbouring points such that each location within a Thiessen polygon is closer to its contained point than any other point (Figure 6.18). This method makes several fundamental assumptions of the data: variation occurs at boundaries, and within boundaries the variation is the same in all directions (isotropic). The limitations of the method are that the size and shape of the resultant areas depend on sample layout, the value of each polygon is estimated from a sample of one, and no allowance is made for the similarity of close points versus distant points.

Global techniques involve the construction of a mathematical model of the spatial pattern using all the data values for a study area (e.g., use of a polynomial function to produce a trend surface). Although such global techniques are easy to use, they possess several limitations: they are only useful when there is reason to expect that the surface can be described by simple polynomial functions, they are very sensitive to boundary effects, and the final surface is smoothed and will rarely pass through any of the original data points. (Burrough, 1986) **Local techniques** estimate a point using only observed values close to it. The most common method of obtaining the interpolated point z_0 is the **distance-weighted average** of a number N of surrounding points z_i:

In a socio-economic context, data are more often only available aggregated over areal or zonal units. The use of aggregated units is problematic (which we will discuss in Section 6.4 below), and you may discover that the areal units for which the data are available are often not those which you may want to study, or you may find that the shape of the aggregated units has changed over time. For example, the boundaries of census enumeration units can change from one census to another, with incompatible zones making between-census comparison difficult. **Area-to-area** interpolation methods are often employed to transfer data from one set of source zones to another set of target zones. There are a variety of methods for area-to-area interpolation. The simplest is the **overlay method** which superimposes target zones over source zones, and estimates target zone values proportionally to the sizes of overlapping regions. The value in the target zone is subject to an areal weighting scheme applied to the value in the source zone. Although

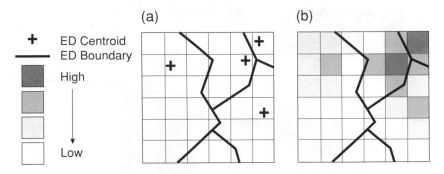

Figure 6.19 (a) Population enumeration district boundaries and centroids, with (b) population surface representation.

simple to implement in a GIS, this method assumes that the variable is uniformly distributed within the source zone, which we know is often an unrealistic model of socio-economic variables such as population distribution. We can obtain more realistic models of population distribution if we can interpolate to a surface in the form of a raster. One such method is the **pycnophylactic method** proposed by Tobler (1979), which is based on the assumption that the variable can be described by an underlying distribution which is continuous, resulting in a smoothed *density surface*. Although this avoids the uniform distribution of the overlay method, the model of continuous/smooth spatial variation may be inappropriate to describe the variation of some human geography variables which are discrete. Finally, **zone centroid techniques** developed by Martin (1989) are based on a distance decay procedure to distribute zone centroid populations to raster cells to create a population surface model (Figure 6.19). Although this method is not strictly an interpolation method, it can be used to good effect to model the distribution of population-based census variables. The results of using this method are displayed in Figure 6.20 which shows the pattern of 200 m cells from individual surfaces modelling the distribution of Protestants and Catholics in the City of Belfast from the 1991 Census.

6.4 Current issues in the use of GIS for socio-economic applications

In this section, we will outline a variety of issues which are important for the use of GIS in a socio-economic context. In the confines of this chapter, we cannot hope to do full justice to these areas of endeavour, many of which are the subject of current research, but in emphasising some of the main points we hope that increased awareness will allow you to reflect upon aspects of your own research project using socio-economic data and GIS. You should aim to read some of the referenced material to obtain a fuller appreciation of the relevant problems and prospects.

6.4.1 Representation

The representation and modelling of socio-economic phenomena in a GIS is both diverse and problematic. Martin (1999) provides an overview of the options available to model phenomena of interest to the human geographer such as population, economic activity and the built environment. Typical spatial objects often used to model such phenomena in a GIS, as well as related GIS-based manipulation and visualisation techniques which could be applicable to each object, are listed in Table 6.1 which has been adapted from Martin (1999).

So how do we choose between these different representations? Where it is possible to represent the same phenomenon in a GIS by different objects, either as a collection of points, a set of lines or zones/areas or a surface, and where there is disagreement over the exact spatial form of the phenomenon, there is no 'correct' method of representation (Martin, 1989). Clearly, if you have a certain type of analysis in mind, you will need to obtain data of the appropriate object type. This is constrained by the availability of sufficient quality georeferenced data. Although a certain amount of transformation between object types can be achieved using some of the GIS tools outlined above, ultimately we are dependent on the supply of georeferenced census data. Although a variety of census

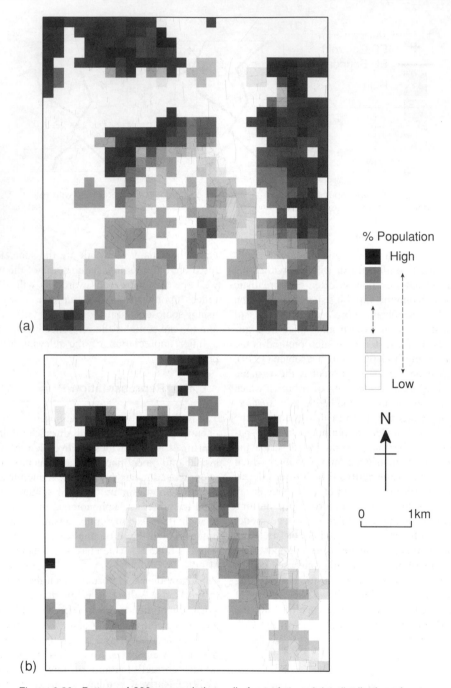

Figure 6.20 Pattern of 200 m population cells for surfaces of the distribution of
(a) Protestants, and (b) Catholics, in the city of Belfast overlaid on enumeration district
boundaries (© Crown Copyright. Permit No 1281).

variables are available in the UK, the finest spatial
resolution for which such data are available is the
enumeration district (ED) (see Section 3.6). Although
census data can be mapped down to the individual

address level in the US, and in the UK with the introduction of ADDRESS POINT referencing, many data
problems exist (e.g., suppression for low-density areas,
under-numeration in high-density areas) which make

Table 6.1   The representation of socio-economic phenomena in GIS.

	Point ■	Line	Area	Surface
Real-world entity	Individual observation	Journey to work	Property ownership	Population density
Digital object	Property or postcode coordinate	Street segment coordinates	Land parcel, census zone boundary	TIN or elevation model
Manipulation techniques	Nearest neighbour analysis, boundary generation, surface generation	Network functions, topological analysis	Areal interpolation, centroid generation, surface generation	Slope analysis, TIN creation, DEM creation, analysis of surface form
Visualisation techniques	Point mapping, multivariate glyphs, convert to 3D	Line mapping, line cartograms, convert to 3D	Choropleth mapping, area cartograms, convert to 3D	Isoline mapping, TIN mapping, grid mapping, convert to 3D

Source: Adapted from Martin in Longley *et al.* 1999.

6.4.2 Modifiable areal units

Many geographical zones represent the results of an arbitrary aggregation of individuals to produce a set of areas. The reporting of census data is an example: here, data collected at an individual household level are only available aggregated up to *larger* zones for which the data are reported. This would clearly not present a serious problem if there were an obvious natural zonal unit which could be used for such aggregation. However, this is not the case: there are many different zonal systems in use. Further, census zones such as enumeration districts, wards and census tracts are often designed simply for ease of enumeration, and may bear little or no relation to the social geography of the people they contain. Such zones can vary in their form and scale: they are **modifiable**. This is important for the geographical (and statistical) analysis of such objects, since changes in the form of a set of zones will alter the resulting pattern of aggregated observations, and any derived bivariate and multivariate statistics. This problem has been referred to as the **modifiable areal unit problem** (MAUP) which possesses two components: a *scale problem*, and a *zoning problem* (also termed an aggregation

problem) (Figure 6.21). The description below draws from reviews by Openshaw (1984) and Wrigley (1995) to which reference should be made for further details.

The **scale problem** describes the variation in results due to the progressive aggregation of smaller zones into larger zones. For example, census data can be aggregated up from enumeration districts to wards (in the UK), or from block to block group (in the US) (contrast Figures 6.22(a) and 6.22(b)). The tendency is for the magnitude of the bivariate correlation coefficient to increase with the size of the areal unit involved. The classic example is from Yule and Kendall (1950) who examined the correlation between yield per acre of wheat and yield per acre of potatoes for 48 counties in England. As the number of areal units decreased to three, the correlation coefficient increased from 0.22 to 0.99. In contrast, the **zoning problem** describes the variation in statistical results due to different arrangements of a fixed set of zones, whilst keeping the scale fixed (e.g., Figure 6.22(c)).

Where the only source of data is in the form of aggregated modifiable areal units, it is extremely dangerous to make inferences about relationships at the individual level. The dubious inference of individual level characteristics from aggregated data is known as the **ecological fallacy problem.** Where individual household-level census data, for example the 2% Sample of Anonymised Records (SAR) from the 1991

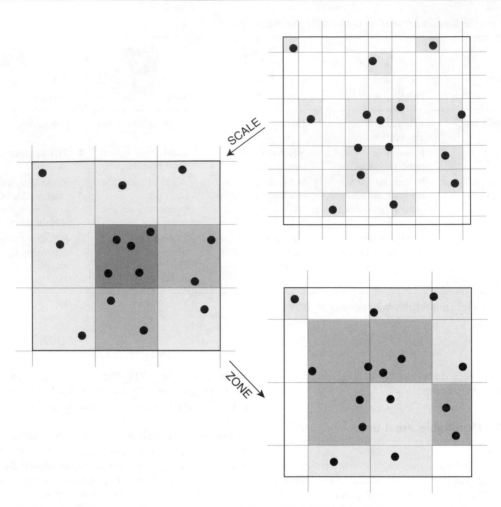

Figure 6.21 The scale and zoning elements of the modifiable areal unit problem.

Census in the UK, are available for comparison with data in modifiable areal units, it has been possible to say that (1) relationships at the areal level, measured by correlation coefficients, are in general *larger in absolute value* than the (unknown) individual household-level correlations, and (2) changes in the sign of the correlation coefficient should be expected (Wrigley *et al.*, 1996). It may be possible to adjust the areal-level correlation coefficients to remove the influence of scale and zoning effects, although such procedures are the focus of current research.

Wrigley *et al.* (1996) noted that the availability of both GIS technology and census/boundary data has led to an increase in data analyses subject to the effects of the MAUP. Indeed, the widespread availability of GIS technology can also avoid the restriction of hav-

ing to use standard areal units such as enumeration districts/blocks in an analysis. For example, implementing the simple area–area interpolation methods covered in Section 6.3 above can allow researchers to design their own zonal systems. Openshaw (1996) has exploited this ability to undertake 'zone carpentry' based on the optimisation of an underlying function. Some of the implications of the MAUP for your own research project are summarised in Box 6.4.

6.4.3 Error and spatial data accuracy

Obtaining some assessment of error and uncertainty is crucial for the accurate manipulation of spatial data using the techniques described in Section 6.3 above. Burrough and McDonnell (1998: 221) have

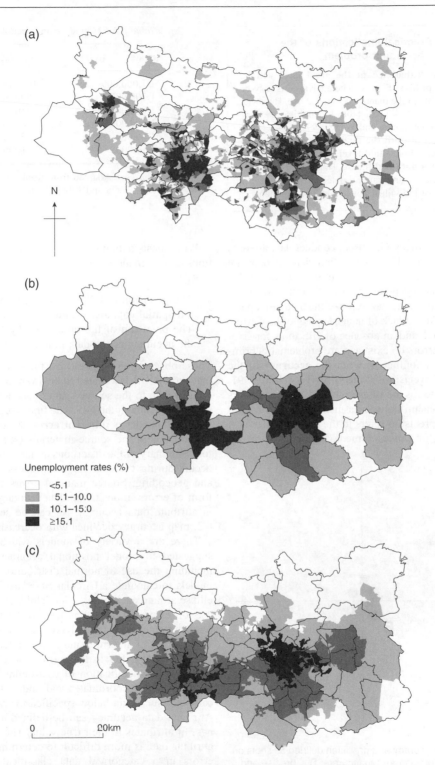

Figure 6.22 1991 Census unemployment rates for Leeds and Bradford in (a) enumeration districts, (b) 63 wards, and (c) 63 equal-area zones (source: Openshaw and Alvanides in press). © Crown Copyright.

Table 6.2   Map scale and accuracy.

Scale	Accuracy
1:1,000,000	1000 m
1:500,000	500 m
1:250,000	250 m
1:100,000	100 m
1:50,000	50 m
1:10,000	10 m

Sources: Adapted from Robinson *et al*. 1995 and Tobler 1988.

> **Box 6.4   Practical implications of the modifiable areal unit problem**
>
> - **Statistics calculated at the areal level will be subject to bias.** For bivariate analysis this may be predictable in the manner detailed above. However, this may be less predictable for multivariate analysis (Fotheringham and Wong, 1991).
> - **Detection of both the magnitude and nature of individual household-level relationships from areal-level data is difficult.** The analysis of areal-level data is an inappropriate surrogate for the analysis of individual household-level problems.

observed that 'many GIS users conduct data analysis . . . under the implicit assumption that all data are totally error free'. They ascribe this incorrect view to various factors: a general absence of information about data quality, the exact concepts of the database operations undertaken, a lack of understanding of how errors are propagated, and an absence of GIS tools for error evaluation. Errors do exist and are problematic, primarily due to a conflict between the **accuracy** of the data and the **precision** of the GIS (Goodchild, 1991, 1993) These terms are often confused: accuracy refers to the relationship between a measurement and reality, whereas precision refers to the degree of detail in the reporting of a measurement (Figure 6.23).

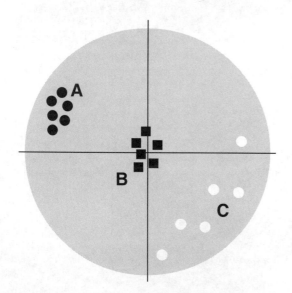

Figure 6.23 Accuracy and precision defined as shots on a target. (A = precise and inaccurate, B = precise and accurate, C = imprecise and inaccurate)

With these definitions in mind, several observations can be made in the context of GIS (Goodchild, 1991):

1 The precision of GIS is effectively infinite.
2 All spatial data are of limited accuracy.
3 The precision of the processing by GIS exceeds the accuracy of the data in almost all cases.

These observations suggest that the sources of error/inaccuracy which we need to be aware of when using a GIS relate to the source spatial data, and the precision of this data in the GIS. The first category of error we will consider is **inherent error**, which is defined as the error in the source materials (see Section 6.2 for a summary of the distortions in maps) or that which occurs during the data capture process of digitising and geocoding. Source material errors can take the form of errors in positional data accuracy and errors in attribute data accuracy. As can be seen in Table 6.2, map accuracy declines with decreasing scale.

There are a variety of models which have been suggested to model positional accuracy in points, including the use of normal distribution probability models to calculate a Circular Standard Error (CSE) ellipse around each point, (Goodchild, 1991) with the use of error bands, often termed **Perkal epsilon bands**, around lines and polygons. (Blakemore, 1984) The concept of an error band is encountered in a GIS such as ARC/INFO in the form of a tolerance: for example the specification of user-defined tolerances for arc lengths, coordinates and nodes, to automatically edit such data below specified tolerance values. Attribute data accuracy can be defined as the closeness of attributes to their true value. The accuracy of attribute data is more difficult to determine; however, errors in a categorical data classification can be determined by the use of an **error/confusion matrix**.

Table 6.3 Image accuracy confusion matrix for Bristol.

CLASSIFIED   REFERENCE DATA

Kappa coefficient	κ = 0.939; n = 250		κ = 0.786; n = 250		κ = 0.737; n = 250			
categories	non-built	built	non-resid.	resid.	low	medium	high	blocks
non-built	86 (96)	0						
built	7	157 (100)						
non-residential		1	47 (89)	2	0	0	0	0
residential		6	6	188 (99)				
low density			0		25 (86)	2	1	0
medium density			2		1	81 (92)	7	2
high density			14		2	5	90 (91)	5
tower blocks			1		1	0	1	10 (59)

Note: Number of correct sample points in **bold** and percentage correct in parentheses; with classification strata shown between bold lines, and 250 sample points in each stratum. 'Low', 'medium' and 'high' refer to residential density categories, and 'blocks' refer to tower blocks.
Source: Mesev 1998.

error/confusion matrix. This requires a measure of **ground truth** which is the actual classification at a given location. Such matrices are used for accuracy assessment of classified, remotely sensed images (Table 6.3), and often employ a measure of classification accuracy termed such as the **percentage agreement** or **Cohen's kappa** (Cohen, 1960).

Percentage agreement is calculated by:

$$\frac{\text{number of agreements}}{\text{number of} \quad \_ \quad \text{number of}} \times 100$$
$$\text{agreements} \quad \text{disagreements}$$

with Cohen's kappa (κ) calculated as follows (Pett, 1997):

$$\kappa = \frac{\text{proportion of} \quad \_ \quad \text{proportion of}}{\text{observed agreement} \quad \text{chance agreement}}$$
$$\frac{}{1 - \text{proportion of chance agreement}}$$

$$= \frac{P_o - P_c}{1 - P_c}$$

where

$$P_o = \frac{\text{number of agreements}}{n}$$

and

$$P_c = \sum \frac{\text{row total} \times \text{column total}}{n^2}$$

The value of kappa is theoretically in the range -1 to $+1$, with a value of $+1$ indicating no agreement by chance and a negative value indicating more agreement

by chance than is observed. The value of kappa can be tested for significance using a similar procedure to the tests in Chapter 5 (see Pett, 1997, for details).

For numerical data, we can employ normal distribution probability models to obtain a measure of uncertainty in the form of a Root Mean Square Error (RMSE, equivalent to the standard deviation) calculated between ground truth observations, and the data values in the data set. A limitation of these methods is that they require higher accuracy data in the form of ground truth which is not always available.

Problems also exist in the use of crisp boundaries to represent spatial variation. The use of zones, or representation of features, in a choropleth map not only assumes a model of spatial uniformity within each zone, but also assumes that we can delineate the boundary between each zone in a precise and meaningful manner. For many geographical phenomena this is simply not the case: natural spatial variation leads to *gradual* change, and the difference between reality and the model can lead to error (Burrough and McDonnell, 1998: 225). Recent progress in the application of **fuzzy set theory** to allow for the imprecision in both attribute assignment (fuzzy membership) and object demarcation (fuzzy boundaries) may reduce such error. Errors can also be introduced in the process of digitising, for example dangling nodes in the form of undershoots and overshoots, and polygon slivers (Figure 6.24).

The second category of error is **operational error**, which accrues as a result of manipulations inside the GIS. Such error can occur during the process of

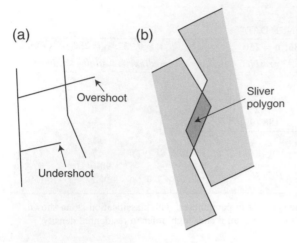

Figure 6.24 Errors in the form of (a) undershoots and overshoots, and (b) polygon slivers.

transformation of data from one data structure to another, as well as in the process of some of the spatial analysis operations considered above. The process of vector to raster conversion can introduce errors, particularly if the cell size of the raster is coarse. This can manifest itself in both geometric terms (the change in shape/area of the resulting objects) and attribute terms (mixed pixels resulting from the integration of fine spatial detail to the size of the cell). There are a variety of statistical procedures which can be used to estimate the errors of vector to raster conversion; these are detailed by Burrough and McDonnell (1998). More important is how the errors due to data manipulation propagate and accumulate through the GIS. A variety of models are available for the modelling of error through GIS operations (e.g., Heuvelink and Burrough, 1993). In this context, two main observations are possible: (1) it is easier to assess the scale of error propagation for some types of analytical operation than for others, and (2) much depends on the degree of spatial auto-correlation in the data. If we consider first uncorrelated variables, we can obtain some simple formula for an operation such as arithmetic overlay. For example, if we want to estimate the error S_C (RMSE) in resultant layer C from adding two layers A and B:

$$S_C = \sqrt{S_A^2 + S_B^2}$$

and for multiplying these two layers:

$$S_C = \sqrt{(S_A^2 \times B^2) + (S_B^2 \times A^2)}$$

An immediate problem is the fact that many of the variables of use in a GIS are likely to be correlated, which complicates the formula (see Burrough and McDonnell, 1998, Chapter 10 for details).

Ideally, in the light of the errors outlined above, all data inputs to a GIS, and products from a GIS, should have a measure of uncertainty attached. This requires suitable measures of uncertainty on all inputs as well as methods for propagating uncertainty through GIS operations. Concern with the quality of spatial data has led to the development of national transfer standards in spatial data, for example the Spatial Data Transfer Standard (SDTS) in the US, and the National Transfer Format (NTF) in the UK. The inclusion of statements on data quality in the form of metadata (*cf.* Chapter 3) are important aspects of such standards.

In order to gauge the potential utility of a data set for GIS analysis, Foote and Hubner (1995) have suggested the set of questions in Box 6.5.

6.5   Sources of digital spatial data

In Chapter 4 we discussed the various sources of digital and analogue non-spatial data which could be used for a quantitative analysis. Most, if not all, of these data could be used in a GIS context if they can be given a geographical reference. For certain data sets this is less of a problem: for example, census

Box 6.6   Example digital spatial data sources in the UK and USA

UK

MIDAS
http://www.midas.ac.uk/

Manchester Information Datasets and Associated Services (University of Manchester) allows computer/Internet access to authorised users for the following spatial data sets:
- Bartholomew digital map data
- 1991 Census digital boundary data
- LANDSAT and SPOT satellite data

OSGB
http://www.ordsvy.govt.uk/

Ordnance Survey of Great Britain. Supplies a great variety of digital map data products which include:
- Boundary-Line: administrative areas for Britain
- ED-LINE: enumeration district boundaries
- ADDRESS-POINT: postal address data
- OSCAR: road networks

THE DATA CONSULTANCY
http://www.datasets.com/

Supplies a wide range of digital data. Digital boundary data sets include:
- Administrative area boundaries
- Census boundaries
- Health authority boundaries
- Postcode boundaries
- Political boundaries
- Drive time boundaries
- Travel-to-work area boundaries

UKBOARDERS
http://borders.ed.ac.uk/

The United Kingdom Boundary Outline and Reference Database for Education and Research Study (part of EDINA national data centre) allows online retrieval for authorised users to the following spatial data sets:
- 1991 Census digital boundary data
- Historical administrative boundaries

USA

USGS
http://mapping.usgs.gov/

United States Geological Survey supply boundary and other spatial data, much online with Internet access. Digital products include:
- Digital Line Graphs (DLGs) of boundary, hydrography and roads
- Digital Elevation Models (DEMs) of topography
- Land Use and Land Cover data (LULC)
- Remote sensing satellite data

US CENSUS BUREAU
http://www.census.gov/

The US Census Bureau supplies a variety of geographic products. Of most use in a GIS context is the TIGER/Line product (Topologically Integrated Geographic Encoding and Referencing) which includes spatial information on:
- Administrative boundaries
- Political boundaries
- Street centrelines
- ZIP codes

data in the US and UK comes with varying amounts of locational information attached; however, such data, particularly where in aggregated form, will often need to be supplemented by digital boundary data. Part of the problem with spatial data is finding out what is available. Examples of some of the products and sources of digital spatial data can be seen in Box 6.6. There are an increasing number of commercial organisations which supply data for the GIS market. For example, in the UK, The Data Consultancy supplies a wide variety of digital datasets which include political, census and health area boundaries as well as postcode and raster maps scanned at a variety of scales. In the US, a Geospatial Data Clearinghouse has been set up as part of the National Spatial Data Infrastructure (NSDI). This is a collection of over 50 spatial data servers which allow a user to search for useful spatial data sets largely covering the US (see http://www.fgdc.gov/Clearinghouse/Clearinghouse.html).

6.6 Planning and implementing analysis using GIS

6.6.1 Using GIS in your research project

We have outlined above some of the main functions and data sources for performing a GIS analysis. But how do you incorporate such material within the context of your own research project? Although the details of your analysis will depend on the specific objectives you have identified, there are some general guidelines which are common to all research projects using an element of GIS. The general model that we will use is outlined in Box 6.7. The points below are based on the ARC/INFO vector GIS model. As indicated above, a vector model is often preferred for socio-economic data. ARC/INFO can also import and manipulate data in a raster form if your analysis requires data in such a format, although our discussion will focus on vector data.

6.6.2 Project objectives

Your research should have produced one or more **problems** to solve or **objectives** to reach for which you are considering the use of a GIS. First, you should try to locate any publications which have made use of a GIS in a similar context to your own project. From this you will be able to estimate the scope of your intended application and the types and quantity of data, and to obtain some idea of the volume of work required, as well as of the types of analysis and potential products. Additional questions might arise – was the GIS used effectively? Was the investment in the GIS analysis justified by the output?

6.6.3 Database construction

Once you have defined the objectives of your GIS application, you should have some idea of your data requirements. The process of database construction is concerned essentially with the design of the database, and the implementation of the design. Essentially, we are concerned with data issues: *data requirements*, *data sources*, *data acquisition*, *input to* and *storage in* the GIS system. Defining the data requirements is concerned with identifying not only a region of interest, but also relevant geographic features and associated attributes for your analysis. For example, your project may be concerned with the use of a GIS to analyse access to health care facilities within the urban district of Toronto. Within this context, you will then need to plan your layers (also termed coverages) of spatial information. You could organise such information by spatial feature (e.g., all point features in one layer, and all line features in another layer) or by attribute (e.g., a transportation layer, a population layer, a health facility location layer). Which you choose will depend on the details of your project and your intended analysis. It may be necessary to adopt a coding scheme to identify the various types of attributes (e.g., major roads defined by a code '10', minor roads by a code '5'). The next step would be to identify sources of appropriate spatial and attribute data. Refer to Section 6.5 above for sources of spatial data, and to Chapter 3, Section 3.5, for sources of attribute data. The data acquisition task is simplified if your data are available in a digital form. Data in an analogue form, or data which require collection as part of a field survey, will be more time consuming to obtain, particularly if you have to do this yourself. Even if digital data are available, they may be prohibitively expensive to obtain, the wrong scale (too small) for your analysis, inaccurate or out of date. You may like to refer to Box 6.6 above for thoughts about data quality.

Box 6.7   GIS project steps

Determine project objectives	What is the nature of your problem?
	What are the final products of the project?
Database construction	Design the database.
	Implement the database design.
Analysis	Perform the required analytical tasks.
Presentation of the results	Output of the GIS as a combination of maps and reports.

Source: Adapted from ESRI 1995.

Table 6.4 Data acquisition.

Type	Spatial	Attribute
Digital	Import	Load in
Analogue	Digitise/scan	Key in
Not available	Field survey	Collect and key in

Source: Adapted from Strachan *et al.* 1993.

Once you have designed your database, you will need to input the data. Attribute data are usually either imported if in a digital format, or keyed in from printed tables (Table 6.4). Spatial data, on the other hand, can be imported if already in a digital format. If you obtain data from an organisation, it will most likely be available in one of a variety of standard formats, for example National Transfer Format (NTF) in the UK, or Spatial Data Transfer Standard (SDTS) in the USA. Other formats include files in the AutoCAD Drawing eXchange Format (DXF), and the United States Geological Survey Digital Line Graph (DLG) format, which can be imported into a variety of GIS systems. Failing this, most GIS systems can import object coordinates (X, Y, Z) in the form of simple ASCII text files.

The process of digitising (for vector data) or scanning (for raster data) is frequently used in student research. **Digitising** is defined as the process of automating the location of geographic objects, by converting their position on a map to a series of X,Y (Cartesian) coordinates. Objects are hand-traced using a digitising table in either **point mode** (entering vertices by hand) or **stream mode** (vertices entered automatically). You should not underestimate the amount of time that digitising even a simple map will take. However, the process can be carried out efficiently if certain steps are followed (Box 6.8).

The alternative, particularly for maps which contain much information in areal form (e.g., an aerial photograph), is to **scan** the information in a raster form. Although high-quality scanners which scan at a micron resolution are increasingly available, the use of such devices may be cost prohibitive. Standard A4 desktop scanners can be used to create images (e.g., in a TIF format); however, the resultant images often consist of non-square cells, which may require re-sampling in the GIS. You may be able to make use of **digital satellite remotely sensed data** (e.g., SPOT or LANDSAT); however, such images are usually in a raw form and will require classification to extract useful information for your GIS. We will concentrate on the steps relating to digitising, as this is the most commonly used method to capture spatial information for the research student. One word of warning, however: the digitising of many commercially available maps (e.g., Ordnance Survey produced maps in the UK) is an infringement of **copyright**. You should make every attempt to ensure that you are not reproducing a map illegally.

Once features have been digitised into the GIS, you will need to make the spatial data usable by constructing topology and editing any errors in the digitised data. For example, overshoots and undershoots will create dangling nodes, which will be incorrect if you have created a coverage consisting entirely of polygons. Most GIS systems will have facilities for the on-screen display and editing of such features, as well as other spatial errors such as slivers (see Section 6.4). The re-creation of topology after editing will reveal any remaining errors which can subsequently be edited.

Box 6.8 Good digitising practice

Use good base maps   Use as large a scale as possible with clear object outlines. You can copy map features onto a stable medium such as a plastic film (MYLAR) to minimise physical map distortion.

Use clear procedures   Be consistent with the level of generalisation, organising codes for different features if necessary. Decide on either point or stream mode for data capture.

Prepare your maps   Mark up features to be digitised on a traced outline of the map in a clear and logical manner. You will need to locate ground control points on your map (also termed tic registration points) for which you know (or can obtain) the actual real-world coordinates. Mark the intersections of lines, and indicate nodes on long arcs and island polygons, ensuring all polygons close have a single label point. It is easier to trim overshoots in subsequent editing than to extend undershoots, so plan to overshoot intersections slightly.

Take regular breaks   Digitising is a very labour-intensive task. You will find that as you become tired, more errors will be introduced to your work, so plan your work and take a break if you are becoming tired.

Source: Adapted from ESRI 1995.

Box 6.9   Entering attribute data in ARC/INFO

Enter your text into an ASCII file	Digital attribute data can come in a wide variety of formats; however, a simple space- or comma-delimited ASCII text file is often the easiest to use, where each column corresponds to a different variable.
Define a database template	This makes use of the database (INFO) part of ARC/INFO. The name, character type, storage and display space are defined here for each variable.
Add data from the file to the template	The data from the ASCII text file is added to the template to create an attribute file. At this stage, the data set is similar in form to a spreadsheet.
Join the attribute file to the spatial data	The final step links the attribute information to the spatial objects.

Source: Adapted from Strachan *et al.* 1993.

Once the spatial data have been entered (and corrected), the **attribute data** can be entered. Similar to spatial data, these may be in a convenient-to-use digital form, or available only in a tabular form or from maps. Although we make use of the ARC/INFO GIS to outline the steps to add attribute data (Box 6.10), the steps are similar to other GIS systems.

6.6.4   Analysis

Once you have reached this stage, you will be able to proceed with the analysis of your data using some of the methods outlined in Section 6.3 above. If possible, *your data requirements should be dictated by the types of analysis you wish to carry out*, although with many sources of secondary data you may well have to make do with whatever data are available. In this context it is worth considering some of the sources of error in Section 6.4 above, in particular noting some of the recommendations in Box 6.9. Analytical results are far more meaningful if you can attach some estimate of uncertainty to the finished product. As with any analytical method, you will need to *understand exactly what is being done with your data*. Although it is possible to treat a GIS as a magical 'black box' (particularly if it is only a minor element of your research), and follow instructions from your supervisor or manual, this will put you in a precarious position when you come to justify and explain your choice of method in the write-up. Again, you may find examples in the literature where similar types of analysis have been performed, which you may be able to adapt or use as a guide for your own.

6.6.5   Presentation of results

The output of your GIS analysis is usually in a map form. Exactly what you include on a map, as well as

the size of the map, is determined by you. For example, the ARC/INFO GIS allows you to control the page units, page size, map size and scale, colour and symbolisation of both objects and text. Although the task of map construction is made easier with a computer, this does not mean that you can automatically construct good maps! Indeed, in our experience, some particularly bad maps can be constructed using a GIS. The principles of map design and map construction for hand-drawn maps apply equally to GIS maps. Although we make some reference to the basis of map construction in Chapter 10, reference to texts such as Robinson *et al.* (1995) and Monkhouse and Wilkinson (1971) will be of considerable use. Again, you can obtain some ideas of what works (and occasionally what doesn't work) by consulting the published literature, although such material is often in a black-and-white form and not the best source for examples. The Environmental Systems Research Institute (ESRI), which makes the ARC/INFO and Arcview GIS packages, publishes an annual map collection, 'ARC/INFO maps', which consists of colour maps contributed by worldwide users of the GIS and which might give you some ideas for your own project. Most GIS systems can export maps in an encapsulated postscript (EPS) format; in this fashion it is possible to export your map into cartographic software such as *Adobe Illustrator* and *Paint Shop Pro*, or directly into your word-processed text.

6.6.6   Some common GIS systems

There are an increasing number of GIS systems available within universities and other institutions of higher education which you might encounter in the process of your research. We list some basic details of some of the more common systems below. Contact details and web addresses for these systems are given in Appendix D.

ARC/INFO GIS software developed by Environmental Systems Research Institute (ESRI). UNIX workstation and PC based, the system supports both vector and raster data structures.

Arcview GIS Desktop mapping and GIS software developed by Environmental Systems Research Institute (ESRI), available on a variety of platforms including the workstation and PC.

IDRISI Grid-based GIS system developed by the IDRISI Project at Clark University; this GIS runs in a Windows-based PC environment.

MapInfo Desktop mapping and GIS software developed by MapInfo Corporation, which runs on a PC platform.

6.7 Spatial statistics

Geographers are often interested in variations of phenomena across space. Spatial statistics allow us to explicitly examine the nature of spatial variation by extending statistical analysis to spatial forms. In general, we can divide spatial forms into four main types: *points*, *lines*, *areas* and *volumes*. It is important to remember that these should not necessarily be viewed as discrete categories. For example, as noted above, an area when viewed at a different scale can appear as a point. Indeed, many point-based statistics are used to examine the spatial distribution of towns and cities. Here we will limit our concern to points, lines and areas as the map objects most commonly examined by geographers. Many spatial statistics are very complicated and often designed for specific purposes. As a consequence we have selected just a few of the more simple and useful spatial statistics. In many ways, the few that we have chosen to illustrate perform an Exploratory Spatial Data Analysis (ESDA) role, similar to the IDA (or EDA) for non-spatial data covered in Chapter 4. As such, they are all descriptive in nature. Taken in isolation, many of the spatial statistics outlined below might seem of limited use. For example, a measure of standard distance has little meaning until compared to another measure of standard distance. Therefore, the real utility of these measures is through cross-comparison. Here, the *results* from a whole series of related measures are compared using the non-spatial statistics outlined in Chapter 5. For a more detailed overview of spatial statistics, see the Further reading section at the end of this chapter, particularly those references by Unwin (1982), Bailey and Gatrell (1995) and Goodchild (1986).

To illustrate each spatial statistic we have followed the format adopted in Chapter 5, first selecting a real-world data set, next outlining how to calculate the spatial statistic by hand, and finally detailing how to find the same result using the GASP computer package. GASP is a basic computer package specifically to teach students how to calculate spatial statistics. GASP can be downloaded, free of charge, from the Web site that accompanies this book (http://www.awl.co.uk/geography). It was co-authored by Mike Bratt and Rob Kitchin.

6.7.1 Description of pattern

Point data: mean centre

Similar to an aspatial distribution (e.g., calculating a normal distribution), some measure of the centre of a spatial distribution is particularly useful. The **mean centre** is perhaps the simplest measure of a point distribution. It represents the 'average' or mean position for a set of points in space. In human geography, the most frequently encountered measure of a mean centre is the **centroid** of population, although for small areas of aggregation the centroid is often approximated by eye. The mean centre can be found simply by calculating the mean of the x coordinates and the mean of the y coordinates, the equations being:

$$\bar{x} = \frac{\sum_{i=1}^{n} x_i}{n}, \quad \bar{y} = \frac{\sum_{i=1}^{n} y_i}{n}$$

In order to illustrate how to calculate the mean centre of a group of points, we have performed the analysis on a data set comprising the cognitive estimates of a group of students undertaking a cognitive mapping exercise (see Table 6.5). Here, respondents were asked to place a series of places onto a bounded space displaying just two known locations. The data shown are the group's estimates of where Mumbles Pier is located (for reference the real location is $x = 5.80$ and $y = 1.00$). By calculating the mean centre we can estimate how accurately, as a group, their estimates were by comparing the mean centre with the real-world location. Through a comparison of this accuracy with the accuracy of other locations we can determine how well the location of Mumbles Pier is known in comparison to other locations within Swansea (Boxes 6.10 and 6.11).

Table 6.5 Cognitive estimate coordinates of a group of students for Mumbles Pier, Swansea.

x coordinate	y coordinate	x coordinate	y coordinate
5.05	0.55	1.70	0.60
1.30	2.90	3.05	1.75
3.70	3.30	2.40	0.75
0.30	1.00	3.45	2.25
4.15	1.85	4.20	3.65
2.95	1.55	4.20	1.30
3.90	3.55	0.90	0.95
3.25	3.60	3.00	2.35
3.90	0.90		

Source: Kitchin 1996.

Box 6.10   Handworked example: mean centre

x	y	x	y
5.05	0.55	1.7	0.6
1.3	2.9	3.05	1.75
3.7	3.3	2.4	0.75
0.3	1	3.45	2.25
4.15	1.85	4.2	3.65
2.95	1.55	4.2	1.3
3.9	3.55	0.9	0.95
3.25	3.6	3	2.35
3.9	0.9	$\sum x = 51.40$	$\sum y = 32.80$

Step 1 Place your data into a table as above.

Step 2 Sum the x and y coordinate values to determine $\sum x = 51.40$ and $\sum y = 32.80$.

Step 3 Divide $\sum x$ and $\sum y$ by the number of coordinate pairs ($n = 17$) to produce $\bar{x} = 3.0235$ and $\bar{y} = 1.9294$.

Step 4 Then plot the mean centre in relation to the cognitive estimates and the real location:

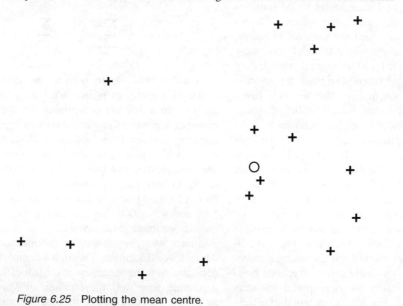

Figure 6.25 Plotting the mean centre.

Point data: standard distance

The **standard distance** (SD) is a single-value spatial equivalent of the standard deviation. It provides a concise description of the spread of points around the mean centre, from which we can determine if the pattern appears to be dispersed or clustered. The larger the standard distance the greater the dispersion of the points. This distance can be shown visually as a circle with the radius equal to the standard distance. It is calculated as follows:

$$\text{SD} = \sqrt{\frac{\sum_{i=1}^{n} (x_i - \bar{x})^2}{n} + \frac{\sum_{i=1}^{n} (y_i - \bar{y})^2}{n}}$$

Points that are far away from the mean centre will exaggerate the standard distance. This is a result of squaring all the distances. In this instance extreme points will have a large squared value and thus distort the summing calculations. In Boxes 6.12 and 6.13 we demonstrate how to calculate the standard distance for the data in Table 6.5.

Box 6.11   GASP example: mean centre

Input data set by loading file 'point.dat' or inserting points using the edit pull-down menu.

Calculate mean centre:

Using pull-down menus:
   ➥ Stats
   ➥ Mean Centre
GASP results:
   x = 3.0235
   y = 1.9294

Point data: standard deviation ellipse

The standard distance summarises the spread of points in just one value. It does not, however, take into account the fact that the spread of points may be different in different directions. The **standard deviation ellipse** summarises dispersion in a set of points in terms of an ellipse, rather than a circle. The ellipse is centred on the mean centre, with its long axis in the direction of the maximum dispersion and its short axis in the direction of the minimum dispersion.

Box 6.12   Handworked example: standard distance

x	x'	x'^2	y	y'	y'^2
5.05	2.03	4.1209	0.55	−1.38	1.9044
1.3	−1.72	2.9584	2.9	0.97	0.9409
3.7	0.68	0.4624	3.3	1.37	1.8769
0.3	−2.72	7.3984	1	−0.93	0.8649
4.15	1.13	1.2769	1.85	−0.08	0.0064
2.95	−0.07	0.0049	1.55	−0.38	0.1444
3.9	0.88	0.7744	3.55	1.62	2.6244
3.25	0.23	0.0529	3.6	1.67	2.7889
3.9	0.88	0.7744	0.9	−1.03	1.0609
1.7	−1.32	1.7424	0.6	−1.33	1.7689
3.05	0.03	0.0009	1.75	−0.18	0.0324
2.4	−0.62	0.3844	0.75	−1.18	1.3924
3.45	0.43	0.1849	2.25	0.32	0.1024
4.2	1.18	1.3924	3.65	1.72	2.9584
4.2	1.18	1.3924	1.3	−0.63	0.3969
0.9	−2.12	4.4944	0.95	−0.98	0.9604
3	−0.02	0.0004	2.35	0.42	0.1764

$\sum x = 51.40$
$\bar{x} = 3.0235$
$\sum x'^2 = 27.41$

$\sum y = 32.80$
$\bar{y} = 1.9294$
$\sum y'^2 = 20.00$

Box 6.12   (cont'd)

Step 1   Place the x and y data in a table as above.

Step 2   Sum the x and y coordinate values to determine Σx and Σy.

Step 3   Divide Σx and Σy by the number of coordinate pairs ($n = 17$) to produce $\bar{x}$ and $\bar{y}$ (as in Box 6.11, mean centre).

Step 4   Subtract $\bar{x}$ from x (e.g., $5.05 - 3.0235$) to produce new column x' and subtract $\bar{y}$ from y to produce new column y'.

Step 5   Square x' and y' to produce new columns x'^2 and y'^2.

Step 6   Sum x'^2 and y'^2 to obtain $\Sigma x'^2$ and $\Sigma y'^2$.

Step 7   Input these values into the formula and work through to calculate SD:

$$SD = \sqrt{\frac{27.41}{17} + \frac{20.00}{17}} = \sqrt{1.61 + 1.18} = \sqrt{2.79} = 1.67$$

Step 8   You can then plot the standard distance by drawing it around the mean centre:

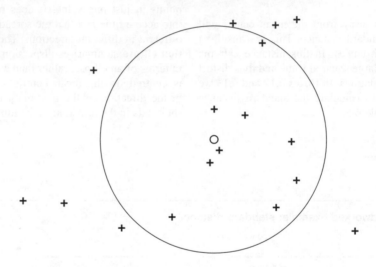

Figure 6.26   Plotting the standard distance.

These axes are always at right angles to each other. The ratio value indicates the strength of the directional bias. The larger the value the more linear the scatter. To calculate the values that describe an ellipse, three values have to be known (Ebdon, 1985):

1   The orientation of the ellipse
2   The length of the shortest axis
3   The length of the longest axis

The equations to calculate these are as follows. First, calculate the angle of rotation through a series of equations:

$$R = \arctan\left(\frac{\left(\sum_{i=1}^{n} x'^2 - \sum_{i=1}^{n} y'^2\right)}{2 \times \sum_{i=1}^{n} x'y'} + \frac{\sqrt{\left(\sum_{i=1}^{n} x'^2 - \sum_{i=1}^{n} y'^2\right)^2 + 4 \times \left(\sum_{i=1}^{n} x'y'\right)^2}}{2 \times \sum_{i=1}^{n} x'y'}\right)$$

where x' and y' represent a transposed coordinates system, obtained by subtracting the mean $\bar{x}$ and $\bar{y}$ coordinate values from the original x and y coordinate values:

$$x' = x - \bar{x}$$

$$y' = y - \bar{y}$$

The angle of rotation represents the rotation needed clockwise from the vertical. Next, calculate the standard deviation along the x axis of the ellipse:

$$\sigma(x) = \sqrt{\frac{\left(\sum_{i=1}^{n} x'\cos R - y'\sin R\right)^2}{n}}$$

Calculate the standard deviation along the y axis of the ellipse:

$$\sigma(y) = \sqrt{\frac{\left(\sum_{i=1}^{n} x'\sin R + y'\cos R\right)^2}{n}}$$

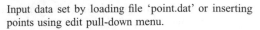

Box 6.13   GASP example: standard distance

Input data set by loading file 'point.dat' or inserting points using edit pull-down menu.

Calculate standard distance:

Using pull-down menus:
➥ Stats
➥ Standard Distance

GASP results:
```
SD = 1.6700
```

The ratio of standard deviation along the x axis as opposed to the y axis is calculated:

$$Ra = \frac{\sigma(x)}{\sigma(y)}$$

To illustrate the calculation of a standard deviation ellipse (Boxes 6.14 and 6.15) we have again used the data in Table 6.5.

Box 6.14   Handworked example: standard deviation ellipse

x	x'	x'^2	y	y'	y'^2	$x'y'$
5.05	2.03	4.1209	0.55	−1.38	1.9044	−2.8014
1.3	−1.72	2.9584	2.9	0.97	0.9409	−1.6684
3.7	0.68	0.4624	3.3	1.37	1.8769	0.9316
0.3	−2.72	7.3984	1	−0.93	0.8649	2.5296
4.15	1.13	1.2769	1.85	−0.08	0.0064	−0.0904
2.95	−0.07	0.0049	1.55	−0.38	0.1444	0.0266
3.9	0.88	0.7744	3.55	1.62	2.6244	1.4256
3.25	0.23	0.0529	3.6	1.67	2.7889	0.3841
3.9	0.88	0.7744	0.9	−1.03	1.0609	−0.9064
1.7	−1.32	1.7424	0.6	−1.33	1.7689	1.7556
3.05	0.03	0.0009	1.75	−0.18	0.0324	−0.0054
2.4	−0.62	0.3844	0.75	−1.18	1.3924	0.7316
3.45	0.43	0.1849	2.25	0.32	0.1024	0.1376
4.2	1.18	1.3924	3.65	1.72	2.9584	2.0296
4.2	1.18	1.3924	1.3	−0.63	0.3969	−0.7434
0.9	−2.12	4.4944	0.95	−0.98	0.9604	2.0776
3	−0.02	0.0004	2.35	0.42	0.1764	−0.0084

$\sum x = 51.40$ $\sum y = 32.80$ $\sum(x'y') = 5.806$
$\bar{x} = 3.0235$ $\bar{y} = 1.9294$
$\sum x'^2 = 27.41$ $\sum y'^2 = 20.00$

Step 1   Place the x and y data in a table as above.

Step 2   Sum the x and y coordinate values to determine $\sum x$ and $\sum y$.

Step 3   Divide $\sum x$ and $\sum y$ by the number of coordinate pairs ($n = 17$) to produce $\bar{x}$ and $\bar{y}$ (as in mean centre)

Box 6.14   (cont'd)

Step 4   Subtract $\bar{x}$ from x (e.g., $5.05 - 3.0235$) to produce new column x' and subtract $\bar{y}$ from y to produce new column y'.

Step 5   Square x' and y' to produce new columns x'^2 and y'^2.

Step 6   Sum x'^2 and y'^2 to determine $\Sigma x'^2$ and $\Sigma y'^2$.

Step 7   Multiply x' with y' to produce a new column $x'y'$.

Step 8   Sum $x'y'$ to determine $\Sigma(x'y')$.

Step 9   Square $\Sigma(x'y')$ to determine $[\Sigma(x'y')]^2$.

Step 10   Input these values into the formula and work through to calculate the angle of rotation, R.

$$R = \arctan \frac{(27.41 - 20.00) + \sqrt{(27.41 - 20.00)^2 + 4 \times (5.806)^2}}{2 \times 5.806}$$

$$= \arctan \frac{7.41 + \sqrt{(7.41)^2 + 4(5.806)^2}}{11.612} = \arctan \frac{7.41 + \sqrt{54.908 + 4(33.709)}}{11.612}$$

$$= \arctan \frac{7.41 + \sqrt{189.744}}{11.612} = \arctan \frac{7.41 + 13.77}{11.612} = \arctan \frac{21.18}{11.612} = \arctan(1.824) = 61.266°$$

Step 11   Next determine $\cos R$ and $\sin R$ and create a new column $x'\cos R$ by multiplying x' with $\cos R$. Repeat to create three more new columns $y'\sin R$, $x'\sin R$ and $y'\cos R$.

Step 12   Create two new columns, $(x'\cos R - y'\sin R)^2$ (by subtracting $y'\sin R$ from $x'\cos R$ and then squaring the result) and $(x'\sin R + y'\cos R)^2$.

Step 13   Sum $(x'\cos R - y'\sin R)^2$ to determine $\Sigma(x'\cos R - y'\sin R)^2$ and similarly determine $\Sigma(x'\sin R + y'\cos R)^2$.

$\cos R = 0.480$ $\quad\quad\quad$ $\sin R = 0.876$

$x'\cos R$	$y'\sin R$	$x'\sin R$	$y'\cos R$	$(x'\cos R - y'\sin R)^2$	$(x'\sin R + y'\cos R)^2$
0.974	−1.209	1.778	−0.662	4.765	1.245
−0.825	0.850	−1.506	0.466	2.806	1.082
0.326	1.200	0.595	0.658	0.764	1.570
−1.305	−0.815	−2.382	−0.446	0.240	7.998
0.542	−0.070	0.989	−0.038	0.375	0.904
−0.033	−0.333	−0.061	−0.182	0.090	0.059
0.422	1.419	0.770	0.778	0.994	2.396
0.110	1.463	0.201	0.802	1.831	1.006
0.422	−0.902	0.770	−0.494	1.753	0.076
−0.633	−1.165	−1.156	−0.638	0.283	3.218
0.014	−0.158	0.026	−0.086	0.030	0.004
−0.297	−1.034	−0.543	−0.566	0.543	1.230
0.206	0.280	0.376	0.154	0.005	0.281
0.566	1.507	1.033	0.826	0.885	3.456
0.566	−0.552	1.033	−0.302	1.250	0.534
−1.017	−0.858	−1.857	−0.470	0.025	5.415
−0.009	0.368	−0.017	0.202	0.142	0.034

$\Sigma(x'\cos R - y'\sin R)^2 = 16.782$ $\quad\quad$ $\Sigma(x'\sin R - y'\cos R)^2 = 30.509$

Step 14   Input the values to calculate the standard deviation along the x axis:

$$\sigma(x) = \sqrt{\frac{16.782}{17}} = \sqrt{0.987} = 0.994$$

Box 6.14   (cont'd)

Step 15   Input the values to calculate the standard deviation along the y axis:

$$\sigma(y) = \sqrt{\frac{30.509}{17}} = \sqrt{1.795} = 1.339$$

Step 16   Calculate the ratio of standard deviation along the x axis as opposed to the y axis:

$$Ra = \frac{0.994}{1.339} = 0.742$$

Step 17   If we wish, we can then plot the standard deviation ellipse by drawing it around the mean centre:

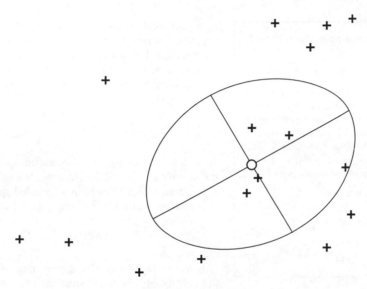

Figure 6.27   Plotting a standard deviation ellipse.

N.B. The results are slightly different from those in the GASP example due to rounding.

Box 6.15   GASP example: standard deviation ellipse

Input data set by loading file 'point.dat' or inserting points using edit pull-down menu.

Calculate standard deviation ellipse:

Using pull-down menus:
   ➡ Stats
   ➡ Std Dev Ellipse

GASP results:
```
angle (rotation) = 61.282
st x (sd(x)) = 0.995
st y (sd(y)) = 1.342
ratio = 1.349 (note that this is σ(y)/σ(x))
```

Table 6.6 Locations of juvenile offenders in one part of a Cardiff housing estate.

x coordinate	y coordinate	x coordinate	y coordinate
66	93	72	81
64	90	74	82
80	95	75	81
79	90	77	88
78	92	80	88
76	92	82	77
71	82	69	80

Source: Bailey and Gatrell 1995.

Box 6.16   Handworked example: density and average area per point

Step 1 Place the data in a table as in Table 6.6.

Step 2 Calculate the area of the bounded space (see Box 6.22). For our purposes, we have used a rectangle defined by the points furthest north (95), east (82), south (77) and west (64) to produce a space to bound the points. The area is thus $(95 - 77) \times (82 - 64) = 18 \times 18 = 324$. The number of coordinate pairs is 14.

Step 3 Input values into the equations:

$$d = \frac{14}{324} = 0.043$$

$$\text{AAPP} = \frac{324}{114} = 23.142$$

Box 6.17   GASP example: density and average area per point

Input data set by loading file 'card.dat' or inserting points using edit pull-down menu.

Calculate mean centre:

Using pull-down menus:
➥ Stats
➥ Density

GASP results:
```
d = 0.043
AAPP = 23.142
```

Point data: density

We can also analyse a set of points in relation to a defined boundary to examine their pattern. **Density**

is the crudest measure which we can calculate to describe the relationship between a set of points and the area that bounds them. The density (d) is simply a measure of the number of points in a unit area and is calculated by dividing the number of points (n) by the area (A):

$$d = \frac{n}{A}$$

To calculate the area of a bounded space (e.g., a polygon) see Box 6.23. We can also calculate the average area (AAPP) within a bounded space that is occupied by a point by reversing the equation:

$$\text{AAPP} = \frac{A}{n}$$

Boxes 6.16 and 6.17 illustrate how to calculate the density of a bounded space and the average area per point by hand and using GASP. The data used (Table 6.6) refer to the locations of juvenile offenders in one part of a housing estate in Cardiff (sub-sampled from Bailey and Gatrell, 1995). Remember, the calculations really become useful only when you compare them to the results for other locations. For example, by calculating the density of offenders in other parts of the city we could determine whether the area detailed has a higher or lower density or average area per point.

Point data: nearest neighbour index

The **nearest neighbour index** describes a point pattern by calculating the mean distance to each point's nearest neighbour. By then adding a density element to the equation, we can determine whether a point pattern is *clustered* (index < 1), *random* (index = 1) or *uniform* (index > 1). The index is always between 0 and 2.15 and is calculated from:

$$NNI = \frac{d_{obs}}{d_{ran}}$$

where d_{obs} is the observed mean nearest neighbour distance (d):

$$d_{obs} = \frac{\sum\limits_{i=1}^{n} d}{n}$$

and d_{ran} is the expected mean nearest neighbour distance for a random arrangement of points and is calculated from:

$$d_{ran} = 0.5\sqrt{\frac{A}{n}}$$

Again using the data in Table 6.6 we can examine the extent to which the location of juvenile offenders is random, clustered or uniform (see Boxes 6.18 and 6.19). Our data give a nearest neighbour index of 1.15, suggesting that the home location is nearly random.

Line data: length

If we possess a set of points in space, we can construct lines between these points. The most fundamental measurement of a line is the **length** (distance) between any two points. This can be defined in a variety of ways (e.g., as a straight line distance, grid (Manhattan) distance, or sinuous distance – see Richardson, 1981; Unwin, 1982). We concern ourselves with straight line distance and sinuosity. Straight lines can consist of one or many line segments. Their length is calculated from Pythagoras' theorem applied between each set of x and y coordinates along the line. For example, to calculate the distance d between two points A and B

$$d = \sqrt{(x_A - x_B)^2 + (y_A - y_B)^2}$$

To find the length of a path consisting of many segments, the calculation is repeated for each section and summed. The sinuosity of a line is its 'wiggliness' or 'curviness'. It can be used to describe a meandering or wandering line. The sinuosity ratio (SR) is calculated from:

$$SR = \frac{\sum\limits_{i=1}^{n} d}{dir}$$

where Σd is the observed path length and dir is the direct length from origin to end.

Box 6.18   Handworked example: nearest neighbour index

Location		Nearest neighbour		Distance
x	y	x	y	
66	93	64	90	3.605
64	90	66	93	3.605
80	95	78	92	3.605
79	90	80	88	2.236
78	92	79	90	2.236
76	92	78	92	2
71	82	72	81	1.414
72	81	71	82	1.414
74	82	75	81	1.414
75	81	74	82	1.414
77	88	79	90	2.828
80	88	79	90	2.236
82	77	75	81	8.062
69	80	71	82	2.828

$\Sigma d = 38.9008$

Step 1   Place the data in a table as above.

Step 2   Plot the coordinates and calculate the distance to each location's nearest neighbour. Judging the nearest point is probably best done by eye. For our data the nearest neighbours are plotted in Figure 6.28.

Box 6.18   (cont'd)

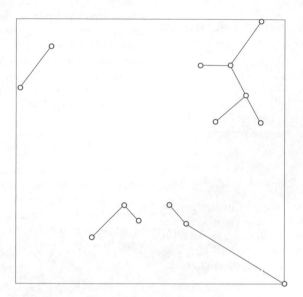

Figure 6.28   Plotting the nearest neighbours.

Step 3   Use the equation in Box 6.20 to calculate the distance between the locations.

Step 4   Sum the distances to determine $\sum d$.

Step 5   Calculate the area of the bounded space (see Box 6.22). For our purposes, we have used a rectangle defined by the points furthest north (95), east (82), south (77) and west (64) to produce a space to bound the points. The area is thus $(95 - 77) \times (82 - 64) = 18 \times 18 = 324$. The number of coordinate pairs is 14.

Step 6   Input the values into the equations:

$$d_{\text{obs}} = \frac{38.9008}{14} = 2.779$$

$$d_{\text{ran}} = 0.5 \times \sqrt{\frac{324}{14}} = 0.5 \times \sqrt{23.142} = 0.5 \times 4.810 = 2.405$$

$$\text{NNI} = \frac{2.779}{2.405} = 1.156$$

Box 6.19   GASP example: nearest neighbour index

Input data set by loading file 'point.dat' or inserting points using edit pull-down menu.

Calculate nearest neighbour:

Using pull-down menus:
➡ Stats
➡ Nearest Neighbour

GASP results:
    NNI = 1.15

The calculation of straight line distance and sinuosity ratio for data representing a student's drawing of the Swansea coastline is illustrated in Boxes 6.20 and 6.21. In the study from which the data have been taken, students were asked to draw the coastline and to locate a number of places (see Kitchin, 1996). By examining the sinuosity ratios of the coastlines drawn it was possible to explore the extent to which the students were aware of the real shape. In a sample of 37, all the student-drawn coastlines had a much lower sinuosity ratio than the 'real' coastline (1.533) demonstrating that they underestimated the degree of curvature in the bay. Figure 6.29 shows the 'real' coastline and the one used in the example.

Box 6.20   Handworked example: path length and sinuosity

Node		Next node		Distance
x	y	x	y	
3.6	1.3	3.4	1.6	0.360555
3.4	1.6	3.3	2.15	0.559017
3.3	2.15	3.4	2.6	0.460977
3.4	2.6	3.7	3.1	0.583095
3.7	3.1	4	3.55	0.540833
4	3.55	4.5	4	0.672681
4.5	4	5	4.3	0.583095
5	4.3	6	4.75	1.096586
6	4.75	7	5.1	1.059481
7	5.1	7.5	5.2	0.509902
7.5	5.2	8	5.25	0.502494
8	5.25	8.25	5.2	0.254951
8.25	5.2			

$$\Sigma d = 7.183667 \qquad dir = 6.068978$$

Step 1   Place the data in a table as above.

Step 2   Use Pythagoras' theorem to calculate the distance between nodes along the path length. For example, the distance between the first two nodes along our path length is:

$$d = \sqrt{(3.6 - 3.4)^2 + (1.3 - 1.6)^2} = \sqrt{(0.2)^2 + (0.3)^2} = \sqrt{0.04 + 0.09} = 0.360555$$

Step 3   Sum the distances to determine Σd.

Step 4   Calculate the distance from the first to the last node (dir) using Pythagoras' theorem.

Step 5   Input the values into the equation to calculate sinuosity:

$$SR = \frac{7.183667}{6.068978} = 1.184$$

Polygon data: area

It is often useful to calculate the **area** A of a polygon. The easiest way of doing this is to input the polygon coordinates into the following equation:

$$A = 0.5 \times \sum_{i=1}^{n} y_i \times (x_{i+1} - x_{i-1})$$

How to use this equation is illustrated in Boxes 6.22 and 6.23, which calculate the area of a generalised polygon representing the Isle of Man (see Table 6.7 and Figure 6.30).

Polygon data: perimeter

The easiest way to calculate the **perimeter** of a polygon is to treat the polygon segments as a path that returns to the same point. In other words, just use the same procedure as in calculating the sinuosity of a line but, instead of dividing the path length by the distance from the first to the last node, add them together.

Box 6.21   GASP example: path length and sinuosity

Input data set by loading file 'line.dat' or inserting points using edit pull-down menu.

Calculate length:

Using pull-down menus:
➥ Stats
➥ Length

Calculate sinuosity:

Using pull-down menus:
➥ Stats
➥ Sinuosity

GASP results:
```
Sin = 1.184
Length = 7.1836
Dist 1st to last = 6.0689
```

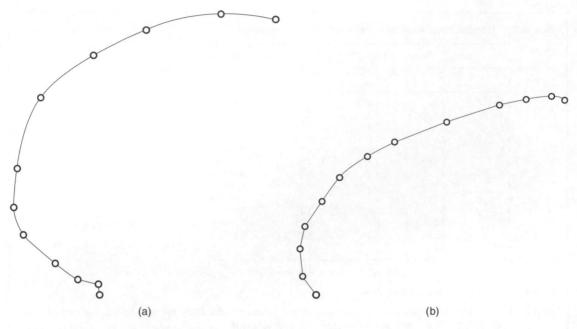

Figure 6.29 Swansea coastline: (a) actual; (b) cognised.

Table 6.7 Generalised outline of the Isle of Man

x coordinate	*y* coordinate	*x* coordinate	*y* coordinate
1.46	8.09	0	6.36
0.91	8.00	0.82	6.45
0.73	7.40	1.28	6.86
0.41	7.31	1.69	7.45
0.18	6.63	1.28	7.68

Polygon data: centre of a polygon

Similarly, the easiest way to find the centre of a polygon is to treat the polygon nodes as a set of points and then just find the mean centre.

Polygon data: shape

All polygons possess a **visual shape** and in certain instances it may be useful to compare the shape of one polygon with another. For example, to continue the theme of students' sketch maps of an area, it may be useful to compare the shape of student drawings within a group and in relation to the 'real' shape (as with sinuosity for a line). Shape, however, is a difficult concept to measure. Most measures compare an area's shape to simple, known forms such as circles, squares and triangles. The four used here are

taken from Ebdon (1985) and compare the shape of a polygon with that of a circle. As such, they examine the compactness of a polygon. The circle is the most compact shape in that it has the smallest possible perimeter relative to the area contained within it. The simplest measure merely calculates the ratio of the perimeter length (P) to the polygon area (A):

$$\text{SI} = \frac{P}{A}$$

The main problem associated with this index is that the index value can alter when the size or units of the polygon vary, even though the shape stays the same. For example, two circles of different sizes will return different shape values, as will one measured in feet and another in metres. Ebdon (1985) presents three alternatives which are a bit more robust (see Figure 6.31).

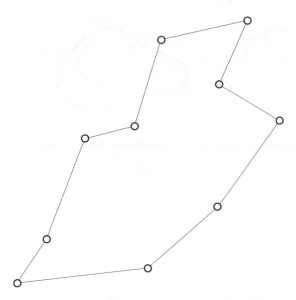

Figure 6.30 Generalised coastline of the Isle of Man.

The second index includes the longest axis (L) across the shape (without leaving the polygon) and is calculated as:

$$SI = \frac{4A}{\pi L^2}$$

The third index uses the diameter of the smallest enclosing circle (D) and is calculated as:

$$SI = \frac{4A}{\pi D^2}$$

D is calculated by finding the longest axis across the polygon regardless of whether the path of this axis leaves the polygon.

The fourth index uses the radius of the smallest enclosing circle (R_1) and the radius of the circle with the same area as the shape (R_2) to calculate the shape:

$$SI = \frac{R_1}{R_2}$$

where

$$R_1 = \sqrt{\frac{A}{\pi}}$$

and

$$R_2 = 2D$$

Again we can illustrate the use of these shape indexes (Boxes 6.24 and 6.25) on the Isle of Man data (Table 6.7).

Box 6.22 Handworked example: area

Node		
x	y	$y_i(x_{i+1} - x_{i-1})$
1.46	8.09	−2.99
0.91	8	−5.84
0.73	7.4	−3.7
0.41	7.31	−4.02
0.18	6.63	−2.71
0	6.36	4.07
0.82	6.45	8.25
1.28	6.86	5.96
1.69	7.45	0
1.28	7.68	−1.76

$$\sum y_i(x_{i+1} - x_{i-1}) = 2.74$$

Step 1 Place the data in a table as above.

Step 2 Work through the data to calculate $y_i(x_{i+1} - x_{i-1})$. x_{i+1} refers to the x coordinate for the next node and x_{i-1} to the x coordinate for the previous node. Remember the data represents a polygon, so x_{i-1} for the first node is the last coordinate pair in the list. For the first node, $y_i(x_{i-1} - x_{i-1}) = 8.09(0.91 - 1.28)$ $= 8.09 \times (-0.37) = -2.99$.

Step 3 Sum $y_i(x_{i+1} - x_{i-1})$ to determine $\sum y_i(x_{i+1} - x_{i-1})$.

Step 4 Put the values into the equation:

$$A = 0.5 \times 2.74 = 1.37$$

Box 6.23 GASP example: area

Input data set by loading file 'area.dat' or inserting a polygon using edit pull-down menu.

Calculate area:

Using pull-down menus:
➡ Stats
➡ Area

GASP results:
```
Area = 1.372
```

6.7.2 Spatial autocorrelation

We can define **spatial autocorrelation** as a measure of the similarity or interdependence of an object with surrounding objects in space. Spatial autocorrelation

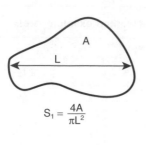

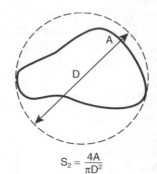

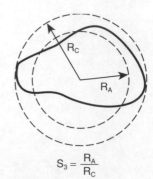

$$S_1 = \frac{4A}{\pi L^2}$$

$$S_2 = \frac{4A}{\pi D^2}$$

$$S_3 = \frac{R_A}{R_C}$$

A = Area
L = Length of longest axis
D = Diameter of smallest circumscribing circle
R_C = Radius of smallest circumscribing circle
R_A = Radius of circle with same area as shape

$$\left(= \sqrt{\frac{A}{\pi}}\right)$$

Figure 6.31   Shape indices (source: redrawn from Ebdon 1985).

Box 6.24   Handworked example: shape indexes

Step 1   Place the data in a table as in Box 6.22.

Step 2   Calculate shape index 1 by inputting area and perimeter values (see above sections):

$$SI = \frac{5.636}{1.372} = 4.108$$

Step 3   Determine the longest axis across the shape (without leaving the polygon). It is easiest to do this by identifying the nodes furthest apart and then using the length equation to determine the distance. In this case, the furthest nodes apart are 1.46, 8.09 and 0.0, 6.36 and the distance (*L*) is 2.26.

Step 4   Calculate shape index 2 by inputting values into the equation:

$$SI = \frac{4 \times 1.372}{\pi \times (2.26)^2} = \frac{5.488}{\pi \times 5.10} = \frac{5.488}{16.048} = 0.341$$

Step 5   Determine the diameter by finding the longest axis across the shape (including leaving the polygon). Again it is easiest to do this by identifying the nodes furthest apart and then using the length equation to determine the distance. In this case, *D* is the same as *L* (2.26).

Step 6   Calculate shape index 3 by inputting values into the equation:

$$SI = \frac{4 \times 1.372}{\pi \times (2.26)^2} = 0.341$$

Step 7   Calculate R_1:

$$R_1 = \sqrt{\frac{1.372}{3.1416}} = \sqrt{0.436} = 0.660$$

Step 8   Calculate $R_2 = D/2 = 2.26/2 = 1.13$.

Step 9   Input values into equation to calculate shape index 4:

$$SI = \frac{0.660}{1.13} = 0.584$$

is intuitively necessary in geographic space, since without it, the distribution of phenomena would be *independent of location*, and we could model patterns as a simple random process. The processes of spatial diffusion and spatial interaction can produce patterns in the human geographical landscape such that events or circumstances at one location are not independent of conditions at surrounding locations (Odland, 1988). However, the fact that many geographic phenomena are autocorrelated provides a problem for the classical statistics considered in Chapter 5, which assume independence of observations, for example the independence of residuals in a regression analysis. Consequently, a number of statistical techniques have been developed to measure and model patterns of spatial dependency in data, which include the **variogram** from regionalised variable theory (geostatistics), and methods of **fractal analysis**. For an overview of regionalised variable theory in the context of kriging and GIS we would recommend Burrough and McDonnell (1998), and for an extensive treatment of the applications of fractal geometry in human geography we would recommend Batty and Longley (1994). We will restrict our concern here to various simple indices of autocorrelation. The key feature of spatial autocorrelation, and the variety of related statistical methods, is that we are dealing with not only the attribute *values* of phenomena, similar to the statistical methods used in Chapter 5, but also the *locations* of these phenomena in space. We can calculate the spatial autocorrelation for interval/ratio data as well as nominal/categorical data.

Autocorrelation for interval/ratio data

Recall from Chapter 5 that for two variables X and Y, Pearson's correlation r is calculated as:

$$r = \frac{\text{covariance of variables X and Y}}{\sqrt{\left(\begin{array}{c}\text{variance of}\\ \text{variable X}\end{array}\right) \times \left(\begin{array}{c}\text{variance of}\\ \text{variable Y}\end{array}\right)}}$$

Autocorrelation is the correlation of a variable with itself: rather than two different variables X and Y, we consider the covariance between the value of a variable at one location with its value at another. Perhaps the easiest way to understand this concept is to consider the formula for the autocorrelation coefficient (also termed the serial correlation coefficient) which we might calculate for an equally spaced series of values, for example as displayed in Figure 6.32.

Such a series is often called a *time series* (if we are using data sampled as a function of time) or a *spatial*

Box 6.25 GASP example: shape indexes

Input data set by loading file 'area.dat' or inserting a polygon using edit pull-down menu.

Calculate shape:

Using pull-down menus:
  ➥ Stats
  ➥ Shape Index

GASP results:
```
Shape Index 1 = 4.108
Shape Index 2 = 0.341
Shape Index 3 = 0.340
Shape Index 4 = 0.583

Perimeter = 5.636
Area = 1.372
L = 2.262
D = 2.264
```

N.B. Because of the difficulty of computing L, GASP will only compute the correct second shape index for regular shaped polygons (ones which have no deep inlets or holes).

---

series (if we are using data sampled as a function of space). Here we are concerned only with the correlation of one variable, which we will term x_t, with itself. Since, clearly in this case, the correlation will be perfect, the resulting coefficient will not be particularly useful, therefore what we do is assign the second variable the value of x_t a certain distance away.

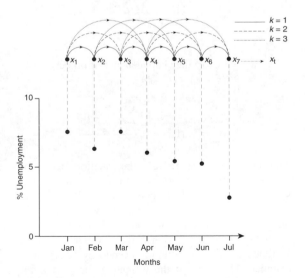

Figure 6.32 Autocorrelation lagged comparisons for a series of points.

In the context of the example in Figure 6.32, we can assemble the data into pairs of objects a certain distance away which we term the **lag distance**, which we will denote as k. For $k = 1$, we produce paired values x_1 and x_2, x_2 and x_3, etc.; for $k = 2$, we obtain x_1 and x_3, x_2 and x_4, etc. The second value of x_t in the pair is treated as the second variable. In this fashion we can modify the equation above to give the autocorrelation function:

$$r_k = \frac{\text{covariance of variables } X_t \text{ and } X_{t+k}}{\text{variance of variable } X_t}$$

The covariance in this case is termed the autocovariance. Where data are in a serial form, we can make use of this equation: a plot of r_k against k is termed a **correlogram**, which is a useful measure of the change in autocorrelation with scale. However, when $k = 1$, this equation forms the basis of the indices of autocorrelation termed **Moran's I** and **Geary's c**.

Moran's I

Moran's Index (Moran, 1948), known as Moran's I, is very similar to the equation above: we divide the covariance between the attribute (or z) values of pairs of n objects by the sample variance to give (Goodchild, 1986):

$$I = \frac{\displaystyle\sum_{i=1}^{n}\sum_{j=1}^{n} w_{ij} \times c_{ij}}{\sigma^2 \times \displaystyle\sum_{i=1}^{n}\sum_{j=1}^{n} w_{ij}}$$

where:

$$c_{ij} = \text{covariance between pair of objects}$$
$$= (z_i - \bar{z}) \times (z_j - \bar{z})$$

$$\sigma^2 = \text{variance} = \frac{\displaystyle\sum_{i=1}^{n}(z_i - \bar{z})^2}{n}$$

The only difference between the equations for I and r_k are the weights w_{ij} which we use as a measure of spatial proximity. These can be defined in various ways. For example, in a simple binary form the weight assigned is 1 if the objects share a common boundary, 0 if not. Alternatively, weights can be assigned as a more complex decreasing function of distance. The form of the weighting function is the most important step in calculating a spatial autocorrelation statistic, since different weights can be used

to represent competing hypotheses in attempts to explain an observed spatial pattern (Odland, 1988). Goodchild (1986) noted that values for Moran's I tend to be positive when the object attributes are similar (positive autocorrelation), negative when the attributes are dissimilar (negative autocorrelation), and $-1/(n-1)$ (which for large n becomes approximately zero) when there is no correlation.

Geary's c

An alternative measure of spatial autocorrelation is Geary's c (Geary, 1968), which is defined as follows:

$$c = \frac{\displaystyle\sum_{i=1}^{n}\sum_{j=1}^{n} w_{ij} \times d_{ij}}{2 \times s^2 \times \displaystyle\sum_{i=1}^{n}\sum_{j=1}^{n} w_{ij}}$$

where:

$$d_{ij} = \text{squared difference in value between pair of objects}$$
$$= (z_i - z_j)^2$$

$$s^2 = \text{sample variance} = \frac{\displaystyle\sum_{i=1}^{n}(z_i - \bar{z})^2}{n-1}$$

Here, the sum of squared differences in value between pairs of objects is used as the measure of covariation. We would expect $c < 1$ in the presence of positive autocorrelation, $c > 1$ for negative autocorrelation and $c = 1$ for an absence of correlation.

To illustrate the calculation of Moran's I and Geary's c we adopt a simplified form of the equations above (primarily to reduce the number of calculations by hand). For Moran's I, we sum the covariance over the number of pairs P of contiguous zones:

$$I = \frac{\displaystyle\sum_{i=1}^{P} c_i}{\sigma^2 \times P}$$

where the c_i are the object-pair covariances defined above. Similarly, for Geary's c:

$$c = \frac{\displaystyle\sum_{i=1}^{P} d_i}{2 \times s^2 \times P}$$

where the d_i are the squared object-pair differences defined above. The denominator in the two equations

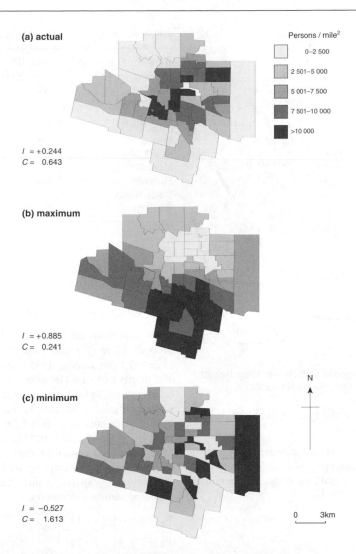

(a) actual

Persons / mile²

☐ 0–2 500

◻ 2 501–5 000

▨ 5 001–7 500

▩ 7 501–10 000

■ >10 000

I = +0.244
C = 0.643

(b) maximum

I = +0.885
C = 0.241

(c) minimum

N

I = −0.527
C = 1.613

0   3km

Figure 6.33 (a) Actual population density pattern in London, Ontario, from the 1971 Census; (b) simulated maximum positive autocorrelation pattern; (c) simulated maximum negative (minimum) autocorrelation pattern (source: redrawn from Goodchild 1986: 15).

above is half that in the full form of the equations, and P is also half the size of the sum of the weights matrix. We can best compare and contrast different values of Moran's I and Geary's c visually. The calculated values of I and c, and the corresponding choropleth of actual population density in the city of London, Ontario, for 1971, are displayed in Figure 6.33, along with the statistics and choropleths for simulated maximum positive and maximum negative patterns of autocorrelation.

As an example (see Box 6.26), we have selected a smaller data set comprising the percentage Catholic for a sample of West Belfast wards as displayed in Figure 6.34 and Table 6.8. The chosen data set is small (10 zones) in order to explain the mechanics of calculation. Any interpretation of the resulting coefficients and their significance should be made with this limitation in mind.

The significance of both Moran's I and Geary's c can be tested. One of two null hypotheses can be

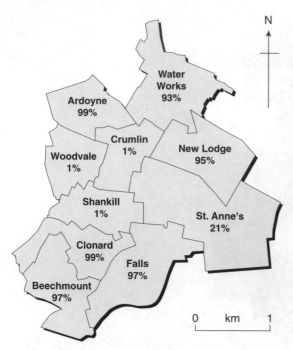

N

Table 6.8 Percentages of Catholics in West Belfast.

Ward	Ward ID	% Catholic
Ardoyne	702	99
Beechmount	707	97
Clonard	718	99
Crumlin	719	1
Falls	721	97
New Lodge	734	95
Shankill	739	1
St Annes	740	21
Water Works	747	93
Woodvale	751	1

Figure 6.34 Percentages of Catholics in West Belfast wards. © Crown Copyright. Permit No 1330.

chosen which relate to the type of sampling procedures chosen:

- *Resampling*: the observed attribute values are a result of random sampling from an infinitely large normal distribution, with each sample containing a different set of values.

- *Randomisation*: the observed attribute values are distributed around a set of spatial objects. All other possible samples would contain the same set of values, but in a different spatial configuration.

The equations for the expected mean and variance of both Moran's I and Geary's c are given below (adapted from Ebdon, 1985, and Goodchild, 1986). The number of terms in some of these equations may be offputting, but the equations require only the careful substitution and simple calculation using the quantities calculated in Box 6.26. We will follow the notation of Goodchild (1986) and use the subscript N to represent the normalisation null hypothesis, and R to represent the resampling null hypothesis. E_N/E_R represents the expected value (mean), and σ_N/σ_R the expected standard deviation.

$$E_N(I) = -1/(n-1)$$

$$E_R(I) = -1/(n-1)$$

$$E_N(c) = 1$$

$$E_R(c) = 1$$

$$\sigma_N(I) = \sqrt{\frac{n^2 P + 3P^2 - n\sum A^2}{P^2(n^2-1)}}$$

$$\sigma_R(I) = \sqrt{\frac{n\left[P(n^2 + 3 - 3n) + 3P^2 - n(\sum A^2)\right] - k\left[P(n^2 - n) + 6P^2 - 2n(\sum A^2)\right]}{P^2(n-1)(n-2)(n-3)}}$$

$$\sigma_N(c) = \sqrt{\frac{(2P + \sum A^2)(n-1) - 4P^2}{2(n+1)P^2}}$$

$$\sigma_R(c) = \sqrt{\frac{(n-1)P\left[n^2 - 3n + 3 - (n-1)k\right] - (n-1)\sum A^2\left[n^2 + 3n - 6 - (n^2 - n + 2)k\right]/4 + P^2\left[n^2 - 3 - (n-1)^2 k\right]}{n(n-2)P^2}}$$

Box 6.26   Handworked example: Moran's *I* and Geary's *c*

Ward	% Catholic	$x - \bar{x}$	$(x - \bar{x})^2$
Ardoyne	99	38.6	1489.96
Beechmount	97	36.6	1339.56
Clonard	99	38.6	1489.96
Crumlin	1	−59.4	3528.36
Falls	97	36.6	1339.56
New Lodge	95	34.6	1197.16
Shankill	1	−59.4	3528.36
St Annes	21	−39.4	1552.36
Water Works	93	32.6	1062.76
Woodvale	1	−59.4	3528.36

$\sum x = 604$ $\sum(x - \bar{x})^2 = 20056.4$ $n = 10$

$\bar{x} = 60.40$ $\sum(x - \bar{x})^2/n = 2005.64$

Step 1   List the wards (zones) as in the table above, and calculate the mean $\bar{x} = \sum x/n = 60.4\%$.

Step 2   The next stage is to calculate the variance for the zonal values. First, calculate the mean difference $x - \bar{x}$ and insert in a new column in the table above.

Step 3   Square each mean difference to obtain $(x - \bar{x})^2$ and insert as another new column above. Obtain the sum of this column (20056.4) divided by the number of values (10) which gives a variance of 2005.64.

Step 4   The next stage is to calculate the covariance between the pairs of zones. We have 18 pairs of zones ($P = 18$) in Figure 6.33.

Pair	x_i	$(x_i - \bar{x})$	x_j	$(x_j - \bar{x})$	$(x_i - \bar{x})(x_j - \bar{x})$	$x_i - x_j$	$(x_i - x_j)^2$
1	93	32.6	99	38.6	1258.36	−6	36
2	93	32.6	1	−59.4	−1936.44	92	8464
3	93	32.6	95	34.6	1127.96	−2	4
4	95	34.6	21	−39.4	−1363.24	74	5476
5	1	−59.4	95	34.6	−2055.24	−94	8836
6	99	38.6	1	−59.4	−2292.84	98	9604
7	1	−59.4	1	−59.4	3528.36	0	0
8	99	38.6	1	−59.4	−2292.84	98	9604
9	1	−59.4	21	−39.4	2340.36	−20	400
10	1	−59.4	1	−59.4	3528.36	0	0
11	1	−59.4	1	−59.4	3528.36	0	0
12	1	−59.4	21	−39.4	2340.36	−20	400
13	21	−39.4	97	36.6	−1442.04	−76	5776
14	97	36.6	1	−59.4	−2174.04	96	9216
15	1	−59.4	99	38.6	−2292.84	−98	9604
16	99	38.6	97	36.6	1412.76	2	4
17	99	38.6	97	36.6	1412.76	2	4
18	97	36.6	97	36.6	1339.56	0	0

$\sum(x_i - \bar{x})(x_j - \bar{x}) = 5967.68$ $\sum(x_i - x_j)^2 = 67428$

Step 5   List the pairs in a column of a table similar to that above. For each value in the pair, calculate the mean differences $(x_i - \bar{x})$ and $(x_j - \bar{x})$, then insert into two new columns.

Step 6   Multiply the mean difference values for each pair to give $(x_i - \bar{x}) \times (x_j - \bar{x})$, and insert into a new column.

Box 6.26   (cont'd)

Step 7   Sum the result $\sum(x_i - \bar{x})(x_j - \bar{x})$ to give 5967.68 and substitute into the following equation to obtain Moran's I:

$$I = \frac{\sum\limits_{i=1}^{P} c_i}{\sigma^2 \times P} = \frac{5967.68}{2005.64 \times 18} = \frac{5967.68}{36101.52} = 0.1653$$

Step 8   To calculate Geary's c, we need to add a further two columns to the table above. First, calculate the difference between the values of each pair $(x_i - x_j)$, and then square the result $(x_i - x_j)^2$ and sum $\sum(x_i - x_j)^2$ to give 67428.

Step 9   Geary's c uses the *sample* estimate of variance, which we can obtain by taking the sum from Step 3 and dividing by $n - 1$ to give 2228.489.

Step 10   Then, substitute into the following equation to obtain Geary's c:

$$c = \frac{\sum\limits_{i=1}^{P} d_i}{2 \times s^2 \times P} = \frac{67428}{2 \times 2228.489 \times 18} = \frac{67428}{80225.604} = 0.84048$$

All the parameters in the above equations except k are from the calculation of Moran's I and Geary's c. Parameter k is the **kurtosis**. This is calculated as follows:

$$k = \frac{\sum\limits_{i=1}^{n}(z_i - \bar{z}_i)^2 \big/ n}{\left(\sum\limits_{i=1}^{n}(z_i - \bar{z}_i)^2 \big/ n\right)^2}$$

The observed values of both I and c can then each be turned into a Z-score, and tested for significance against critical values of the Z distribution. We have chosen to illustrate the calculation of these quantities (Box 6.27) for the values of Moran's I and Geary's c for the percentage of Catholics in West Belfast obtained above in Box 6.27.

Autocorrelation for categorical/nominal data: join count statistic

This is a measure of spatial autocorrelation used for nominal data, and refers to the number of adjacent objects on a map (Goodchild, 1986). The simplest example is a binary map, i.e., comprised of two classes which we can term **black objects** (B) and **white objects** (W). We can obtain three possible adjacent combinations of these objects: BB, BW and

WW. To obtain the observed counts for each of these, it is simply a matter of counting the number of BB, BW and WW joins. Mathematically we can define these calculations as follows (from Odland, 1988):

$$BB = \frac{1}{2}\sum_{i=1}^{n}\sum_{j=1}^{n} w_{ij} z_i z_j$$

$$BW = \frac{1}{2}\sum_{i=1}^{n}\sum_{j=1}^{n} w_{ij} z_i z_j$$

where the binary variable z_i is assigned to each region, such that $z_i = 1$ if a region is black, and $z_i = 0$ if a region is white, where w_{ij} is the weighting function. The WW statistic can be obtained by reversing the assignments such that $z_i = 0$ if a region is black and $z_i = 1$ if a region is white.

Similar to the autocorrelation statistics for interval and ratio data considered above, we can test the significance of the number of joins. We will restrict our attention to the significance of BW joins using the equations below, where $E_{BW(N)}$ and $E_{BW(R)}$ represent the expected value (mean) and $\sigma_{BW(N)}$ and $\sigma_{BW(R)}$ the expected standard deviation for the resampling and normalisation null hypotheses respectively. You should consult Goodchild (1986) for details of the calculation of the significance of BB and WW joins.

Box 6.27   Handworked example: significance of Moran's *I* and Geary's *c*

Here we use data from Table 6.7 and Box 6.26.

Step 1   Substitute into the equations below:

$$E_N(I) = -1/(10 - 1) = -0.111$$

$$E_R(I) = -1/(10 - 1) = -0.111$$

$$E_N(c) = 1$$

$$E_R(c) = 1$$

$$\sigma_N(I) = \sqrt{\frac{(10^2 \times 18) + (3 \times 18^2) - (10 \times 137)}{18^2 \times (10^2 - 1)} - \frac{1}{81}} = \sqrt{\frac{1402}{32076} - \frac{1}{81}} = 0.177096$$

$$\sigma_R(I) = \sqrt{\frac{10[18(10^2 + 3 - (3 \times 10)) + 3 \times 18^2 - 10 \times 137] - 1.25[18(10^2 - 10) + 6 \times 18^2 - 2 \times 10 \times 137]}{18^2(10 - 1)(10 - 2)(10 - 3)} - \frac{1}{81}}$$

$$= \sqrt{\frac{8130}{163296} - \frac{1}{81}} = 0.193497$$

$$\sigma_N(c) = \sqrt{\frac{(2 \times 18 + 137)(10 - 1) - 4 \times 18^2}{2(10 + 1)18^2}} = \sqrt{\frac{261}{7128}} = \sqrt{0.0366} = 0.191353$$

$$\sigma_R(c) = \sqrt{\frac{(10 - 1)18[10^2 - (3 \times 10) + 3 - (10 - 1)1.25] - (10 - 1) \times 137 \times}{[10^2 + (3 \times 10) - 6 - (10^2 - 10 + 2)1.25]/4 + 18^2[10^2 - 3 - (10 - 1)^2 1.25]}{10(10 - 2)18^2}}$$

$$= \sqrt{\frac{5852.25}{25920}} = \sqrt{0.2258} = 0.475164$$

Step 2   Then calculate the *Z*-scores:

N hypothesis Moran's *I*: $Z = (0.1653 + 0.111) / 0.177096 = 1.56$
R hypothesis Moran's *I*: $Z = (0.1653 + 0.111) / 0.193497 = 1.428$
N hypothesis Geary's *c*: $Z = (0.84048 - 1) / 0.191353 = -0.834$
R hypothesis Geary's *c*: $Z = (0.84048 - 1) / 0.475164 = -0.335$

Step 3   At a 0.05 significance level, the two-tailed critical value of *Z* is 1.96, therefore the null hypothesis cannot be rejected, and our arrangement is not significantly different from a random pattern, which we might expect with such a small set of zones.

$$E_{BW(N)} = 2Jp_B p_W$$

$$E_{BW(R)} = 2Jn_B n_W / n(n - 1)$$

$$\sigma_{BW(N)} = \sqrt{\left[2J + \sum A(A - 1)\right]p_B p_W - 4\left[J + \sum A(A - 1)\right]p_B^2 p_W^2}$$

$$\sigma_{BW(R)} = \sqrt{E_{BW(R)} + \frac{\sum A(A - 1)n_B n_W}{n(n - 1)} + \frac{4\left[J(J - 1) - \sum A(A - 1)\right]n_B(n_B - 1)n_W(n_W - 1)}{n(n - 1)(n - 2)(n - 3)} \quad E_{BW(R)}^2}$$

Box 6.28   Handworked example: join count statistics

Ward	% Catholic	Predominance	A	$A(A-1)$
Ardoyne	99	C	2	2
Beechmount	97	C	6	30
Clonard	99	C	3	6
Crumlin	1	P	3	6
Falls	97	C	4	12
New Lodge	95	C	4	12
Shankill	1	P	5	20
St Annes	21	P	2	2
Water Works	93	C	3	6
Woodvale	1	P	3	6

$$\sum A(A-1) = 102$$

Step 1   Count the respective number of PP, CP and CC zones from Figure 6.32. In this case PP = 5, CP = 8 and CC = 5, and the total number of paired zones $J = 18$.

Step 2   For each zone, obtain the count of contiguous zones and enter in a new column in the table above. Then obtain $A(A-1)$ for each zone, and sum to give $\sum A(A-1) = 102$.

Step 3   Substitute into the equations below, where p_p = probability of a Protestant zone ($n_p/n = 4/10 = 0.4$) and p_c is the probability of a Catholic zone ($n_c/n = 6/10 = 0.6$).

$$E_{CP(N)} = 2Jp_c p_p$$

$$E_{CP(R)} = 2Jn_c n_p/n(n-1)$$

$$\sigma_{CP(N)} = \sqrt{[2J + \sum A(A-1)]p_c p_p - 4[J + \sum A(A-1)]p_c^2 p_p^2}$$

$$\sigma_{CP(R)} = \sqrt{E_{CP(R)} + \frac{\sum A(A-1)n_c n_p}{n(n-1)} + \frac{4[J(J-1) - \sum A(A-1)]n_c(n_c-1)n_p(n_p-1)}{n(n-1)(n-2)(n-3)} - E_{CP(R)}^2}$$

Substituting in:

$$E_{CP(N)} = 2 \times 18 \times 0.6 \times 0.4 = 8.64$$

$$E_{CP(R)} = 2 \times 18 \times 6 \times 4/10(10-1) = 9.6$$

$$\sigma_{CP(N)} = \sqrt{[2 \times 18 + 102] \times 0.6 \times 0.4 - 4[18 + 102] \times 0.6^2 \times 0.4^2} = \sqrt{5.472} = 2.339$$

$$\sigma_{CP(R)} = \sqrt{9.6 + \frac{102 \times 6 \times 4}{10(10-1)} + \frac{4[18(18-1) - 102] \times 6(6-1)4(4-1)}{10(10-1)(10-2)(10-3)} - 9.6^2}$$

$$= \sqrt{2.925714} = 1.7105$$

Step 4   Calculate the Z-scores:

N hypothesis Catholic–Protestant joins $Z = (8 - 8.64)/2.339 = -0.27359$
R hypothesis Catholic–Protestant joins $Z = (8 - 9.6)/1.7105 = -0.93541$

Step 5   At a 0.05 significance level, the two-tailed critical value of Z is 1.96, therefore the null hypothesis cannot be rejected, and our arrangement is not significantly different from a random pattern, which we might expect with such a small set of zones.

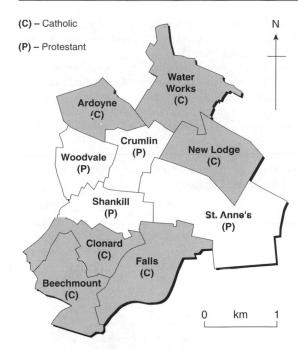

(C) – Catholic

(P) – Protestant

N

Ardoyne (C)

Water Works (C)

Crumlin (P)

Woodvale (P)

New Lodge (C)

Shankill (P)

St. Anne's (P)

Clonard (C)

Falls (C)

Beechmount (C)

0   km   1

Figure 6.35 Binary classification of Catholic and Protestant wards in West Belfast. © Crown Copyright. Permit No 1330.

where p_W = probability that a cell is white, p_B = probability that a cell is black, J = number of joins, A = number of joins between a zone and its contiguous zones, and n_B and n_W are the respective number of black and white zones. Although the probabilities p_W and p_B should be assigned based on what we know about the occurrence of the zones, we can make use of the empirical data and obtain p_W and p_B based on the relative number of white and black zones. The results can be tested for significance similar to Moran's I and Geary's c above.

To illustrate this technique (Box 6.28), we will re-express the zones in Figure 6.34 as either predominantly Protestant (P) or predominantly Catholic (C) to give Figure 6.35.

6.8 Summary

After reading this chapter you should:

- understand the power of maps as media of communication;
- understand the basic functionality of a GIS;
- understand some of the conceptual issues to be conscious of when using spatial data;
- be able to obtain secondary spatial data and assess their merits and limitations;
- be able to perform analysis using basic spatial statistics.

In this chapter we have introduced spatial data and explored some of the avenues open for spatial analysis. Maps and the presentation of spatial data can provide a powerful medium of communication. Spatial analysis builds upon this visual power to provide statistical support. However, like the aspatial data we have discussed in the previous three chapters, spatial data need to be pre-processed and checked before use. There are also a number of conceptual issues, such as the Modifiable Areal Unit Problem, that you should keep in mind when undertaking your analysis. Again, the watchword here is care – you should take care at all stages of your analysis to ensure a valid piece of research. In relation to our treatment of GIS, we have limited our discussion to basic functions and conceptual issues that are important to consider whichever package you end up using. If you wish to undertake analysis using GIS you should refer to one of the books detailed in the Further reading section of this chapter. Similarly, we have only introduced basic, descriptive spatial statistics and for a fuller treatment you are referred to more specialised texts such as that by Bailey and Gatrell (1995). In the next chapter, we turn our attention away from quantitative data to examine how to generate qualitative data.

6.9 Questions for reflection

- *Why is an understanding of map distortion important in the context of GIS?*
- *What types of data are stored in a GIS?*
- *Can a GIS be used for data transformation, and if so how?*
- *Is the representation of socio-economic information in a zonal form problematic for analysis, and if so why?*
- *What is the difference between 'inherent error' and 'operational error' in a GIS context?*
- *What options do you have for the organisation of spatial data for input to a GIS?*
- *Are geographic phenomena autocorrelated?*

Further reading

Bailey, T. and Gatrell, A.C. (1995) *Interactive Spatial Data Analysis*. Longman Scientific and Technical, Harlow.

Burrough, P.A. and McDonnell, R. (1998) *Principles of Geographical Information Systems*. Oxford University Press, Oxford.

Goodchild, M.F. (1986) *Spatial Autocorrelation*. Concepts and Techniques in Modern Geography 47, Geo Books, Norwich.

Longley, P., Goodchild, M.F., Maguire, D. and Rhind, D. (1999) *Geographical Information Systems: Principles, Techniques, Management, Applications*. John Wiley, New York.

Martin, D. (1996) *Geographic Information Systems: Socioeconomic Applications*. Routledge, London.

Monmonier, M.S. (1996) *How to Lie with Maps*, 2nd edition. University of Chicago Press, Chicago.

Robinson, A.H., Morrison, J.L., Muehrcke, P.C., Kimerling, A.J. and Guptil, S.C. (1995) *Elements of Cartography*. John Wiley, New York.

Unwin, D. (1982) *Introductory Spatial Analysis*. Methuen, London.

CHAPTER 7

Producing data for qualitative analysis

This chapter covers

7.1 Introduction
7.2 Qualitative approaches
7.3 Primary data production
7.4 Specific approaches to producing qualitative data
7.5 Secondary sources of qualitative data
7.6 Summary
7.7 Questions for reflection

7.1 Introduction

As we discussed in Chapter 2, choosing a research method is not just a case of picking the one that seems the easiest but picking the most appropriate relative to the knowledge you require. Qualitative data consist of words, pictures and sounds and are usually unstructured in nature. As such, qualitative data are not easily converted into a numeric format and, in general, need to be analysed using a different set of techniques from quantitative data. Studies which utilise, generate, analyse and interpret qualitative data are quite complex to design. You should be under no illusions – the production and analysis of qualitative data is *no easier* than for quantitative data. Admittedly, data produced using qualitative techniques are generally not analysed using dreaded statistics. However, qualitative studies require just as much planning as quantitative studies and in many ways are more demanding. It is therefore important that you plan your project carefully (Chapter 2) and read this chapter and Chapter 8 before you try to undertake a project which utilises qualitative data. This chapter is designed to provide you with a knowledge of how to produce qualitative data, the different approaches concerning the use of qualitative techniques, and appropriate sources of secondary data.

7.2 Qualitative approaches

Just as there are many different schools of thought concerning how research in human geography should be conducted (see Section 1.4), there are a variety of opinions concerning how qualitative research should be undertaken. As such, there is more to undertaking qualitative-based data production than just conducting interviews, observing people, or undertaking secondary analysis of archival sources. There are also a multitude of ways in which to approach these tasks, depending upon your perspective and purpose. Tesch (1990), for example, identifies a multiplicity of labels that researchers in the social sciences have used to define their qualitative research (Table 7.1). She suggests that the main problem associated with these labels is that each is defined in a different way. For example, some labels refer to the perspective that a researcher adopts (e.g., naturalistic, clinical), others to the field in which they are based (e.g., phenomenology), others to the research approach used (e.g., case study, discourse analysis) and some to the data type, method or research location (e.g., field research, participant observation). To go through each of these labels and approaches would take a whole book in itself and might be a fruitless exercise when so many of them overlap in nature, or are synonyms for others,

Table 7.1 Different approaches to qualitative research.

Action research	Focus group research
Case study	Grounded theory
Clinical research	Hermeneutics
Cognitive anthropology	Heuristic research
Collaborative inquiry	Holistic ethnography
Content analysis	Imaginal psychology
Conversational analysis	Intensive research
Delphi study	Interpretative evaluation
Descriptive research	Interpretative interactionalism
Dialogical research	Interpretative human studies
Direct research	Life history studies
Discourse analysis	Naturalistic enquiry
Document study	Oral history
Ecological psychology	Panel research
Educational connoisseurship and criticism	Participant observation
Educational ethnography	Participant research
Ethnographic content analysis	Phenomenography
Ethnography	Phenomenology
Ethnography of communication	Qualitative evaluation
Ethnomethodology	Structural ethnography
Ethnoscience	Symbolic interactionalism
Experiential psychology	Transcendental realism
Field study	Transformative research

Source: Tesch 1990: 58.

and have more relevance to other social sciences than to human geography. The fact that all these perspectives have been grouped together as qualitative in nature suggests that they share common attributes. Tesch (1990) herself divided these labels into three more manageable groupings: language-orientated approaches concerned with the use of language; descriptive/interpretative approaches concerned with experience and meaning; and theory-building approaches concerned with identifying the connections between phenomena.

Tesch's classification is by no means the only attempt to pigeon-hole different studies. Other researchers have also tried to classify qualitative research. This book, however, is not the place to become enmeshed in trying to explain the differences between all the approaches to qualitative research. Needless to say, the differences are often tied to wider theoretical debates, as discussed in Chapter 1, and to specific areas of research (see Cresswell, 1998, for an extended discussion on five different qualitative approaches at various stages of the research process). In our discussion, we therefore concentrate on the qualitative techniques themselves (interviewing and observation), how to undertake qualitative analysis in relation to human geography, and three different approaches to qualitative analysis which have gained some currency

in geographical enquiry (ethnography, action research and case studies).

7.3   Primary data production

When designing your study there are two important considerations: first, '*How am I going to produce data?*', and second, '*How am I going to approach data production?*'. Researchers can produce qualitative data from primary sources in a number of different ways. For the purpose of our discussion we have classified qualitative techniques into two generic classes: **interviewing** and **observation**. Within the accounts detailed below, interviewing and observation are considered from a traditional scientific perspective. That is, it is assumed that the researcher is an objective scientist producing data in a neutral fashion and for no other purpose than to increase understanding of a particular phenomenon. We have chosen to frame our account in such a fashion for ease of description. Later in this chapter (Section 7.4), three other approaches to interviewing and observation are detailed. These are **ethnography**, where the researcher attempts to understand the world as seen through the eyes of the participants,

action research, where the researcher is seeking, along with the participants, to explicitly alter a situation, and the use of **case studies**, where research is directed towards specific cases in real-life settings.

7.3.1 Interviewing

The interview is probably the most commonly used **qualitative technique**. It allows the researcher to produce a rich and varied data set in a less formal setting. Like its cousin the questionnaire (see Chapter 3), the interview comes in many forms, ranging from the highly structured to the completely unstructured. The interview differs from the questionnaire in the nature of its questions and its manner of presentation. While the questionnaire is useful for asking very specific questions concerning quantifiable information such as age, income and sex or for converting general information into a closed form through rating or ranking, the interview allows a more thorough examination of **experiences**, **feelings** or **opinions** that closed questions could never hope to capture. Dey (1993) characterises this difference such that questionnaires concern numbers or facts and interviews concern meanings or beliefs, although this difference is not mutually exclusive. The interview is also more informal in nature than the questionnaire and cannot be self-administered. While a questionnaire has a very formal question-then-answer structure, interviews are often described as entering and maintaining a conversation. This is not to deny that some interviews do follow a highly structured path, but the nature of this path does differ in the nature of presentation. Oppenheim (1992) suggests that the interview is really a precursor to a larger questionnaire survey, with the interview providing the basis for the closed-ended questions on the questionnaire. Used together, the interview provides a pilot for formulating relevant questions and the questionnaire ensures a larger sample size and data that can be analysed quantitatively. While interviews and questionnaires can be used in this way, many researchers would disagree with such a strategy for two reasons. First, some would argue that an interview alone will provide a sound data set. Second, some would argue that the closed questions on a questionnaire filter out meaningful information. As such, they would contend that questionnaire surveys are fundamentally flawed as it is meaningless or pointless to try and measure and analyse experiences, opinions and attitudes in such a manner.

Interviews can provide rich sources of data on people's experiences, opinions, aspirations and feelings. There is, however, more to interviewing than simply asking a participant questions. Interviewing can be a complex social encounter. As a researcher, before you undertake any interviews, you need to understand the dynamics of interviewing and the various different interviewing strategies, and be aware of both the strengths and limitations of interviewing.

Types of interview

There have been various different classifications of interviews. We will use a variation of Patton's (1990) classification. Patton identifies four different interview strategies:

- The *closed quantitative interview*
- The *structured open-ended interview*
- The *interview guide approach*
- The *informal conversational interview*
- To this we will add *group discussion*

These categories are not mutually exclusive and two or more interviewing strategies can be used during the same interviewing session depending on what sort of information is sought. For example, we might combine a closed quantitative interview designed to gain background facts with an interview guide approach designed to elicit a greater depth of information about specific topics. The differences between these categories essentially concern the degree to which the interviewer controls the conversation and the degree of standardisation in the questions asked. Effectively, the **closed quantitative interview** is discussed in Chapter 3 and is essentially a verbal questionnaire. All the questions are predetermined and the questions closed so that respondents have only limited choices for a response.

Structured open-ended interview

Within a structured open-ended interview the conversation is highly controlled by the interviewer. Like the closed quantitative interviews, the questions are also highly structured and standardised. However, rather than consisting of closed questions which would transform this type of interview into a questionnaire, a series of **open-ended questions** are asked. Open-ended questions mean that the interviewee's responses are not constrained to categories provided by the interviewer; respondents can give whatever answer they wish. It is hoped that open-ended questions better reflect a person's own thinking. Using this strategy all the interviewees are asked the same basic questions

---

Box 7.1   An example of a structured open-ended interview

Preamble

I am interested in problems of access for disabled people in urban environments. I am going to ask you a series of questions relating to this topic. Your answers to these questions will be treated in strictest confidence.

Questions

1   If you have experienced difficulty gaining access to buildings, can you please describe some situations?
2   Similarly, if you have experienced access problems in a public space, can you please describe the various problems encountered?
3   If you have experienced access difficulties accessing public transport, please can you describe the difficulties experienced?
4   Can you think of any short-term solutions to access problems within the urban environment?
5   What do you think long-term solutions to access problems might consist of?

Box 7.2   An example of an interview guide approach

Preamble

I am interested in problems of access for disabled people in urban environments. I am not going to ask you specific questions but rather just want to discuss with you some general topics. Your answers to these questions will be treated in strictest confidence.

Topics to be covered (for the eyes of the interviewer only)
1   Access to buildings.
2   Access to public spaces.
3   Access to transport.
4   Short-term solutions?
5   Long-term solutions?

---

in the same order. The exact wording and sequence of the questions is determined before any of the interviews are conducted. This structured strategy is meant to try and increase the comparability of responses and ensure responses to all questions for every interviewee. It is also thought that this strategy might reduce interviewer effects and biases introduced through free conversation. Furthermore, this structured approach provides a 'natural' basis of organisation for analysis of data. However, this strategy does have its weaknesses. Because the interview is so highly structured and standardised, it allows little flexibility in relating the interviews to particular individuals or circumstances (it removes individuality) and may also constrain and limit the naturalness and relevance of questions and answers (the question might not be relevant to an interviewee but requires a response, and a particularly interesting response cannot be followed up in more detail). Box 7.1 provides an example of a structured open-ended interview (note how this differs from the questionnaire questions in Box 3.4). The respondents for the study outlined in Box 7.1 are known to have experienced problems of disabled access.

Interview guide approach

An **interview guide approach** is less structured than that taken in a standardised open-ended interview.

Topics and the issues to be covered are specified in advance in an outline form but the interviewer can vary the wording of the questions and the sequence in which the questions are tackled. As a result, the interviewer has much greater freedom to explore specific avenues of enquiry, and logical gaps within the data can be anticipated and closed. The interview also takes on a more conversational feel while ensuring that all the topics of interest are explored. Because the interview is more **free-form** there is, however, the possibility that specific topics may be inadvertently omitted. Furthermore, because of the flexibility in sequencing and wording, the questions posed to interviewees may vary, thus reducing the comparability of the responses. Interview guide approaches therefore require the interviewer to have the ability to keep the conversation based around specific topics, within a more informal interview style, and not to let the conversation take off on wild tangents. Box 7.2 provides an example of a interview guide approach.

Informal conversational interview

An **informal conversational interview** is generally considered to lack any formal structure. The questions the interviewer asks are meant to emerge from the immediate context of the conversation and are asked in the natural course of a discussion. Similarly, there is meant to be no predetermination of question topics or wording. With little or no direction from the interviewer the respondents are encouraged to relate their experiences, describe events that are significant to them, and reveal their attitudes and opinions as they see fit. The great strength of such an approach is that the interviewees can talk about any issue in

Box 7.3   An example of an informal conversational interview

Preamble

I am interested in problems of access for disabled people in urban environments. I would like to discuss this subject with you in detail. Your answers to these questions will be treated in strictest confidence.

No predetermined agenda (questions are asked in the context of the conversation).

any way they feel, thus challenging the preconceptions of the researcher. The unstructured format allows respondents to talk about a topic within their own 'frame of reference' and thus provides a greater understanding of the interviewees' point of view. Thus, this method is said to increase the salience and relevance of questions as they emerge within the natural flow of the conversation. Furthermore, the interviewer is also given a great deal of freedom to probe various areas and to raise specific queries during the course of the interview. Informal interviews can vary in nature from discussing specific topics to constructing life history or biographical accounts (see Cresswell, 1998, for discussion). While this method of interviewing does provide a very detailed and rich source of information, the data produced can vary substantially from one respondent to the next. Further, some aspects that may be of interest to the respondent might not arise naturally. As a result, comparability across interviewees is more difficult. Data produced in this fashion can also be difficult to organise and analyse. While an informal conversational interview produces the 'richest' source of data, it also demands a relatively high degree of interviewer skill. You must be able to keep the conversation flowing naturally and show a strong set of interpersonal skills to try to elicit deeper insights into the interviewees' thoughts or experiences. Box 7.3 provides an example of an informal conversational interview format.

Group interview

A group discussion can sometimes be a useful alternative or supplement to one-to-one interviews. A group discussion generally consists of a set of three to ten individuals discussing a particular topic under the guidance of a **moderator** who promotes interaction and directs the conversation. The dynamics of a group discussion often bring out feelings and experiences that might not have been articulated in a one-to-one interview. For example, in a group discussion workers may be more likely to express grievances against management, as they feel 'safer' within a collective environment. The group dynamics might also work in a negative way as well, with some participants reluctant to voice an opinion through shyness or fear of embarrassment. As a result, a group discussion might produce different perspectives to one-to-one interviews concerning the same issues. The challenge for the interviewer is to keep the conversation flowing and to try to involve all members of the group. The selection of people for the group is also crucial. The individuals chosen should generally be from the same background and have the same characteristics. For example, there is no point in trying to run a group discussion on racial abuse on a housing estate with members of both the abused and (potential) abusers present, as both may be reluctant to speak freely in each other's company.

Conducting interviews

Robson (1993) suggests that the interview is a commonly used method because it seems a relatively straightforward and non-problematic way of finding things out. However, from the discussion so far it should be clear that *there is more to interviewing than just talking to people*. An interview is a complex social interaction in which you are trying to learn about a person's experiences or thoughts on a specific topic. You must remember that the interview is not just a passive means of gathering information but is also a social encounter. As with all social encounters the interview is rule-guided with both parties bringing expectations about the content and role they may play. Interviewing requires a high level of interpersonal skills such as putting the interviewee at ease, asking questions in an interesting manner, an ability to listen to the responses and act accordingly, recording the responses without upsetting the conversational flow, and giving support without introducing bias (Oppenheim, 1992). In general, you must also be able to try to balance the establishment and maintenance of a **rapport** with the interviewee, so that a trusting relationship is developed, while maintaining a neutral position about the topic under discussion so that you can be objective in your analysis. You should be aware that the adoption of neutrality depends upon your position as a researcher. If you subscribe to a feminist approach you may reject the concept of researcher neutrality and try to adopt a more emancipatory or empowering approach (see the section on critically appraising interviewing below). Without these skills the data set you produce may be underdeveloped or lacking in depth because the respondent may be unwilling to impart information to you.

Compounding the interview process are a series of personal characteristics such as age, skin colour and accent which may affect how the interviewee will react or 'take' to you. Clearly there is little you can do to alter these characteristics. You can, however, dress accordingly and by treating the respondent with respect try to gain their confidence. In other words, you must dress and act in an appropriate manner to the situation and interviewee. An interviewer who turns up at an 'eco-warrior' or new-age traveller camp wearing a suit and tie is not likely to get very forthcoming responses to their questions. Remember to try to listen more than you speak and to look as if you are enjoying the conversation. If you look bored the interviewee is likely to be less forthcoming with their views.

Choosing an interview style

The first decision you have to make is which style of interview you are going to use. As discussed, all have their advantages and limitations. One of the main considerations at this point might concern your **interviewing ability**. To be able to conduct informal conversational interviews requires a great deal of interviewing skill. You must be able to keep a conversation flowing while probing the interviewee for relevant information. If the interview concerns a sensitive issue, using an informal conversational or interview guide approach means that there is more scope for causing offence through a badly phrased question. These approaches also require you to listen carefully and be able to think quickly 'on the go'. If you are a novice interviewer it may be wise to start with a standardised open-ended interview or an interview guide approach as you develop your interviewing technique. If time is of the essence then a standardised open-ended interview may be a more suitable interview method to use, as the data are more formally organised and easier to analyse. Remember, interviews are **time consuming** to undertake and even more time consuming to analyse.

Choosing an interview medium

The second main concern relates to the medium in which you are going to conduct the interviews. There are three basic choices. The most common interview medium is a **face-to-face meeting**. This sort of meeting has distinct advantages, not least of which is that it is more personal in nature and that you can more easily gauge the interviewee's reaction to a specific topic through their body language and facial expression. A second option might be to conduct the interview using a **telephone**. Telephone interviews might be a viable option if it is difficult or expensive to meet face-to-face. Alternatively you might require a random sample of the general public which may be drawn from a phonebook, or speed might be the essence (telephone interviews take less time than face-to-face because there is no travelling). While less personal in nature, a telephone interview does allow you to pick up voice inflections. A final option may be to conduct an interview via **e-mail** or **Internet Relay Chat**. The Internet's growing use has led to several studies which have utilised it as a medium for research. For example, Correll (1995) conducted several interviews with users of an online lesbian 'café' via the Internet. In this case the medium was particularly useful, firstly because the respondents were widely scattered across the United States, and secondly because some of the interviewees were initially reluctant to be formally identified: the Internet did allow them a degree of anonymity. Many studies of online interaction have found that this anonymity leads people to express themselves more openly and honestly (see Turkle, 1996).

Recruitment

Interviewing requires quite a **large commitment** from the interviewee, as they are not only giving up their time but also imparting significantly more information than they might otherwise do in another medium (e.g., questionnaire). This information is more likely to be personal in nature and may be sensitive. Therefore, you need to be able to persuade people to take part in your research project. This can be achieved in two ways. First, you need respondents to feel that taking part will be pleasant and satisfying. If you are dealing with a particularly sensitive topic that may arouse deep emotions, this might be more difficult. Second, you need interviewees to feel that your study is worthwhile. Convincing potential respondents that a student project is worth contributing towards your project mark might be quite difficult. Our advice is to prepare in advance a summary of why you have chosen that research topic. These two methods of persuasion can be achieved by formally **briefing** respondents when first contacted. This briefing should consist of four parts:

1 Tell the respondent who you are and who you represent.
2 Tell the respondent what you are doing in a way that will stimulate their interest.
3 Tell the respondent how he or she was chosen.
4 Try to create a rapport between yourself and the respondent.

In terms of finding interviewees to take part in the study, the **sampling strategies** discussed in Section 3.4 are still applicable. However, some might have more relevance than others. In general, respondents are not drawn from the whole population but from very specific groups of people. In these cases, individuals might be contacted through institutions (sampling strategies still apply) or through snowballing.

Asking questions

Whichever type of interview you decide to undertake, through whichever medium, how you phrase your questions is of critical importance. 'Good' interviews are those which ask 'good' questions. A good question is generally one which is clear, concise and easy to understand. The questions throughout the interview should vary in format to try to keep the conversation flowing and minimise boredom. For example, in Box 7.1 we could have phrased the first three questions in the same manner, just changing the last few words. Instead, we have tried to vary how these essentially similar questions have been phrased. Questions which start with *Who? Why? What? Where? When? How?* help to establish the basic framework. Mikkelsen (1995), however, suggests that *Why?* questions be used sparingly because they can put the interviewee on the defensive. Also, care must be taken to ensure that questions are open-ended in format to ensure explanation rather than 'yes' or 'no' answers. The more specific a question to the interviewee's situation, the 'better' the information that can be expected. In other words, try to **personalise** the questions. For example, rather than ask 'what motivates people to move house?', ask 'what motivated *you* to move house?'.

In an informal conversation or an interview guide approach, **probing questions** (cross-checks) are also important to establish the depth of feeling and validity of a statement (that is, whether the opinion tallies both times). This should be done by further exploring answers (e.g., *'In your answer you mentioned . . . ; do you mean . . . or were you referring to . . . ?'*; *'Could you go over that again?'*; *'Anything more?'*) and should not amount to a cross examination that will put the interviewee ill at ease. Remember, you are not

trying to interrogate your participants. Probe questions might also be used to get the respondent to elaborate further on a specific subject. For example, in the following passage the interviewer has repeated a phrase in order to try to get the interviewee to elaborate further on a point. The probe has resulted in a great deal more information.

INTERVIEWEE: The main problem I encounter is steps. Being in a wheelchair basically means that anywhere that has steps at the entrance is inaccessible. Public buildings are the worst.

INTERVIEWER: Public buildings are the worst?

INTERVIEWEE: You know, churches, town halls, sports centres and so on . . . I mean, they are meant to be built so that everyone can get in them. But they aren't. Most of them are just little oases for the able-bodied. In my local sports centre they haven't got a lift so I can't visit the first floor without somebody carrying me up there. The weight-room, where I want to go to train, is up there. I've tried complaining but it's the same old story – no money. What can you do?

Another way to get an interviewee to explore a situation might be use **prompts**. Here, you offer the interviewee a few more alternatives and ask them to comment on them. For example, you might ask: 'You mentioned the rate of pay as being important, how about other factors such as work conditions or work hours?'. When using prompts, however, you must be careful not to bias the questions. Refer to Box 3.5 for a general guide to asking questions; Box 7.4 lists common interviewing mistakes.

Box 7.4 Common interviewing mistakes

- Failing to listen carefully.
- Repeating the questions.
- Helping the interviewee to give an answer (e.g., 'You mean like when . . .').
- Asking vague questions (e.g., 'What's your opinion on research?').
- Asking insensitive questions (e.g., 'It must be really awful being disabled?').
- Failing to judge an answer (e.g., asking a question that is out of context given the previous answer).
- Failing to explore an interesting answer (e.g., leaving an interesting topic to move on to the next topic).
- Asking leading questions (e.g., 'Do you not think that . . .', or 'Is it not likely that . . .').
- Letting the interview go on too long.
- Boring the interviewee.
- Failing to adequately record the interview.

Source: Adapted from Mikkelsen 1995.

Recording the interview

It is essential that the interview is **adequately recorded**. Failure to adequately capture the interview discussion will lead to problems of analysis and weakens the validity of the study (remember that full capture might be difficult in some situations). There are a number of ways that interviews might be recorded depending upon the nature of the interview. A face-to-face interview can be recorded by either jotting down comprehensive notes, tape recording the discussion, or video filming the interview. If the person is deaf you may need to video their hands to transcribe the signing at a later date or get a third person to transcribe the interview for you if you are also signing. Telephone interviews can be recorded by taping the discussion or making notes. The Internet interview can also be recorded by capturing the written dialogue to a file or by saving the e-mails. Each medium of recording has its merits and limitations.

Audio recording of an interview allows you to accurately record an interview word-for-word with a minimum amount of effort. Recording the interview does allow you to concentrate fully upon the discussion rather than trying to balance conversation and note-taking. While it does provide a rich data set, you should be aware that there are problems associated with its use. For example, Hester Parr (in press), in her study of the geographical worlds of people who are mentally ill, illustrates that taping interviews is not always an easy process:

Taping an interview was also problematic in my study as some interviewees often experienced states of mind which incorporated paranoid thoughts about being monitored, recorded and bugged. The geography researcher who uses recording equipment in this context has to negotiate and be sensitive to these meanings. At times it was appropriate that I abandoned the tape, abandoned my quest for 'order', and was content to hold halting conversations that could take hours, over several cups of tea and periodic silences. This of course does not apply to all people with mental health problems and not to all of the people who I interviewed, and some were happy to speak with the tape recorder running, be that in short and sporadic bursts.

Some interviewees may be uncomfortable knowing that they are being recorded. Also, while the data recorded on tape will be rich in detail you must also remember that its analysis will be time consuming. General estimates place full transcribing time for a one-hour interview at about 6 to 9 hours. There might also be a tendency for you to lose concentration and not to listen fully to the respondent because you know that everything is being faithfully recorded. One way to retain concentration is to take short notes – key words and phrases – to keep account of what has been said and what still needs to be covered. These notes can also act as a backup if your tape recording has failed. Remember that the tape recorder will not record body language or prompts such as nodding of a head or gesturing with hands. Therefore you might have to articulate responses for the benefit of the tape or jot down certain responses. When using a tape recorder always make sure you have spare tapes and batteries. Position the tape recorder directly in front of the interviewee and try to use an external microphone. Be aware that background noise can make your job of transcribing much harder, so try to conduct the interview in a quiet (or quietest) location.

Video recording provides additional information to the spoken word. When reviewing the recording the interviewer can note body language that may have been missed at the time of interviewing. Again, interviews captured using video filming might also suffer because interviewees are self-conscious about being recorded. With film, a great degree of anonymity is also lost. When using the video, position the camera so that it includes both yourself and the interviewer 'in shot'. If possible, use an external microphone.

Note taking can provide a rich description of an encounter when well written. However, in the course of an interview where the discussion often moves at a pace much quicker than can be transcribed, the note-taker must be skilled in identifying and jotting down the most important aspects of the discussion. At the same time as noting down the interview you must be able to concentrate on what has been said and to keep the conversation flowing. The notes taken will probably be piecemeal and disjointed with just key concepts and phrases recorded. As a result, if you do take notes after the interview it is wise to go back through them and add in extra comments to give key comments context. This ensures that when you come to look at the notes at a later date they make sense.

Pilot study

Similar to the administering of a questionnaire in Chapter 3, given the range of issues concerning the interview style, medium and method of recording, it is advisable to undertake a pilot study to try to iron out any difficulties that might be encountered in a larger study. There is no point in starting your project to discover at some later date that there is a significant flaw in your data generation or analysis. For

Box 7.5   Interviewee evaluation of pilot study

Issues to be raised:

- Did you understand why I am conducting the research?
- Were my instructions clear?
- Were any of the questions unclear or confusing?
- Did you object to answering any of the questions?
- In your opinion, was any major, relevant topic omitted?
- Any other comments?

Source: Adapted from Barrat and Cole 1991.

example, you might find that some of the questions you thought were perfectly reasonable cause offence or that people interpret them incorrectly. You can assess your pilot study in two ways. First, you can self-evaluate the effectiveness of the strategy used in terms of response rates, answers given, ease of analysis, etc. Second, you can ask your interviewees to evaluate the interview for you (see Box 7.5).

Critically appraising interviewing

From the discussion above it should be clear that there are advantages and disadvantages to interviewing. The interview can provide a fuller and richer data set than might otherwise be gained through highly structured, closed questions. This allows the interviewee to explain further their experiences, attitudes and opinions. Further, the less structured the interview the greater the flexibility the researcher has to direct the conversation and to explore specific issues in depth. On the negative side, interviews are more costly to undertake and analyse because of their time-consuming nature. There might also be problems relating to a greater likelihood of interviewer bias introduced through prompting and question phrasing. As with questionnaires, there is also the possibility that the respondent might try to predict what the interviewer wants to hear, and rather than articulating their own views, instead might forward an alternative they feel that will satisfy the interviewer.

Feminist commentators have been critical of textbook (cookbook) guides to undertaking interviews. They suggest that the description rarely conforms to the actual event. In textbook descriptions of an interview the interviewer is usually totally in control of the situation, asking the questions and responding to the answers. 'Normal' conversations do not take this form. In 'normal' conversations both parties ask and

respond to questions. In most interviews the interviewee will also ask the interviewer questions about specific issues. Feminists argue that in such a case, if the interviewer refuses to answer an opinion-based question, they are compromising the rapport and trust of the interviewee. If the interviewer does respond to such a question, however, then they are breaking the notion of neutrality or objectivity. Furthermore, feminists would argue that the interview strategy described in most methods books represents a masculinist view of research. This strategy seeks to exploit the knowledge of the respondent rather than to empower them. When interviewing people from marginalised sections of the community, feminists would argue that traditional interview methods maintain and reinforce current social **power relations**. Feminists suggest that the power relations within an interview must be renegotiated and that the interviewer must recognise that they cannot be neutral and objective. **Genuine trust** must therefore exist between the interviewer and interviewee, and the outcome of the research must also be genuinely empathetic and empowering. The interviewer then must develop a genuine rapport with the interviewee based upon shared concerns. To achieve this rapport the researcher must become engaged in the life of the interviewee. In other words, if you are interested in homelessness, rather than just interviewing homeless people you need to either work in an organisation devoted to helping homeless people or actually live among them. Only in these ways can you hope to build a genuine trust between yourself and your homeless interviewees. Your experiences will also help you to put your research into context. Clearly, this may be impractical for a student project. Nonetheless, feminists would argue that every effort should be made to become involved with the research participants beyond an interview. Dyck (1993) suggests that researchers should be more reflexive in their approach. This means that you reflect fully on your assumptions, and your part in the research process. Before we discuss alternative approaches to qualitative research, however, we will examine the other component of qualitative study, observation.

7.3.2   Observation

Wolcott (1995) suggests that the difference between interviewing and observation is that in observation you watch as events unfold whereas with interviews 'you get nosy'. Interviews are self-reports of experiences, opinions and feelings, whereas observation relies on the observer's ability to interpret what is happening

Table 7.2   Types of observation.

	Straight observation	Participant observation
Overt	Researcher does not engage with the group under study but makes no attempt to conceal fact of observation e.g., observing how children learn a map in a classroom setting.	Researcher joins a group as a participant in an event but does not hide fact that (s)he is observing them e.g., observing domestic labour relations by working as a cleaner where fellow cleaners know you are a researcher.
Covert	Researcher does not engage with the group under study and does not reveal to the group that they are being studied e.g., observing farming practices by walking down country lanes.	Researcher joins a group as a participant in a situation without telling them that s(he) is observing them e.g., observing tourist behaviour on a coach journey by posing as another tourist.

and why. Observation then 'entails the systematic noting and recording of events, behaviours, and artefacts in a social setting' (Marshall and Rossman, 1995: 79). Data can relate to observing conversations and overt behaviours. In particular, observation focuses upon people's behaviour in an attempt to learn about the meanings behind and attached to actions. Observation then assumes that people's behaviour is purposeful and expressive of deeper values and beliefs. Observation does not, however, need to be confined to observing people in contemporary settings. Much research within traditional cultural geography uses observation to study the cultural landscape. By observing human structures and practices upon the landscape, researchers try to determine the socio-cultural basis of society within certain time frames.

In general, observation is an **inductive method** of data generation (see Chapter 1). It works from data to refine hypotheses and produce a theory. This means that as observations are made the hypotheses of the study alter through a process of **negative case analysis**. When entering the observation phase a researcher should have a general hypothesis as to what will be observed. As the study progresses the researcher looks for cases which do not fit the hypothesis and seeks to find a new hypothesis that will include these cases. In such a fashion a hypothesis that explains the observations is found. Theories thus emerge from data analysis rather than theories being tested by data analysis. Such a strategy is said to circumnavigate the problem of the researcher deciding upon what is important (e.g., what questions should be asked to test a theory).

Frankfort-Nachmias and Nachmias (1996) suggest that the major advantage of observation as a technique is its directness. Rather than asking people

about their views and feelings, you watch what they do and listen to what they say. This directness provides a degree of validity as it concentrates upon what people really *do* as opposed to what they *say* they will. In an interview or when answering a questionnaire it is easy for a participant to claim a particular response. For example, it is easy for somebody to claim that they are not racist when completing a questionnaire. As studies have demonstrated, however, this does not always follow in real life. Saying something and doing something are not equivalent. Observation also allows you to record the lives of people as they live it rather than asking them to reflect critically upon their actions in an artificial social encounter such as an interview. This is particularly useful if the group who you are studying are unable to articulate themselves meaningfully or are not given to introspection, for example, young children and mentally impaired people. Observation then is generally a **naturalistic** (not necessarily naturalist) technique (although it is possible to observe people in an artificial situation created by the researchers). There are two sorts of observation: **straight observation** and **participant observation**. These can be divided into overt and covert types (see Table 7.2).

Straight observation

In straight observation the researcher is a visible and **detached observer** of a situation. For example, if we were studying children's ability to learn how to use maps we might undertake a classroom-based observation study. The researcher would be clearly identified as somebody who was sitting in the class to observe how the children learnt from a map. The researcher would make no effort to undertake a particular role within the classroom and would seek to be tolerated as an **unobtrusive observer**. By remaining in the background the researcher hopes that children will forget about her or his presence and act normally. By watching the children's actions and listening to their conversations the researcher seeks to understand the process by which a map is understood.

Marshall and Rossman (1995) note that this process of straight observation can range from a highly structured, detailed notation of behaviour guided by **checklists** (coded schedules) to a more **holistic description** of events (narratives). In the process of research they suggest that it is usual to progress from the latter to the former, although this is not always the case. For example, in our mapping example the researcher might enter the classroom with a broad area of interest but with no predetermined categories or checklists. After observing the behaviour and discovering recurring patterns of behaviour, the observer might start to construct a more context-sensitive checklist based upon the field notes (see coding data, Section 4.2.1). This checklist is then used to monitor behaviour over time. Some researchers would reject this structuring of the observation process, preferring instead to record their observations in a less constricting manner. Data generated using a formal framework such as checklists impose a large amount of structure on what is to be observed and consequently the data produced. This will increase the ease by which the data can be analysed, but only at the expense of complexity and completeness. Data generated using an **informal framework** such as note-taking and diary-keeping are relatively complex and more difficult to analyse. It is also possible to undertake straight observation by observing a situation without the observed community knowing or realising the true purpose. For example, you could observe the way farmers harvest their crops by watching over a gateway. In this scenario, your presence may not be suspicious, the farmers are not aware of the reasons for your observation, you remain firmly detached from the harvesting and do not seek to engage the observees in any way. Robson (1993) terms such a role as **marginal observation**.

Participant observation

In participant observation the researcher seeks to observe events and the behaviour of people by taking part in the activity themselves. This involves going beyond just being present at the same event to include sharing life experiences, becoming a member of the observee's social world. This might entail learning social conventions, the verbal and body language used, and establishing a social role within a community. The philosophy here is, 'if you want to know what it's like being an eco-warrior, then rather than just asking them, *become* an eco-warrior'. In living and acting out the life of the observed community the researcher becomes the **research instrument**. The observations made are usually recorded in a diary in a narrative form (unstructured). They will usually be accompanied by on-the-spot analysis and interpretation. Because the researcher is living with the researched community it may be impossible to take a break from the data generation to analyse data before going back into the field (as with our mapping example).

There are two forms of participant observation. The first is the **participant as observer**. In this form the observer reveals their intentions to the observed group from the beginning. The observer then attempts to build trusting relationships with the host community. This stance has distinct advantages because the observer can ask the community members to explain certain events to them and record events as they happen. If trusting relationships are not developed then the project will suffer greatly. It is important that the observees know why you are conducting the research and that they are happy to let you into their community willingly. If you are not accepted you will be marginalised from the community. The great danger with such an open approach is that the community will modify its behaviour for your benefit. If you are already a member of the group you are seeking to study then there may be problems relating to objectivity. Also if the community has a hierarchical structure then the problem of power relations applies. If you are a lowly member then senior members may be reluctant to confide in you or allow you to observe them. If you are a senior member, lowly members may resent intrusive observation.

In the second form, observation is undertaken **covertly**. The researched community has no knowledge that they are being studied. This strategy of research raises ethical questions concerning deception (see Section 2.5.2). It is usually justified, however, by arguing that the group would not have agreed to take part otherwise or would have acted differently if they had know about the researcher's presence. Covert research is relatively rare in geographical studies. One example is Moss's (1995) study of the oppression and exploitation in domestic waged labour, and in particular domestic cleaners. She took a job as a cleaner and while working chatted to the other staff about their experiences of the domestic labour market. In general, covert participation observation is frowned upon. It raises serious ethical questions and some researchers have questioned the extent to which the researcher can remain objective and neutral. Others suggest that postponing recording until one is safely alone may lead to incomplete and selective accounts.

It is generally assumed that the observer, although immersed in the world of their observed community, will remain objective and neutral. As such, they will record events as an **impartial observer**. Participation observation then can be considered 'good science' (Robson, 1993). By gaining access to the private life-world of the observed community the researcher can objectively validate their hypotheses with greater accuracy. As we have discussed elsewhere, some researchers would disagree with such claims to objectivity. They would argue that the research process is inherently subjective. An approach which explicitly recognises the subjective nature of observation is **ethnography**. This approach is discussed in the following section.

Participation observation requires a large commitment from the researcher both in terms of time and at a personal level. It may mean cutting off your own social relations in order to go and live with another community for an extended period of time. It may also mean getting involved in activities you might otherwise avoid. You should think carefully before committing yourself to such a venture.

Conducting observation studies

Like interviewing, observation can seem deceptively simple. However, *there is more to observation than simply watching people in a particular situation*. As noted, if you are to observe people in an everyday setting (or laboratory) then a trusting relationship must be developed, unless you occupy a marginal position. At first you may find the process of building relationships difficult and frustrating and you should not expect immediate results. Some experienced researchers might spend months developing an environment conducive to their research. There are differences in opinion amongst researchers as to how much preparation work you should have undertaken before proceeding into the field. Some would argue that the theory should arise out of your observations. However, what you observe and why you are observing is framed within the context of wider knowledge. We suggest that it is best at least to have a good grounding in the relevant literature concerning the

phenomenon you are observing, even if your exact theory is not fleshed out.

As we have discussed, there are two main ways of generating and recording your observations in the field. The first is to use a structured approach consisting of **coding schemes** (checklists). Coding schemes consist of predetermined categories for recording observations. In many ways a coding scheme is similar in format to a questionnaire but responses are interpreted and recorded by the observer. Great care must be taken when designing a coding scheme. The scheme must be **exhaustive**. That is, all possible **relevant behaviours** must be available for recording, in the format required (e.g., if timings of each behaviour are required then there should be a space to record these). Here, the keyword is *relevant*. In most cases not all behaviour is being recorded, only that which is relevant to the particular phenomenon or situation being studied. In general, you are recording specific events and their timings. Box 7.6 details four different sample coding schemes relating to traffic flows. It is extremely unlikely that you will be able to identify observation categories from the literature alone. It will therefore take much piloting time to construct a useful coding scheme. Robson (1993) provides a useful guide to designing a coding scheme (Box 7.7). The main advantage of this approach is that, once a scheme has been designed, it is fast and efficient and it minimises recording error by being highly structured and minimising interpretation. The approach is, however, limited to straight observation and provides a more constricted and limited set of data.

The second way of generating and recording observational data consists of a more **holistic account**. This approach, although less structured in nature than checklists, consists of more than just random note-taking. Here, the observer records observations using detailed notes rather than just ticking relevant boxes. This provides a richer and more detailed account. These notes should consist of more than random observation, and using a structured format may help you to record all the necessary information you might need later when analysing your records. Robson (1993) suggests structuring each entry into two accounts. The first is a descriptive account relating to the place, time, date, who is there, actions, etc. (see Box 7.8). The second is a narrative account which uses the elements within the descriptive account to construct the 'story' being told and a theory to explain what is being observed. The narrative then extends beyond the descriptive to detail your opinions and hypotheses as to what is happening.

Box 7.6 Sample coding scheme

Coding schemes for recording traffic flows

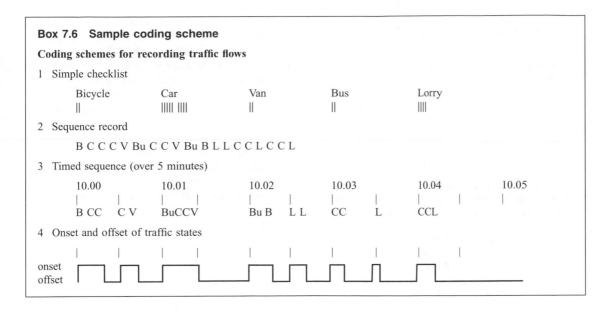

1 Simple checklist

Bicycle	Car	Van	Bus	Lorry
‖	‖‖‖‖ ‖‖‖	‖	‖	‖‖‖‖

2 Sequence record

B C C C V Bu C C V Bu B L L C C L C C L

3 Timed sequence (over 5 minutes)

10.00	10.01	10.02	10.03	10.04	10.05

B CC   C V   BuCCV   Bu B   L L   CC   L   CCL

4 Onset and offset of traffic states

onset
offset

Box 7.7 Considerations in developing a coding scheme

- **Focusing:** The categories on the coding scheme should be relevant to your research questions.
- **Objective:** The recording should require little interpretation by the coder.
- **Non context-dependent:** The categories to be recorded should not vary greatly over different contexts (unless it is part of the research aims).
- **Explicitly defined:** The categories should be easily identified.
- **Exhaustive:** All possible behaviours are available for coding.
- **Mutually exclusive:** Where possible, each observation should have only one code (e.g., just code e rather than codes e and f).
- **Easy to record:** For example, just ticking a box rather than any calculation.

Source: Adapted from Robson 1993.

Box 7.8 Descriptive observations

- *Where and when*: place, time and date
- *Space*: layout of setting, rooms and outdoor spaces
- *Actors*: names and details of people involved
- *Activities*: various activities of the actors
- *Objects*: physical elements; furniture
- *Acts*: specific individual actions
- *Events*: particular occasions
- *Timings*: sequences of acts or events
- *Purpose or goals*: what were the motivations behind act or events
- *Feelings*: emotions in particular contexts

Source: Adapted from Robson 1993: 200.

Your observations should be recorded using a research diary. Remember, recording in a more holistic manner is time-consuming. Where possible, you should record your observations there and then (even just in an abbreviated form to act as memory jogs at a later date). If this is not possible, record your entry at the nearest opportunity (visiting a restroom will provide you with a couple of minutes to jot down notes). You might find it helpful to get into a regular diary entry routine, setting aside a time every day to

flesh out your descriptive account and work on your narrative account. You should try to write a full account of your observations within 24 hours of the events while they are fresh in your mind. After this period, reflection and subsequent events are likely to cloud your memory and you may misinterpret some of your notes. If you are observing a situation covertly you may find it difficult to find the time and space to note down what you have observed and make attendant notes. In such a situation, note down your observations in your research diary whenever an opportunity arises. This might mean writing down snippets at irregular intervals. An alternative to a written diary might be a taped diary which can be transcribed at a later date. Harry Wolcott (1995) provides

Box 7.9   Some useful tips for undertaking observation studies

- Remember there is no such thing as 'just observing'. A lens can have a focus and a periphery but it must be pointed somewhere.
- Review constantly what you are looking for and whether or not you are seeing it or are likely to see it.
- Remember that you cannot sustain your attention indefinitely. Observation is necessarily an 'averaging out' process.
- Try to assess what you are doing, what you are observing and what you are recording in terms of the kind of information you will need to report rather than the kind of information you feel you ought to be putting together.
- Reflect critically on your note-taking and subsequent writing-up practices as part of your fieldwork. Are you writing too much or not enough? What are you recording and why?

Source: Adapted from Wolcott 1995: 96–101.

some useful tips when undertaking an observation study (Box 7.9).

Critically appraising observation

Observation has its merits and limitations. On the plus side is the directness and openness of the method. Observation is sensitive and receptive to the individuals within the study and also allows a certain amount of growth and progression of ideas and focus during the course of the study. While this might be limited in a structured format, ideas can progress during the piloting stage. As such, some lines of enquiry will wither and die while others will be followed up and observed more closely. Observation is the only methodology available to study what people actually do rather than what they say they may do. As such, its value to human geography research is great, although at present it remains relatively under-utilised for two principal reasons: first, observation is time consuming and difficult to instigate; second, it is not a trouble-free technique.

There are two major doubts concerning the use of observation. The first doubt concerns whether deeper meanings can be attached to overt behaviour. Here, we can get drawn into an age-old debate as to whether all behaviours are conscious acts. The second concerns the effect of the researcher upon the behaviour of the observees. Clearly in covert observa-

tion this effect is negligible as the observees do not know they are being observed. However, where the observer is known to be recording the actions of the observees, there is a question as to the degree to which the observees might be acting for the benefit of the observer rather than acting as usual. This is a difficult problem to resolve with no way of testing without covert observation. Here, we get drawn into the ethical problems of observing people covertly.

Observation is also very time consuming. This is particularly so for participant observation where strong trusting relationships often need to be developed. There are also methodological doubts concerning the nature of observer recording. It has long been documented that observational techniques suffer from a number of biases. When we observe people or a situation we do so from a **selective position**. Our position as a researcher but also our individual feelings and experiences affect what we see and also our interpretation of the events unfolding. As such, we have **selective attention**. Accompanying this is **selective encoding**. This is where our judgement concerning an action may be clouded and we pigeon-hole an observation into the category where we think it should go rather than the one where it should be placed. The risk of incompleteness or inaccuracy is increased with time and the dangers of a selective memory.

7.4   Specific approaches to producing qualitative data

7.4.1   Ethnography

Ethnography seeks to understand the world as it is 'seen through the eyes' of the participants. The aim of ethnography is not to deceive or exploit your respondents but to empathise with them and to gain an understanding of their lives through a genuine trusting relationship. In general, it is a naturalistic approach where you engage with your respondents in their everyday life-world rather than an artificial setting or encounter. As such, ethnography is an **empathetic** approach that combines aspects of informal conversational interviewing with straight and participant observation. As in participant observation, ethnographers look, listen and enquire about certain aspects of people's lives, noting down and recording encounters. Unlike participant observation, the researcher is trying to describe a situation from the participant's perspective rather than just recording events and interpreting the social processes leading to an event from a neutral position. This is a crucial difference. Like most other approaches ethnography consists of many variants. For example, many ethnographers would use the interviewing and observational methods in the scientific way we have just described. However, an emerging strand of ethnographic work argues that it is impossible to assume an objective role within the research process. Rather it is recognised that the ethnographer is a subjective agent within the research process and therefore cannot adopt a neutral position. Within this alternative ethnography, the use of covert observation is also generally rejected. As such, the researcher exposes themselves in their true role. While this might make the process of developing trust more difficult it also allows a freedom to record data more openly. An exploration of the **social politics** of the research process is often incorporated explicitly within the research.

7.4.2 Action research

Action research has been a distinctive study approach since the mid-1940s. Action research is formulated upon the basis of trying to *change* a social system at the same time as generating knowledge about it. As such, rather than trying to produce new knowledge by tackling scientific questions, action research aims to create new knowledge through the solving of practical problems. Clearly, remaining objective within action research remains difficult, as a particular action is sought. In general, action research is aimed at practical and technical problems and has a stronger history within physical geography than in human geography. Action-orientated approaches with political and social consequences are generally confined to neo-Marxist and feminist approaches within human geography and are more abstract in nature than action research projects.

7.4.3 Case studies

Case studies involve studying a phenomenon within its real-life setting. Rather than studying a phenomenon in general, a specific example within time and space is chosen for study. This allows a particular issue to be studied in depth and from a variety of perspectives. There are different types of case study approach (see Box 7.10). No one data generation method is used and quite commonly a number of techniques are employed. We have included case studies within the qualitative approach because, in

---

Box 7.10 Types of case study

- **Individual case study**. Detailed account of one person.
- **Set of individual case studies**. Several inter-related accounts of particular individuals.
- **Community studies**. Studies of one or more local communities (area defined).
- **Social group studies**. Studies of people belonging to a particular social group (occupation, activity defined).
- **Organisation and institutional studies**. Studies of people within particular working units.
- **Studies of events, roles and relationships**. Focuses upon specific events or encounters.

Source: Adapted from Robson 1993: 147.

---

the main, case studies are qualitative in nature, using observation and interviewing as methods of data generation. However, case studies can also be quantitative in nature or use a mix of both sorts of data. Often some of the data will be secondary in nature, consisting of summary statistics relating to the phenomenon or historical accounts relating to a phenomenon. It is therefore possible to undertake a case study which is also ethnographic in nature or action research-led.

7.5 Secondary sources of qualitative data

Just as with quantitative studies, there are a whole series of archival, qualitative data which you can utilise in your research. These can range from the formally recorded, such as an historical inventory, to the more informal such as letters and photographs. In this section we outline briefly some of the main sources of secondary qualitative data you might use. There are different approaches to analysing secondary data. For example, there are **biographical approaches** that seek to reconstruct the life histories of people to understand their life-worlds; **content analysis** which seeks objectively and quantifiably to identify patterns within the text; and **deconstructive** or **hermeneutic approaches** which seek to tease out the wider meanings held within the sources. We do not detail these approaches here but rather just present possible secondary sources of data. As such, we do not present different methods of how to use secondary sources. It is suggested that observation and interviewing analysis strategies be applied to the secondary sources (see next chapter). These recognise the need to 'read' the text and to focus upon noting and interpreting what is said or displayed. It should be realised that these data sources are often used together rather than exclusively.

7.5.1   Classifications of secondary sources

May (1993) reports that secondary sources of qualitative data have been classified in three main ways. First, sources can be *primary*, *secondary* or *tertiary* in nature:

- **Primary sources** refer to materials that have been recorded by those who actually witnessed an event. They represent knowledge by acquaintance.
- **Secondary sources** refer to materials recorded after the event from second parties. The recorder, therefore, has no personal experience of the event itself.
- **Tertiary sources** enable us to locate other sources and consist of indexes, bibliographies, abstracts, etc.

Second, sources can be **public** or **private** sources. This refers to the ownership and regulation of a source. This is important in a study as this indicates whether you might gain access to read or observe the source. Those sources in private hands are generally inaccessible without permission. Access to sources in public ownership can also be regulated. This has led Scott (1990) to categorise access into four classes:

- **Closed** (unobtainable)
- **Restricted** (special permission needed)
- **Open-archival** (no permission needed but sources are archived at one site)
- **Open-published** (freely available)

If you cannot secure access to a source you may have to reconsider your project objectives. Lastly, sources can be **solicited** or **unsolicited**. Solicited sources refer to those sources which have been created with the researcher in mind. An example might be a diary kept by a participant on the instructions of the researcher. Unsolicited sources refer to those that were not deliberately created for the purpose of the researcher. For the purpose of our discussion we have grouped the sources into three classes.

Diaries, letters, autobiographies and biographies

For those researchers interested in the life histories of certain figures or events, diaries, letters, autobiographies and biographies are valuable sources of information.

Diaries and letters, in particular, provide a rich source of information concerning the personal feelings, opinions and experiences of the writer. Historically, many 'people of note' wrote regular journal entries and letters to friends and many of these journals are held in public or private collections. This is particularly the case with fairly recent historical 'geographical' figures who kept logs of their explorations and whose exploits were sponsored by organisations such as the Royal Geographical Society. By linking the diaries and letters of several individuals together, rather than focusing upon a particular person, a particular event can be studied. Accompanying diaries and letters, many 'key figures' were also members of 'high society', regularly in and out of the press. **Press clippings**, particularly from the letters pages and gossip columns, are also sources of information. Many figures have also converted their diaries into more formal, edited autobiographical accounts of their lives. These accounts may contain retrospective narratives concerning certain events. **Biographies**, where a third party was writing at the time or from an historical perspective, may also shed important light on some figures or situations. These accounts benefit from the biographer drawing together several sources to document certain key stages within a life history. Historical geographers have long used such sources. In a recent example, McEwan (1996) studied the travel diaries and biographies of nineteenth-century female travellers to Africa. Using these sources, she has demonstrated the way in which colonial histories have been written almost exclusively by men. Similarly, Royle (1998) has studied the diaries, letters and press clippings of Alice Stepford Green, an aid worker in a Boer War prisoner-of-war camp on St Helena. It should be noted that these sources are necessarily subjective and in the cases of letters and autobiographies will most often be written in such a way as to flatter or argue the case of the writer. Diaries, if private, may allow a more detailed insight into the true, rather than public, opinions of the writer. Biographies similarly may not be objective studies of a person but may be written out of admiration or loathing, or for a fee.

Literary sources and 'official' documents

As well as more personal accounts, documentary sources provide a wealth of secondary data. These can range from literature to government and quasi-government reports, academic studies, minutes of meetings, and the 'unofficial' reports of interested parties. While of particular interest to historical geographers, the importance of these sources for contemporary analysis should not be understated. For those interested in studying contemporary society, documentary data can provide valuable insight into the structures and mechanisms of socio-spatial thinking and practice. Literary sources are particularly useful for detailed accounts of the **geographical imagination** of the writer and her or his characters within the context of an area.

In contrast, more 'official' documents provide detailed accounts of how individuals and institutions thought and reacted to certain geographical contexts. For example, Chris Philo (1987) has studied 'official' documents relating to the siting and running of mental institutions. He is particularly interested in ideas and proposals forwarded in the *Asylum Journal*, a quasi-academic journal concerning mental health institutions. By sifting through back issues and analysing the articles and editorials contained within, he provides a detailed historical account of asylums in nineteenth-century Britain. When you use official sources it is important that you take account of who wrote the document and ask why it might have been written rather than just studying the content. All documents are subjective and represent a particular viewpoint. For example, Barrat and Cole (1991) note that the 1985 review of social security was undertaken by four committees, with a total of 18 members. Of these 18 committee members half were government (Conservative) ministers; the others were from private industry and right-wing organisations. There were no trade unionists or representatives from any of the leading poverty agencies. Is this report likely to be an objective assessment of the needs of those on social security benefits?

Paintings, photographs, films and sound recordings

As well as textual sources of secondary information there are visual and auditory sources. Visual data can be classed into two general themes: stills and action. Stills are snapshot recordings of a particular scene. A painting would be an interpretative still, capturing a scene through the 'painter's eye'. A photograph, in contrast, would be a more faithful recording of a scene (although what is photographed and how is also subjective). This is not to say that a photograph should be preferred over a painting, particularly if you are interested in the geographical imagination of a certain period. Each medium records the same sorts

of information but in different ways. In recent years, geographers have been increasingly recognising the importance of paintings and photographs as sources of information. For example, a number of studies within cultural and historical geography have analysed landscape paintings (e.g., Daniels, 1988). Visual data are not confined to stills. Films, home videos and television programmes can also provide useful observational data and an insight into the geographical imagination of the camera operator and prospective viewers. Although film studies are still relatively rare in geography, Aitken (1990) has studied the films of the Scottish director Bill Forsyth. Auditory data can also be used as a secondary source of information. Recordings might consist of the taped conversations of another researcher's study or of music and lyrics. McLeay (1995), for example, has studied the sense of place within the music and lyrics of U2.

7.5.2 Assessing secondary sources

Whatever the information source, Scott (1990) suggests that before you go ahead and use the data you should assess its usefulness in the context of four things:

1 **Authenticity**. Is the source correctly attributed?
2 **Credibility**. Credibility relates to the accuracy of the source and the sincerity with which it was recorded (e.g., did the recorder believe in what was recorded?)
3 **Representativeness**. Is the source representative of opinion at that time and place? Are there other sources which might also be used to check representativeness? If there are only a few surviving sources, how much weight should be attached to them?
4 **Meaning**. Should the source be used in a literal sense or is some level of interpretation needed to fully understand the purposes of the recorder?

In conjunction these four factors relate to the validity of the data – the confidence with which you can use it to draw valid conclusions about a situation or phenomenon. When you analyse and interpret data from secondary sources, you should consider carefully these four factors and keep them in mind throughout the project. In addition, you should also define some criteria by which you select sources and which parts of a source will be used in your research. These criteria should follow the same boundary-making principles that you might follow when deciding upon questions for an interview. Essentially you need to be able to identify what is, and what is not, relevant.

7.6   Summary

After reading this chapter you should:

- understand the basic principles underlying interviewing;
- understand the basic principles of observation;
- be aware of the different sorts of secondary data available;
- be able to generate qualitative data relevant to your research project.

In this chapter we have explored how to generate primary qualitative data and to identify and locate secondary sources of data. We have discussed a number of themes, such as strategies for data generation, recording medium and recruitment, that you will need to consider when generating primary qualitative data. Each aspect of the data generation phase needs careful thought and planning. You should only proceed with the generation of data when you are satisfied that you have considered thoroughly the options available to you. In the following two chapters, we will examine how to analyse and interpret qualitative data, first using a handworked example and second using a qualitative data analysis package (NUD-IST).

7.7   Questions for reflection

- *What are the merits and limitations of interviewing and observation?*
- *What are the merits and limitations of the different interview media?*
- *What ethical questions are associated with covert observation?*
- *How does straight observation differ from participant observation?*
- *How do the qualitative approaches detailed in Section 7.4 differ from the main approach used to structure the discussion throughout the rest of the chapter?*
- *Why are secondary sources of qualitative data particularly useful to historical geographers?*

Further reading

Cresswell, J.W. (1998) *Qualitative Inquiry and Research Design: Choosing Among Five Traditions*. Sage, London.

Dey, I. (1993) *Qualitative Data Analysis: A User Friendly Guide for Social Scientists*. Routledge, London.

Eyles, J. and Smith, D.M. (1988) *Qualitative Methods in Human Geography*. Polity, Cambridge.

Marshall, C. and Rossman, G.B. (1995) *Designing Qualitative Research*, 2nd edition. Sage, London.

Robson, C. (1993) *Real World Research: A Resource for Social Scientists and Practitioner-Researchers*. Blackwell, Oxford.

Scott, J. (1990) *A Matter of Record: Documentary Sources in Social Research*. Polity, Cambridge.

CHAPTER 8

Analysing and interpreting qualitative data

This chapter covers

8.1 Introduction
8.2 Description, classification, connection
8.3 Transcribing and annotation
8.4 Categorising qualitative data
8.5 Splitting and splicing
8.6 Linking and connecting
8.7 Corroborating evidence
8.8 Analysing qualitative data quantitatively
8.9 Summary
8.10 Questions for reflection

8.1 Introduction

As detailed in the last chapter, there are different ways to approach the task of producing qualitative data. Similarly, there are different ways to approach the analysis of the data produced. For example, Patton (1990) uses an **interpretative approach** which emphasises the role of patterns, categories, and basic descriptive units; Strauss and Corbin (1990) use a **'grounded theory' approach** which emphasises different strategies of coding data; Miles and Huberman (1992) use a **quasi-statistical approach** which seeks to minimise interpretative analysis and introduce an objective, prescriptive approach to analysis. As Silverman (1993) notes, these approaches are often utilised in analysing different sorts of data. When writing this chapter we had to make a decision as to how to proceed. There were two very different routes open to us. We could choose to discuss either methods to analyse interview, observational and secondary data separately (as with Cresswell, 1998), or a **universal approach** which can be applied to study all types of qualitative data. We have chosen the latter. As Dey (1993) argues, despite differences in emphasis, the various approaches to qualitative analysis all seek to make sense of the data produced through categorisation and connection. In our discussion of qualitative analysis we follow Dey's (1993) approach which seeks to combine different aspects of various other approaches to gain a deeper understanding of qualitative data. It must be appreciated that different commentators might, therefore, approach the data we present in a different manner and you are encouraged to read beyond our prescription. This is not to say that we have not drawn from alternative sources. Although the method of analysis we present follows Dey's prescription, we do incorporate the advice of other researchers.

Some researchers might be unhappy about following a prescriptive approach to qualitative data analysis. They would argue that the analysis of qualitative data is an **art** rather than something that can be undertaken through prescription (as with quantitative analysis, where you generally put in your data and get out a result). There is a degree of truth in this statement; qualitative data are not as rigidly defined as quantitative data and analysis lacks the formal rigour of standardised procedures. In an introductory text such as this, however, we feel that a prescriptive approach is the most appropriate. It provides you with a clear set of guidelines for analysing your data. Trying to practise qualitative data analysis as an art before you know how to draw is, in our view, a recipe for disaster. We therefore recommend that, until you are comfortable with the basics of qualitative analysis, you follow our prescriptive approach. While fairly systematic in nature we have endeavoured not to slip into the quasi-scientific approach to qualitative data analysis as practised by Miles and Huberman (1992) and Yin (1989). In other words, while we recognise that qualitative data analysis is largely an inductive, open-ended process that is not easily captured by a mechanical process of assembly-line steps (Lofland and Lofland, 1995), formal guidelines are useful in a learning context. Our main concern is to give you a practical insight in how to analyse qualitative data for the purposes of your study. Once you are familiar with these processes we encourage you to explore different approaches.

Dey (1993) describes his approach to qualitative data using an omelette analogy. He suggests that just as you cannot make an omelette without breaking and then beating the eggs, you cannot undertake data analysis without breaking down data into bits and then 'beating' the bits together. He suggests that the core of qualitative analysis consists of the **description** of data, the **classification** of data, and seeing how concepts **interconnect**. As we noted for quantitative data, analysis is more than just describing the data – we want to be able to interpret meanings, and explain or understand the data generated. Description lays the basis for this deeper analysis, but then we need to progress beyond this stage to try to determine the interconnections and relationships between data, and to tease out the ways in which to reconceptualise our constituent bits. Each section of this chapter is designed to guide you through qualitative data analysis using Dey's approach. While the following prescription has been written as if all the data are generated before any analysis starts, it can equally be used to start making sense of data as they are generated. That is, you can start to describe, classify and connect data before you have a full data set. This will help you make sense of the data being generated and to also guide further data generation. In this chapter, we describe the process of analysis and interpretation by hand. In the next chapter, we describe the analysis of qualitative data using a computer.

In order to illustrate how to analyse and interpret qualitative data, four example data sets will be used (Boxes 8.1–8.4). These data were generated as part of a large international study which concerned how blind people interact with and learn urban environments (see Kitchin *et al.*, 1997). The study consisted of three different stages:

1   Respondents were interviewed about their spatial competence and their experiences in interacting with an environment.
2   Respondents described a familiar route. This route was then walked and various spatial tasks such as pointing to other locations, constructing a model of the route and estimating distances were completed.
3   Respondents learned a new route and completed the same tasks as in stage 2.

The data presented were taken from the interview in stage 1. The interview consisted mainly of closed questions designed to elicit quantitative data concerning levels of blindness and mobility. These closed questions were complemented by questions styled in an interview guide approach, asked after the questionnaire was completed. The data presented were recorded during the interview. The interviews were conducted by Dan Jacobson between March and May 1997 in the Belfast Urban Area. Interviewees were paid £10 for their time. People and place names have been changed. A number of factors should be noted. First, these data are the results of Dan's first-ever interviews. Second, Dan is not visually impaired. Third, prior to the interview both he and the interviewee had never met before. Therefore, a strong research relationship had not yet been forged. Fourth, the interviews were recorded on video as well as tape. As part of his own PhD project, after the interviewees had completed the various stages of the research, Dan reinterviewed some of the respondents. By this stage he had spent quite a long time in the company of each respondent (at least two full days interviewing and retracing routes). In the second set of interviews he found that the respondents were more open and frank about their lives and their experiences of interacting with urban environments. His interviewing skills had also developed and he found that he could probe the respondents more effectively.

While these data are not from a purely qualitative study they are sufficient to demonstrate how qualitative data can be analysed. It should be possible for you to substitute whatever data you have produced, whether from primary or secondary sources, and undertake a similar analysis. For example, you should be able to analyse notes from an observation study or to perform an analysis of secondary, documentary data using this approach. The example data have been chosen for two main reasons. First, one of us has personal experience of the whole study. Second, the data are likely to reflect those generated by first-time interviewers with missing data, leading questions (see if you can

Box 8.1   Interview with respondent A

Male
Age: 43
Age since blind: 19
Location of interview: Respondent's home.
Spatial context: Urban estate, privately owned (formerly council), semi-detached.
Date/Time of interview: */*/97; 4 pm.
Who was present: Dan Jacobson and Respondent A
Interview was recorded on tape and video.
Notes: Interviewee was cautious but reasonably forthcoming.

INT: *How do you approach and resolve problems encountered en route?*
RES: Such as?
INT: *Well, how about detours?*
RES: I'm OK with them. But I got lost once. It tends to knock you sideways a bit. I crossed one too many streets and I ended up going in the opposite direction. Well, I would first of all – if I didn't know the area – I would ask someone. Or I would go back on myself and retrace, and maybe get a taxi or some other transport. Or maybe I would go to a friend's house nearby.
INT: *How about if you found yourself off-route?*
RES: I don't really like finding myself in this situation. If I was caught in a security scare or something like that, I would wait until someone came along, or I would go to the door of a house. Or I would retrace.
INT: *Do you come across any specific hazards?*
RES: Sometimes you have to be very wary, if there's a lorry parked on the pavement, or some awkward street furniture blocking the way. When two or three things

happen at once, it can be really confusing. The only thing you can do is to backtrack.
INT: *Do you think environmental modifications might make navigating easier?*
RES: There is a conflict between dropped kerbs and flush pavements. I can't tell the difference. If you go from Castle Street, from Castle Court, the place where the pavement should be, it just isn't there. There is nothing to tell you where it is, it is all flush. Those bobbly tiles – I don't like those.
INT: *How do you find out about new routes?*
RES: By asking people. I haven't been down in the town for a while, but I knew there was work done on this part of the road. I didn't realise the Westlink went under the Falls Road. Nobody had ever told me that. I was walking across towards Divis Street, and suddenly I heard the roaring of a heavy thundering truck. 'Shit, this one's going to get me', I thought. Then it went underneath and I was OK.
INT: *How often do you learn a new route?*
RES: Very rarely, once or twice a year, most places I go, I go regular.
INT: *How do you prepare for learning a new route?*
RES: If I can get the information in advance I may go over it beforehand. If I can get detailed instructions, 'cross three roads, turn left, on the up kerb, turn right', it's a great help.
INT: *Have any route learning strategies proved to be particularly successful or disastrous?*
RES: No.

---

spot them) and a lack of probing. It should be clear from these interviews that some people are a lot more forthcoming than others! A paper discussing analysis of all the interviews is Kitchin *et al.* (1998).

8.2 Description, classification, connection

As noted, Dey's approach, like most other qualitative analysis methodologies, consists of the description, classification, and making of connections between the data. Before we detail how you might go about undertaking these tasks, each will be discussed briefly. While presented in turn, it should be noted that the three do not always follow in a strict linear

order. The process is more **iterative** than linear (Dey, 1993). That is, while classification cannot precede description, and the making of connections cannot precede classification, we can go back to modify the previous task and to take a new route into the next task (see Figure 8.1 on p. 235).

8.2.1   Description

Description is central to any study, whether using qualitative or quantitative data. It permeates all levels of enquiry. Description concerns the portrayal of data in a form that can be easily interpreted. For example, a description might be a verbal or written account of the data or a graphical illustration. Description can be either **thin** or **thick** in nature. In describing quantitative data we were seeking what has been termed 'thin'

Box 8.2   Interview with respondent B

Female
Age: 42.
Age since blind: 0.
Location of interview: Respondent's home.
Spatial context: Suburban environment, privately owned, semi-detached.
Date/Time of interview: */*/97; 2.30 pm.
Who was present: Dan Jacobson and Respondent B.
Interview was recorded on tape and video.
Notes: Interviewee was very outgoing and forthcoming.

INT: *How do you approach and resolve problems encountered en route?*
RES: Say I arrived in town and I wanted to go to a department store, I would get off the bus at the appropriate stop to find that the road had been dug up. The noise of the men working lets me know that they are working, but it is then off-putting as you can't find other sound clues to get around that. I would, perhaps . . . if in town, I try to follow the flow of people – use them as guides even though they don't know it. The same is true for crossing roads, rather than asking each time, I would find somebody who I think is reasonably competent and then follow them.

INT: *How about off-route?*
RES: Generally I don't like this. I can get anxious. Sometimes it is OK though, but especially . . . if I'm not familiar with it I may panic a bit and wonder 'am I where I want to be?', 'Is this where I want to be?' This can happen. One day a few weeks ago I was in big hurry, I was taking my daughter to the opticians, and we had to come back in half an hour. We had a few other things to do so we left the shop to do those. They were in another part of town and then we had to get back to Castle Court shopping centre. Our normal route would be up Royal Avenue and Castle Court would then be on your left. But we found ourselves around Queen Victoria Street and I found myself thinking, 'I wonder if I can cut through these streets?', 'will it bring me out at Castle Court?' Very much a guessing game to head in the right direction. We did get there in the end, but I had my sighted child with me. If I had been there on my own I probably would have gone the long way around.

INT: *How do you cope with being lost?*
RES: I don't like it, I hate being 'out of control'. It is not a good feeling. It can be an anxious feeling. It's also quite a confidence downer. All of a sudden you are not where you think you are and you feel a bit stupid. You think any other adult could manage this and maybe you can't so you feel inadequate in some way. How I would resolve it would be to ask somebody. This has, believe it or not, taken me nearly 42 years to reach the point where

I can ask somebody. You get a whole range of answers that you need emotional strength for, like the 'are you blind?' question.

INT: *Are there specific hazards that you encounter?*
RES: A lot of shops now have huge amounts of pavement clutter. Sometimes I can see these, pick them out OK. But if it is bright and there are lots of people then I will bump into things.

INT: *How do you learn a new route?*
RES: By trial and error and by talking to friends. If I am going alone I try to remember the way I have been and to follow it back.

INT: *Do you learn many new routes?*
RES: One a month. I would be open to trying out new routes.

INT: *How do you prepare for a new route?*
RES: Friends, an A to Z with a magnifying glass, and generally try to get as much information as possible to be prepared before I go. Last year I was told I was going away for the weekend but I wasn't told where I was going, it was to be a surprise, and the person I was going with was completely blind. Then I panicked totally. 'How will we ever find our way around?', 'What will we do?', 'How will we manage?' Just total panic really. We flew to Edinburgh and by basically asking people where to get the bus etc. we made our way to the town. Once in town we asked for the street with the hotel on, where we were staying. The first thing we did was to get a map of Edinburgh, I was able to see some of this with my magnifier, although I had trouble translating that into my surroundings. The person I was with could translate this much easier, by not looking for visual cues as he can't see.

INT: *Have any route learning strategies proved to be particularly successful or disastrous?*
RES: The trip to Edinburgh, it was just one weekend, at the time we came back I felt fairly confident about finding my way. This was news to me, as if somebody beforehand had told me you would go to Edinburgh and be able to find your way about and that you'll manage to do all the things that we did, I would not have believed them. It was a confidence booster and it did prove successful. Another one was unsuccessful. I went to Westport for the day. It was such a small place compared to Edinburgh, but I didn't manage too well. It was built on a square, but I never knew what section of the square I was at. I could have been three-quarters of the way around, when I needed to only go back a quarter, but I would go around. There was a river and a bridge and everyone seemed to go by these, but they didn't mean anything to me. It was frustrating and I was quite disorientated.

Box 8.3   Interview with respondent C

Male:
Age: 35.
Age since blind: 7.
Location of interview: Respondent's home.
Spatial context: Inner city, privately owned, semi-detached.
Date/Time of interview: */*/97; 2 pm.
Who was present: Dan Jacobson and Respondent C.
Interview was recorded on tape and video.

INT: *How do you approach and resolve problems encountered en route?*
RES: I do have a cane, and I suppose I would use it at night, but not during the day. 'Watch that puddle, watch that kerb, watch that car'. If I could watch the kerb I wouldn't need you to tell me to watch the kerb. I've had 3 months of O&M [orientation and mobility] training with the long cane about 2 years ago for 2 hours a week. If Alison is there, it is a lot easier, picking out the gaps in the wall, or going into shops. When I do go into shops everything just brights out for a moment. So you stand there and look like a wally, everybody looking at you. So it is a lot easier with someone to guide me.
INT: *How do you cope with more specific problems, such as crossing roads?*
RES: I tend to cloak people, wait on somebody that's going across and then shadow them.
INT: *How about attempting detours?*
RES: They can be awkward. Generally Alison, my partner, would be with me. If it was left to me I would be in for a lot of trouble, so I would probably ask.
INT: *How about if you were off-route or lost?*
RES: I would ask somebody. A few times I've been, not lost, but mislaid. Recently, when I was on a course I got the wrong bus one morning. There were two buses leaving Lisburn, one was an express to the back of the City Hall in Belfast and the other one went to Newry – somewhere. I soon knew the bus I was on wasn't going to the

right place and I started sort of panicking, and then it was more really of a mental panic; 'I'm in deep shit'. Although in the end I recovered easily enough.
INT: *Do you come across many hazards?*
RES: I keep bumping into a lot of them such as when I come across them outside shops and the like. Very often I'd be kicking things. And people tend to think, what's up with him, it is a bit early in the day to be drinking.
INT: *Do you think improved training would help?*
RES: Really it's not the training, the training I had was fine, but for me it was a bit of a let down. I was expecting something a little different, like failsafe, or a guide dog. What you think it is about and what it actually is are two different things. My expectations and reality didn't match.
INT: *Do you think environmental modifications might make navigating easier?*
RES: Some changes are definitely needed. Those wiggly pavement bobbles are good for totals, but for me they are a menace – it throws you. Some places in the shopping centres you have to go up and grope the shopfront looking for the door. You don't know if it is a window or a door. There are just so many things we have to put up with. But for me definitely the likes of kerbstones being painted helps.
INT: *How do you find out about new routes?*
RES: I don't think I would get maps or anything like that. I would just learn by walking around with a friend. I only need to do something a couple of times and then I've got it. Probably a couple of times a year.
INT: *How do you prepare for learning a new route?*
RES: Every time, if I was going by myself, I would always have it mentally well worked out. For the first couple of times I would go with my wife. That helps me build up a lot of confidence.
INT: *Have any route learning strategies proved to be particularly successful or disastrous?*
RES: No, not really.

description – a clear statement of facts; how many, in what form, etc. The description of qualitative data is usually more 'thick' in nature. Thick description seeks to provide a more thorough and comprehensive description of the subject matter. It includes information concerning the situational context, the intentions and meanings associated with an act, and the process in which the situation is embedded. Essentially the thick description consists of the descriptive observations listed in Box 7.8 (in Chapter 7).

A description of the **situational context** consists of a detailed account of the *social settings* (e.g., home), *social context* (e.g., paid interview), and *spatial arena*

(e.g., place) within which an action or phenomenon occurs, and the time frame within which the data were generated. This information is important because it provides a detailed account in which to situate the analysis. It is well known that the social, spatial and temporal context can all significantly affect the data generated. We must, therefore, be able to take into account any factors which may have influenced the nature of the data when conducting our analysis. This information also allows us to compare data generated under different circumstances.

The intentions and meanings of the actors (i.e., people being interviewed) similarly provide us with

---

Box 8.4   Interview with respondent D

Male
Age: 31.
Age since blind: 17.
Location of interview: Respondent's home.
Spatial context: Satellite village, private home, semi-detached.
Date/Time of interview: */*/97; 3 pm.
Who was present: Dan Jacobson and Respondent D.
Interview was recorded on tape and video.
Notes: Interviewee was cautious and guarded.

INT: *How do you approach and resolve problems encountered en route?*
RES: Such as?
INT: *Detours, for example?*
RES: Well, most obstacles are small and in areas that I'm familiar with it certainly wouldn't be a problem – I would just take an automatic route to bypass them. For example, I was on the Antrim Road, lines of shops, traffic and there was CableTel digging holes. There were pneumatic drills going, it was really disorientating – I couldn't use the traffic noise to work out which way I was facing. So I turned, took the dog to the kerb to reorientate myself, turned back and walked through the parallel streets that I know.

INT: *How about if you were off-route?*
RES: It doesn't usually happen, but if I was unsure I would retrace my steps. I've never been totally lost, but disorientated, for sure. One time another dog came up and we got entangled and spun around, there was no traffic to orientate myself so again I had to take my dog to the kerb and get my bearings.
INT: *Do you think environmental modifications might make navigating easier?*
RES: Yes. Tactile pavements need to be standardised and there need to be more of them. Shopfronts with baskets and stalls are a nightmare, but you can't outlaw that.
INT: *How do you find out about new routes?*
RES: With some difficulty, but mainly when I come across them.
INT: *How often do you learn a new route?*
RES: Not very often, but for example I've learned a new route for a course I am attending.
INT: *How do you prepare for a new route?*
RES: I'd walk over it with someone from Guide Dogs – 'down kerb to up kerb, keeping the traffic on my right'.
INT: *Have any route learning strategies proved to be particularly successful or disastrous?*
RES: No.

---

contextual information. However, unlike situational context factors, these data are more difficult to record and unless stated by the interviewee require some interpretation. In our data, the intentions and meanings of the actors when navigating an environment are recorded in some of their answers. In an observational study it would be difficult to insinuate (imply) some of their responses purely on the basis of observing their actions. However, it must be remembered that what is said in an interview is not always the same as what the interviewee actually believes or what they would actually do. As noted in Chapter 7, an interviewee will sometimes give the answer they think you want to hear, rather than their own opinion. We therefore cannot always rely on interviewees giving truthful and rational statements concerning their intentions or meanings.

Qualitative data generation rarely consists of snapshot studies. Interviews and observation are often the product of several encounters over a period of time. As such, data generation is usually part of an ongoing process of social relations. Another type of description then concerns the extent to which situational context, intentions and meanings remain stable or change over various encounters. Further, a focus on

process allows us to examine the consequences of action and subsequent alterations in thought or behaviour. This descriptive data can relate to both the people or phenomena under study but also to the researcher. As the research has progressed, so will the researcher's thinking and possibly her/his approach. The researcher's own field notes can therefore become a useful source for contextualising the progress and process of the research. These field notes should be analysed in the same manner as described below.

8.2.2   Classification

Classification is where you move beyond data description to trying to interpret and make sense of data. By undertaking **interpretative analysis** you are seeking to understand the data generated more fully and make the data meaningful to others. The classic way to start to make sense and interpret qualitative data is to 'break up' the data into constituent parts and then place them into similar categories or classes. In this process we start to identify which factors are important or more salient; to draw out commonalities and divergences. By classifying the data we can start to make more effective comparisons

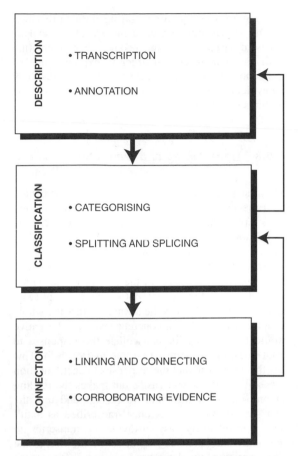

Figure 8.1 Description, classification and connection.

between cases. While we naturally or implicitly classify objects in our daily lives (e.g., types of shop or transport), classification within the research process is more systematic and explicit. **Implicit classifications** help us make sense of our world. In contrast, **systematic classifications** help the researcher understand the thoughts and actions of several people. As such, different classification classes do not have to be defined by the actors or their actions within the research but can be imposed by the researcher. For example, it might seem logical to classify our visually impaired interviewees into:

- those who were blind from birth and those who lost their sight later in life;
- those with residual vision and those with none;
- those who use guide dogs and those who do not;
- those who have had orientation and mobility training and those who have had none;

- those who have lived at their address for more than two years as opposed to those who have recently moved.

Alternatively we could use something less tangible like interviewees' answering style, whether they are positive or negative in their responses, etc. In other words, classification is more than the identification of the most logical or implicit classes within the data. Which classification we choose to impose is dependent upon what we wish to examine. Clearly, there are some data that do seem to classify themselves. The majority, however, need to be purposively sorted and a carefully selected order imposed. Undertaking classification, effectively 'breaking up' your data and collecting like-with-like against some standard, means that your data lose their original format (e.g., interview transcript). What is gained is an organisation which will aid further analysis and the process of interconnection. We will examine the process of classification in more detail in Section 8.5.

8.2.3 Connection

Classification is concerned with identifying coherent classes of data. Connection is concerned with the identification and understanding of the **relationships** and **associations** between different classes. This consists of more than just identifying the similarity or difference between categories. Instead we are more interested in the **interactions** between classes. Dey (1993), using a building analogy, suggests that whereas classification concerns putting all the bricks, frames, glass panes and beams in separate places (and classifying these according to type, size, etc.), making connections concerns how they fit together and relate to one another as a structure such as a house. In other words, our data generation may lead to a building yard full of material. Our analysis consists of describing the yard's material so that we know what we have and why, classifying the various data forms into relevant building materials, and connecting the classes together to construct a coherent and stable structure. Clearly if analysis is occurring as an ongoing process our structure might undergo several modifications and extensions as the building material is refined and classed more effectively. If you have started from the position of testing a theory then the structure built should generally conform to some sort of plan (theory). If, however, you are inductively building a theory then your final construction will provide the blueprint of your theory.

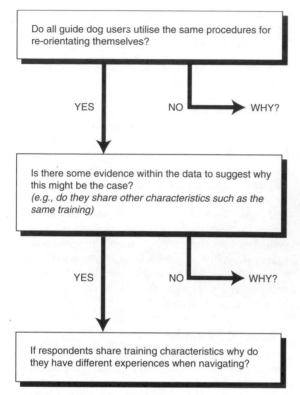

Figure 8.2 Interrogating data.

As we will discuss in Section 8.6, making connections is not always a simple process. One common method is to search for **recurring patterns** within the data. This is achieved by identifying whether individuals with a particular characteristic also possess other common features. For example, do all guide dog users utilise the same procedures for reorientating themselves? The question now, if the answer is 'yes', is why? Is there some evidence within the data to suggest why this might be the case? Do they share other characteristics such as the same training? Again, if 'yes', does this lead to similar experiences when navigating? If the answer is 'no', then why? If these interviewees share so many characteristics in terms of training and procedure when navigating, why do they have different experiences when navigating? Again are there any clues within the data? At this point we are shifting from examining the relationships as if they are external and contingent. That is, we are moving beyond identifying whether there are relationships between data to trying to find a reason why any relationship apparently exists. A useful way to do this is to visualise the process as a flowchart

(see Figure 8.2). By **interrogating data** in this way we are slowly building up a picture of the experiences of visually impaired people when interacting with an environment. In order to explain some of the connections it might also be necessary to draw on evidence from beyond your data, for example the findings of others.

8.3 Transcribing and annotation

The first stage after you have generated your data is to **transcribe** your notes or interviews into coherent transcripts. This should be done as soon after data generation as possible. If the data are from a secondary source the relevant sections should be transcribed. There are two main methods of transcribing your data. The first is to transcribe all the data provided within one data generation session onto a single script (e.g., as in Box 8.1). This has the advantage that the whole script can be read as a coherent text. An alternative would be to transcribe and collate the responses to each question separately (e.g., as in Box 8.5). This has the advantage that the responses to each question can be viewed together easily, but makes the reading of one interview difficult. We recommend that the first strategy be used, as once transcribed as a full text, it is relatively easy to divide the transcript up into individual question responses. In addition to the transcription of the data, it is important that you also fully transcribe the 'thick' description to accompany the data. In our example, we have transcribed interview data. If you have undertaken an observational study or examined secondary sources you should simply substitute your entries for ours.

There are also two main ways in which you can transcribe your data. First, as we have done, you can just copy out what was said using a minimum of codes (e.g., '. . .' means a pause). An alternative is to use a much more rigorous method of transcription. In general, the more rigorous method is applied only to interview data, although it could be applied to other data. We suggest that unless necessary (e.g., you are interested in speech patterns) you transcribe your data using a minimum selection of **codes** (see Box 8.6). This will save unnecessary data preparation and valuable time. Remember, the transcription of data is very time consuming. A one-hour interview can easily take up to 6–9 hours to fully transcribe and annotate. In order to try to minimise time and effort we recommend that you follow some general rules (see Box 8.7).

Box 8.5   Transcribing each question or theme separately

(Details of each interviewee held on master cards)

CATEGORY: OFF-ROUTE/LOST

RESPONDENT A

INT: *How about if you found yourself off-route?*
RES: I don't really like finding myself in this situation. If I was caught in a security scare or something like that, I would wait until someone came along, or I would go to the door of a house. Or I would retrace.

RESPONDENT B

INT: *How about off-route?*
RES: Generally I don't like this. I can get anxious. Sometimes it is OK though, but especially . . . if I'm not familiar with it I may panic a bit and wonder 'am I where I want to be?', 'Is this where I want to be?' This can happen. One day a few weeks ago I was in big hurry, I was taking my daughter to the opticians, and we had to come back in half an hour. We had a few other things to do so we left the shop to do those. They were in another part of town and then we had to get back to Castle Court shopping centre. Our normal route would be up Royal Avenue and Castle Court would then be on your left. But we found ourselves around Queen Victoria Street and I found myself thinking, 'I wonder if I can cut through these streets?', 'will it bring me out at Castle Court?' Very much a guessing game to head in the right direction. We did get there in the end, but I had my sighted child with me. If I had been there on my own I probably would have gone the long way around.
INT: *How do you cope with being lost?*
RES: I don't like it, I hate being 'out of control'. It is not a good feeling. It can be an anxious feeling. It's also quite a confidence downer. All of a sudden you are not where you think you are and you feel a bit stupid. You think any other adult could manage this and maybe you can't so you feel inadequate in some way. How I would resolve it would be to ask somebody. This has, believe it or not, taken me nearly 42 years to reach the point where I can ask somebody. You get a whole range of answers that you need emotional strength for, like the 'are you blind?' question.

RESPONDENT C

INT: *How about if you were off-route or lost?*
RES: I would ask somebody. A few times I've been, not lost, but mislaid. Recently, when I was on a course I got the wrong bus one morning. There were two buses leaving Lisburn, one was an express to the back of the City Hall in Belfast and the other one went to Newry – somewhere. I soon knew the bus I was on wasn't going to the right place and I started sort of panicking, and then it was more really of a mental panic, 'I'm in deep shit'. Although in the end I recovered easily enough.

RESPONDENT D

INT: *How about if you were off-route?*
RES: It doesn't usually happen, but if I was unsure I would retrace my steps. I've never been totally lost, but disorientated, for sure. One time another dog came up and we got entangled and spun around, there was no traffic to orientate myself so again I had to take my dog to the kerb and get my bearings.

Effective transcribing consists of more than just accurately writing down an interview or observation. While transcribing you should be thinking carefully about what is being transcribed, trying to get a feel for the data, and starting to develop ideas about specific lines of enquiry. As such, while transcribing your interviews you should jot down **ideas** and **memos** relating to the transcription. You should try to distinguish between ideas and memos. Ideas relate to your own thoughts about the data. Memos are notes about the data. It is here that you have started the process of *description*. You should jot down notes immediately, before your inspiration disappears. Once you have finished the complete transcription you should then thoroughly annotate the transcript. Start **annotating** your data immediately after transcribing it, while both the interview and transcription are still fresh in your mind. In this way your annotations will be fully situated within the context of the data. While transcribing and annotating might seem like a boring chore, they have their utility. Transcribing requires you to study each interview, and annotating forces you to think about the data. Your annotations, in particular, open the data up and start the process of analysis in earnest. Annotating your data now will make subsequent categorisation and connecting easier. Annotations are also extremely useful guides to future data generation.

There are a number of strategies you can use to aid the process of annotating your transcripts. Box 8.8 provides a description of each. Each strategy makes you think about your data in a slightly different way. Using a combination of these strategies will allow you to examine different possibilities and will shed light on your data that you might not otherwise have thought of. To illustrate the transcription process Box 8.9 (on p. 240) details the initial annotation notes for

---

Box 8.6 Transcription codes

We recommend that you use only a sample of these codes (the first five listed). The ones we would avoid, unless necessary, are those which concern speech patterns and timings. While these codes might, at first, seem tedious to use, they are quick to learn and soon become natural additions to your transcriptions. Adding these codes will provide a richness to your transcript that will aid subsequent analysis.

(.)	RES: Yeah (.) I've done	*A dot in parentheses indicates a slight pause. Use three dots for a slightly longer pause.*
( )	RES: Yeah, I've done ( )	*Empty parentheses indicate that the transcriber could not hear or make out what was said.*
(word)	RES: Yeah, I've done (that)	*Parenthesised words are possible hearings.*
(( ))	RES: Yeah, I've done ((laughs))	*Double-parenthesised words are author descriptions.*
WORD, *word* word	RES: YEAH, I'VE DONE RES: Yeah, *I've done* RES: Yeah, I've done	*Capitals indicate loud sounds.* *Italics indicate a change in pitch.* *Underlining indicates phrases that are stressed.*
[	INT: Have you [ever RES: [yeah	*Left bracket indicates the point at which a current speaker's talk is overlapped by another's.*
=	INT: Have you ever = RES: = Yeah, I've done	*Equals signs, one at the end of a line and one at the beginning, indicate no gap between talk.*
::	O::kay	*Colons indicate prolongation of the immediate prior sound.*
.hhhh	Yeah .hhhh I've done	*.hhhh indicate breathing patterns. A row of h's prefixed by a dot indicate inbreath. No dot indicates an outbreath. The number of h's indicates length of in- or outbreath.*
(.4)	Yeah (.4) I've done	*Numbers in parentheses indicate elapsed time in seconds.*

Source: Adapted from Silverman 1993: 118.

Box 8.7 General rules for transcription

If you have access to a computer
- Where possible type your transcriptions into separate word-processed files. There are two benefits to this: the data will be amenable for analysis using a suitable qualitative data analysis package; if you have no access to such a package, 'manual' analysis will be easier because you will be able to cut-and-paste sections of transcriptions easily.
- Save the file you are working on at regular intervals.

If you do not have access to a computer
- Divide the paper you are transcribing onto into two halves. In the left-hand half write out the transcription. The right-hand half is for annotations and memos.
- Make at least two photocopies of the transcription.
- Keep the original transcription safe and use the copies for annotating and cutting-and-pasting.

Respondent A's transcript. Box 8.10 (on p. 241) illustrates the full annotation notes once transcription has been completed. The full annotated notes for the other

three data sets are listed in Appendix B. Note how the initial annotation has been used as pointers to aid full annotation. We have also detailed whether the annotation is a memo (**m**) or idea/thought (**i**) to illustrate the difference between the two. Information in square brackets is presented only to help you understand abbreviations or specialised terms. These annotated notes (and the notes accompanying the other transcriptions) will now be used to help focus analysis and identify categories. Equally they could be used as a reference for subsequent data generation.

Once your data are transcribed and you have decided specifically what to focus your analysis upon, you need to start categorising and then connecting your data. It is to this that we now turn.

8.4 Categorising qualitative data

In many respects categorising quantitative data is relatively straightforward. Numeric data are easily grouped into ordinal, nominal, interval and ratio categories

Box 8.8   Strategies to aid annotation

- **The interrogative quintet**: Who? What? When? Where? Why? These are exploratory questions that can be easily applied to your transcriptions to open lines of enquiry. Another useful question is 'so what?'.
- **The substantive checklist**: Do the data correspond to any of the substantive issues which the research concerns? (e.g., a checklist developed from our reading of the literature for our example study might be: spatial confusion; situational confusion; coping strategies; learning strategies).
- **Transposing data**: This concerns asking 'what if' questions. For example, by substituting in members of a different social group would the answers still apply? These questions allow you to explore and identify the uniqueness or generalities of the data, and to situate the data within a context.
- **Making comparisons**: Transposition is a form of comparison. You can also make more formal comparisons to other data sets you might have transcribed (e.g., does this happen elsewhere?).
- **Free association**: This is the jotting down of all the other things that you might associate with a particular section of the data: in a sense, a brainstorming session.
- **Shifting focus**: We can think of the data at different levels (e.g., at a universal level or specific points). Shifting focus concerns looking at the data at these different levels and trying to 'see' the relationship between the 'small' and 'large' pictures.
- **Shifting sequence**: By reading the transcription in a different order, varying the sequence of entries, you may spot things you may have missed when read linearly.

Source: Adapted from Dey 1993: 83–88.

and the relationships between data in these categories are logical (e.g., 1 < 2). Categorising qualitative data is not so simple. Because the data consist of non-numeric information there can be few logical relationships between the data. Placing qualitative data into **meaningful categories**, then, can be a tricky task. Your annotated notes, however, should make this task much easier. This is because, in part, your annotations represent **informal coding strategies**.

The task now is to log the data within the transcripts into **formal categories** to aid further analysis. While your annotations might have provided you with a pretty good feel for your data, categorisation will allow further and deeper insights. The main question you need to ask yourself at this point is: 'on what basis are data going to be assigned to categories?'.

Generally speaking, you are seeking to identify the similarities and differences between data. This is not always an easy task and is part of a **creative process**. As we described in Section 8.2, there are any number of ways we can categorise the same data. We can demonstrate this with a small example. Sears Tower (Chicago), the Grand Canyon, the River Nile and the Pyramids might be classified into three different classifications (Figure 8.3 on p. 242).

All three classifications lead to different pairings. The decision as to which scheme to use really centres upon what you interested in and the quality of your data. Remember that there are many different ways of seeing the same data. None are right or wrong, but some are more appropriate than others.

The easiest way of classifying your data, if you have carried out a systematic interview or observation study, is to group the answers to specific questions or observations. While this might seem logical, it does not always make the best line of enquiry. Our suggestion is to think carefully about the focus of your analysis – what you really want to know about – and then using your annotated notes devise a series of **master categories**. If, after you have started to code your transcripts, you change your mind about the categories you wish to use, you can always go back one or several stage(s) and modify or extend a category (or categories) or devise a new one (or set). Creating categories is a continual process of development as you refine your ideas. When starting and throughout the process, remember you are seeking to group like-with-like, creating a series of category levels that work from the general to the specific. To do this effectively each category must be easy to identify and clearly distinguishable from others. That is, you must have a well-developed set of criteria for placing data into different categories. This does not mean that all categories must be mutually exclusive – in some cases the same data can belong to more than one category. These criteria should be **conceptually and empirically grounded**. That is, criteria should relate to the overall focus of analysis but should also have some empirical basis (the data can be easily classed in this way). In general, if you have to force your data into categories then you need to think of a new classification. Dey (1993) thus suggests that categories should have an **internal aspect** (they must be meaningful in relation to the data – i.e., they should not be arbitrary) and an **external aspect** (they must be meaningful in relation to other categories). Categories then should not be created in isolation from other categories within the analysis. They should relate to each other in meaningful ways.

Box 8.9   Initial transcription

Respondent A's transcript	Initial thoughts
INT: *How do you approach and resolve problems encountered en route?* **RES:** Such as?	
INT: *Well, how about detours?* **RES:** I'm OK with them. But I got lost once. It tends to knock you sideways a bit. I crossed one too many streets and I ended up going in the opposite direction. Well, I would first of all – if I didn't know the area – I would ask someone. Or I would go back on myself and retrace, and maybe get a taxi or some other transport. Or maybe I would go to a friend's house nearby.	**(m)** Multiple coping strategies
INT: *How about if you found yourself off-route?* **RES:** I don't really like finding myself in this situation. If I was caught in a security scare or something like that, I would wait until someone came along, or I would go to the door of a house. Or I would retrace.	**(m)** Particular hazard – terrorism. **(i)** A Northern Ireland factor?
INT: *Do you come across any specific hazards?* **RES:** Sometimes you have to be very wary, if there's a lorry parked on the pavement, or some awkward street furniture blocking the way. When two or three things happen at once, it can be really confusing. The only thing you can do is to backtrack.	**(m)** Situational confusion
INT: *Do you think environmental modifications might make navigating easier?* **RES:** There is a conflict between dropped kerbs and flush pavements. I can't tell the difference. If you go from Castle Street, from Castle Court, the place where the pavement should be, it just isn't there. There is nothing to tell you where it is, it is all flush. Those bobbly tiles – I don't like those.	**(m)** Conflict in planning. **(i)** What is planning policy in BUA [Belfast Urban Area]? **(m)** But bobbly tiles are for VI [visually impaired] people!
INT: *How do you find out about new routes?* **RES:** By asking people. I haven't been down in the town for a while, but I knew there was work done on this part of the road. I didn't realise the Westlink went under the Falls Road. Nobody had ever told me that. I was walking across towards Divis Street, and suddenly I heard the roaring of a heavy thundering truck. 'Shit, this one's going to get me', I thought. Then it went underneath and I was OK.	**(m)** Spatial confusion – lack of information
INT: *How often do you learn a new route?* **RES:** Very rarely, once or twice a year, most places I go, I go regular.	**(m)** Sticks to known routes
INT: *How do you prepare for learning a new route?* **RES:** If I can get the information in advance I may go over it beforehand. If I can get detailed instructions, 'cross three roads, turn left, on the up kerb, turn right', it's a great help.	**(m)** Propositional learning [spatial language – left, right, etc.]
INT: *Have any route learning strategies proved to be particularly successful or disastrous?* **RES:** No.	

Box 8.10 Full transcription

Respondent A's transcript	Full annotation
INT: *How do you approach and resolve problems encountered en route?* **RES:** Such as?	
INT: *Well, how about detours?* **RES:** I'm OK with them. But I got lost once. It tends to knock you sideways a bit. I crossed one too many streets and I ended up going in the opposite direction. Well, I would first of all – if I didn't know the area – I would ask someone. Or I would go back on myself and retrace, and maybe get a taxi or some other transport. Or maybe I would go to a friend's house nearby.	**(m)** Spatial confusion – miscounting streets. **(i)** Is this common? Do sighted people do this? – literature. **(m)** Self-recognition of confusion. **(m)** Multiple coping strategies: ask; retrace; get somebody else to navigate (taxi); evasion.
INT: *How about if you found yourself off-route?* **RES:** I don't really like finding myself in this situation. If I was caught in a security scare or something like that, I would wait until someone came along, or I would go to the door of a house. Or I would retrace.	**(m)** Fear of disorientation or being lost. **(m)** Particular hazard – terrorism. **(i)** A Northern Ireland factor? Compare with other respondents. Does this respondent live in a 'high' terrorism area? If so, how will this affect behaviour – less likely to travel alone? **(m)** Multiple coping strategies: stay put, seek shelter(?), retrace. **(i)** What is advice of O&M [orientation and mobility] training? What is most effective strategy?
INT: *Do you come across any specific hazards?* **RES:** Sometimes you have to be very wary, if there's a lorry parked on the pavement, or some awkward street furniture blocking the way. When two or three things happen at once, it can be really confusing. The only thing you can do is to backtrack.	**(m)** Self-awareness of fear (expressed other places in interview). **(m)** Multiple hazards: pavement parking; street furniture. **(m)** *Situational confusion.* **(i)** Any situation particularly hazardous? How often does this occur? **(m)** Single hazard solution: retrace. **(i)** Could there be others? **(i)** What is advice of O&M [orientation and mobility] training?
INT: *Do you think environmental modifications might make navigating easier?* **RES:** There is a conflict between dropped kerbs and flush pavements. I can't tell the difference. If you go from Castle Street, from Castle Court, the place where the pavement should be, it just isn't there. There is nothing to tell you where it is, it is all flush. Those bobbly tiles – I don't like those.	**(m)** Conflict in planning. **(i)** What is planning policy in BUA [Belfast Urban Area]? Has this changed because of the DDA (1996) [Disability Discrimination Act]? **(m)** Another hazard: flush pavements. **(i)** But bobbly tiles are for VI [visually impaired] people! Why disliked?
INT: *How do you find out about new routes?* **RES:** By asking people. I haven't been down in the town for a while, but I knew there was work done on this part of the road. I didn't realise the Westlink went under the Falls Road. Nobody had ever told me that. I was walking across towards Divis Street, and suddenly I heard the roaring of a heavy thundering truck. 'Shit, this one's going to get me', I thought. Then it went underneath and I was OK.	**(m)** Single spatial learning strategy: asking people. **(m)** Spatial confusion again: lack of information – new road layout. **(m)** Fear. **(i)** Seems to be a strong link between spatial confusion and fear.
INT: *How often do you learn a new route?* **RES:** Very rarely, once or twice a year, most places I go, I go regular.	**(m)** Sticks to known routes. Little new learning. **(i)** Why is this? – terrorist area? Doesn't need to? Is one or two routes all that is needed?
INT: *How do you prepare for learning a new route?* **RES:** If I can get the information in advance I may go over it beforehand. If I can get detailed instructions, 'cross three roads, turn left, on the up kerb, turn right', it's a great help.	**(m)** Single learning strategy: propositional learning. **(i)** Is this the only strategy used, or just most common?
INT: *Have any route learning strategies proved to be particularly successful or disastrous?* **RES:** No.	**(m)** No details! **(i)** What are the different route learning strategies?

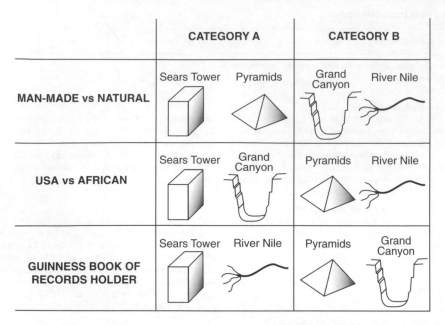

	CATEGORY A		CATEGORY B	
MAN-MADE vs NATURAL	Sears Tower	Pyramids	Grand Canyon	River Nile
USA vs AFRICAN	Sears Tower	Grand Canyon	Pyramids	River Nile
GUINNESS BOOK OF RECORDS HOLDER	Sears Tower	River Nile	Pyramids	Grand Canyon

Figure 8.3 Categorising data.

After studying our data (Boxes 8.1–8.4), we decided that the three things we were most interested in were:

- Spatial confusion
- Situational confusion
- Learning strategies

We have chosen these three categories as they most directly relate to the main aspects of the overall study. In our project, we were interested in how visually impaired people interact with and learn urban environments. These then are our master categories. They are clearly defined and easily distinguishable (see Box 8.11). However, as they stand, these categories are very broad in nature. Just assigning the relevant portions of data to these categories will not provide sufficient disaggregation to make analysis easier. This would especially be the case if we were categorising the data from all 20 interviews rather than just four.

We therefore need to **disaggregate** (refine) our data further by devising some sub-categories. This means examining the data more closely by carefully sifting through the transcripts to identify possible sub-categories. One possible way to make this process easier is to undertake the next two steps (assigning codes and sorting data) and then follow the process through once more, treating each of the four main categories as separate data sets. Following this iterative process, however, we must be careful to ensure that data from within the main categories is still coded in any appropriate sub-categories that may have grown out of another main category. In other words, this process is no substitute for sifting through data. **Category development**, if done properly, requires you to become thoroughly familiar with the data. Taking short-cuts will lead to deficiencies in your analysis. Admittedly this may be necessary, given time constraints, but be aware of potential shortcomings.

Box 8.11   Criteria of inclusion in master categories

Spatial confusion: Include data about spatial behaviour, getting lost, confused or disorientated.
Situational confusion: Include data about specific hazards or situations.
Learning strategies: Include data about how people learn new routes, specialised training.

Note how these categories relate to each other.

There is no limit to the number of categories you can develop. However, while the list needs to be comprehensive, you need to find a balance between depth and breath. With not enough categories, the data will be difficult to compare. With too many categories, it will be difficult to assign data to categories because of similarities and overlaps. It will also be difficult to relate categories and develop links further on in the analysis. We suggest progressing with categorisation up to the point where the data within categories seem to sit naturally with one another and no further division is immediately obvious. The categories you do determine must be **exhaustive**. All useful data must be able to reside in an appropriate category. In addition, we recommend that you are also careful to record why you have chosen your category choices over others. This will be useful at a later stage if you are trying to remember why you have assigned data in certain ways. Figure 8.4 illustrates all the categories identified for the purpose of our study and how they relate to each other.

Once you have decided how the data are to be categorised you need to assign data within your transcripts to specific categories. This process is known as **coding**. The easiest way to code your transcripts is, using a photocopy, to work your way through each of the transcripts, placing a specific code next to each relevant piece of data, termed a databit. A relevant section of data might consist of a phrase, or a sentence, or even a whole passage. As we have discussed, this is an ongoing process as you refine your choice of categories. The codes you use should be recorded on a **master sheet**. Remember that your data may well belong to more than one category. It should be coded as such, otherwise a salient point might be missing from some categories. When assigning codes to your data you should also do the same with your memos and ideas. Linking together relevant data and annotation through coding will help further analysis. Although it is probably sensible to work through the data in a sequential manner, you should by this stage be sufficiently familiar with the data to use a selective path of coding. Remember, you are sorting your data only into the categories at the bottom of the category tree (see Figure 8.4). The upper levels exist only to show you the pathway and links between the categories. In terms of analysis these are redundant as there is no point in replicating the data at all different levels. Box 8.12 illustrates the codes assigned to Respondent A's transcript. To aid your analysis we suggest that you adopt a similar format. The coded transcripts of Respondents B, C and D are provided in Appendix B.

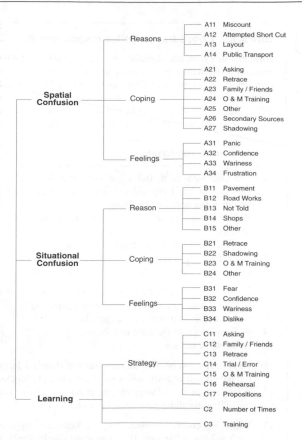

Figure 8.4　Category codes for example data.

When you have completed the coding process on your data, the next task is to sort the data into the relevant categories. This consists of a cut-and-paste process. When completing the task using a word-processor or by literally cutting up and pasting photocopies of your transcripts together, good **data management** really comes to the fore. Before you cut-and-paste you must devise a suitable **filing system**. If you cut up your transcripts and paste them back together again without labelling your pastings with master codes relating to the source of the data, you will soon become confused as to where a piece of data originated. Similarly you should keep a master record of where all the pastings of a source have been copied to. This **housekeeping**, while time consuming and tedious, will improve the efficiency and level of analysis at a later date. It will also stop your analysis from falling into disarray. Box 8.13 is an example of a **sorted category**.

Box 8.12   Coded transcript for respondent A

Note that this is an edited transcript – only the data that have been coded are present. The annotations have also been removed. The transcript has been placed into blocks and numbered phrases. Each block represents what was a passage in the original transcript. Each numbered phrase is a piece of coded data (databit). The relevant code is listed on the same line. To help you understand the coding system, we have provided a summary code rather than just an abbreviation. Usually you would just use an abbreviation, referring to your master sheet for full details. In places we have provided brief context annotation within the transcript (e.g., [off-route]).

Respondent A's transcript	Category code
1 I'm OK with them. But I got lost once. It tends to knock you sideways a bit. I crossed one too many streets and I ended up going in the opposite direction.	A11 [Sp Con – reason – miscount]
2 Well, I would first of all – if I didn't know the area – I would ask someone.	A21 [Sp Con – coping – asking]
3 Or I would go back on myself and retrace,	A22 [Sp Con – coping – retrace]
4 and maybe get a taxi or some other transport.	A25 [Sp Con – coping – other]
5 Or maybe I would go to a friend's house nearby.	A25 [Sp Con – coping – other]
6 I don't really like finding myself in this situation [off-route].	A33 [Sp Con – feel – wariness]
7 If I was caught in a security scare or something like that,	B15 [Sit Con – reason – other]
8 I would wait until someone came along,	B24 [Sit Con – coping – other]
9 or I would go to the door of a house.	B24 [Sit Con – coping – other]
10 Or I would retrace.	B21 [Sit Con – coping –retrace]
11 Sometimes you have to be very wary, if there's a lorry parked on the pavement, or some awkward street furniture blocking the way.	B33 [Sit Con – feel – wariness] & B11 [Sit Con – reason – pave]
12 When two or three things happen at once, it can be really confusing. The only thing you can do is to backtrack.	B21 [Sit Con – coping – retrace]
13 There is a conflict between dropped kerbs and flush pavements. I can't tell the difference. If you go from Castle Street, from Castle Court, the place where the pavement should be, it just isn't there. There is nothing to tell you where it is, it is all flush. Those bobbly tiles – I don't like those.	B11 [Sit Con – reason – pave] & B34 [Sit Con – feel – dislike]
14 By asking people.	C11 [Learn – strategy – asking]
15 I haven't been down in the town for a while, but I knew there was work done on this part of the road.	B12 [Sit Con – reason – road works]
16 I didn't realise the Westlink went under the Falls Road. Nobody had ever told me that.	B13 [Sit Con – reason – not told]
17 I was walking across towards Divis Street, and suddenly I heard the roaring of a heavy thundering truck. 'Shit, this one's going to get me', I thought. Then it went underneath and I was OK.	B31 [Sit Con – feel – fear]
18 Very rarely, once or twice a year,	C2 [Learn – No. of times]
19 most places I go, I go regular.	A41 [Sp Con – Sp Beh – set pattern]
24 If I can get the information in advance I may go over it beforehand.	C16 [Learn – strategy – rehearsal]
25 If I can get detailed instructions, 'cross three roads, turn left, on the up kerb, turn right', it's a great help.	C17 [Learn – strategy – propositions]

8.5   Splitting and splicing

At this point, your data should be in two different forms: the original transcripts and as sorted categories. While our attention, up until now, has focused upon annotating, categorising and coding the original transcripts, we are now about to shift our focus to the sorted categories. Splitting and splicing is concerned with reassessing the organisation and data management

---

Box 8.13   A sorted category

Note that this contains both data and annotations. Also, there are two identifying labels, e.g., RES A (respondent A) and (8) (number of line in transcript the databit came from), so that all the data and annotations can be traced back to their original sources. Remember that these data are drawn from only four interviews. With more interviews the richness of this category would improve.

CATEGORY B11 [Situational confusion – reasons – pavement]:

Databits

i) **RES A:** (11) Sometimes you have to be very wary, if there's a lorry parked on the pavement, or some awkward street furniture blocking the way.

ii) **RES A:** (13) There is a conflict between dropped kerbs and flush pavements. I can't tell the difference. If you go from Castle Street, from Castle Court, the place where the pavement should be, it just isn't there. There is nothing to tell you where it is, it is all flush. Those bobbly tiles – I don't like those.

iii) **RES B:** (12) A lot of shops now have huge amounts of pavement clutter. Sometimes I can see these, pick them out OK.

iv) **RES C:** (11) I keep bumping into a lot of them [hazards] such as when I come across them outside shops and the like. Very often I'd be kicking things.

v) **RES C:** (14) Those wiggly pavement bobbles are good for totals, but for me they are a menace – it throws you.

vi) **RES D:** (7) Tactile pavements need to be standardised and there need to be more of them.

Annotations

i) (**m**) Self-awareness of fear (expressed other places in interview). (**m**) Multiple hazards: pavement parking; street furniture. (**m**) Situational confusion. (**i**) Any situation particularly hazardous? How often does this occur?

ii) (**m**) Conflict in planning. (**i**) What is planning policy in BUA [Belfast Urban Area]? Has this changed because of the DDA (1996) [Disability Discrimination Act]? (**m**) Another hazard: flush pavements. (**i**) But bobbly tiles are for VI (visually impaired) people! Why disliked?

iii) (**m**) Situational hazards: pavement clutter.

iv) (**m**) Situational confusion: street obstacles.

v) (**i**) Planning design for VI people poor?

vi) (**m**) Planning: needs improvement and standardisation. Tactile pavements inadequate. (**i**) Need new laws to aid disabled people's interaction with environments?

---

of data within sorted categories. Just as we carefully sorted through our original transcripts, we now turn to sorting through our reorganised data. This stage then has two purposes. First, we consider ways of refining or focusing analysis in the light of any revelations while categorising the data. 'Playing' with the data in various ways (annotating, categorising, sorting) will probably have led to insights that just reading the transcripts alone will have failed to highlight. Second, this acts as a cross-checking phase. Some of our categories may be large and cumbersome, needing further refinement, and some, while seeming valid and important during categorisation, are small and need to be merged and integrated to be viable.

During the process of creating, modifying and extending our categories we have created a number of **databits** (coded pieces of text). These have been assigned to sorted categories (see Box 8.13). By categorising the data and placing them into sorted categories we have removed the data from their original context. In return, we have gained a new way to think about our data. It is now easy to compare data relating to a similar theme (e.g., pavement-based hazards). In order to refine our analysis and also compare

data within a category, Dey (1993) suggests that the data should be categorised further. **Splitting** refers to the task of refining the analysis of the data by sub-categorising databits within a sorted category (Figure 8.5). Clearly, we have already 'split' our data several times to categorise the original data. However, in the last stage (categorisation) we divided up our transcripts into separate databits which were then placed into categories. Here, we are not trying to create new databits, just to categorise existing databits into relevant sub-categories. These sub-categories should be **internally consistent** (e.g., the data within each should refer to the same things), be **conceptually related** to each other (e.g., all categories are variations on a theme), and be **analytically useful** (e.g., they relate to the aims of the study). The aim here is to identify the various dimensions of the data and to start establishing the associations and relationships between data. Not all sorted categories will merit splitting. In some, the data will be relatively consistent, in others the data might be more complex. Also some categories will be of more interest than others. For example, in our analysis we may have become particularly interested in confusion caused by situational hazards. We

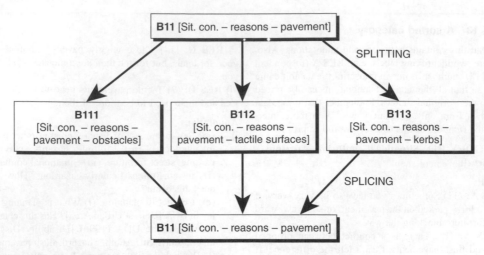

Figure 8.5   Splitting and splicing data.

therefore might decide to further analyse data within those categories by seeing whether we can split the data further. Box 8.14 illustrates the splitting of sub-category B11 [pavement-based confusion]. By splitting the data within this category we are seeking to identify particular hazards or annoyances that visually impaired people encounter when navigating along pavements.

Splicing, alternatively, concerns the interweaving of related categories (Figure 8.5). While we split categories for greater resolution and detail, we splice categories for **greater integration** and scope. Splicing categories together is a search for understanding how different themes intertwine and relate to each other. In its simplest form, splicing just takes the form of re-merging categories together. For example, if we had originally classified our pavement sub-categories as full categories (e.g., B111 [Sit Con – reason – pavement – obstacles], B112 [Sit Con – reason – pavement – tactile], B113 [Sit Con – reason – pavement – kerbs]) we might want to splice these together to get an integrated picture of the role of pavements in situational confusion. Similarly we might want to splice together situational coping strategies or learning strategies to try to gain a greater understanding of these processes. Alternatively we might splice together categories in different parts of our category schema, for example all the data concerning coping strategies regardless of whether they relate to spatial or situational confusion. Dey (1993) suggests that splicing provides a greater intelligibility and coherence to analysis through the informal linking of associated data. Splicing should also be seen as a method of integrating, and not

ignoring, relevant data that sit in more marginal (less central) categories. That is, splicing is used to re-sort data in a process of bringing marginal or small categories to the centre through integration. This creates a smaller number of stronger and more central categories. As we can see by looking through the categories we assigned data to, many of them are small in size. By merging them together we ensure that the relevance of these data is not ignored through a concentration on the 'richer' categories. For an example of splicing, return to Box 8.14. Instead of identifying sub-categories, imagine a reversal of the process so that these sub-categories are merged to create a universal pavement category (our original B11 category).

Splicing, like splitting, is time consuming and should be undertaken only after careful consideration of which data are to be spliced together and whether the resultant category will be strongly related to the central themes of the study. The processes of splitting and splicing should not, however, be avoided as they do provide quite powerful tools to gain further insight into your data. It should be noted that while splitting and splicing provide further depth to the process of categorisation and allow better comparison between data, they do not identify the nature of the relationships between data. It is to this that we now turn.

8.6   Linking and connecting

The tasks of categorising, sorting, splitting and splicing data have allowed us to analyse our data through

---

Box 8.14   Splitting sorted data

B11 [Situational confusion – reasons – pavement]:

Databits

i) **RES A:** (11) Sometimes you have to be very wary, if there's a lorry parked on the pavement, or some awkward street furniture blocking the way.

ii) **RES A:** (13) There is a conflict between dropped kerbs and flush pavements. I can't tell the difference. If you go from Castle Street, from Castle Court, the place where the pavement should be, it just isn't there. There is nothing to tell you where it is, it is all flush. Those bobbly tiles – I don't like those.

iii) **RES B:** (12) A lot of shops now have huge amounts of pavement clutter. Sometimes I can see these, pick them out OK.

iv) **RES C:** (11) I keep bumping into a lot of them [hazards] such as when I come across them outside shops and the like. Very often I'd be kicking things.

v) **RES C:** (14) Those wiggly pavement bobbles are good for totals, but for me they are a menace – it throws you.

vi) **RES D:** (7) Tactile pavements need to be standardised and there need to be more of them.

New sub-categories

Obstacles

RES A: (11) Sometimes you have to be very wary, if there's a lorry parked on the pavement, or some awkward street furniture blocking the way.

RES B: (12) A lot of shops now have huge amounts of pavement clutter. Sometimes I can see these, pick them out OK.

RES C: (11) I keep bumping into a lot of them [hazards] such as when I come across them outside shops and the like. Very often I'd be kicking things.

Tactile surfaces

RES C: (14) Those wiggly pavement bobbles are good for totals, but for me they are a menace – it throws you.

RES D: (7) Tactile pavements need to be standardised and there need to be more of them.

Kerbs

RES A: (13) There is a conflict between dropped kerbs and flush pavements. I can't tell the difference. If you go from Castle Street, from Castle Court, the place where the pavement should be, it just isn't there. There is nothing to tell you where it is, it is all flush. Those bobbly tiles – I don't like those.

While the initial category (B11 is small) and the sub-categories contain only a couple of entries each, remember that these data have come from only four short interviews. If all 20 interviews had been included each of these sub-categories would increase in size.

---

comparison. When undertaking these tasks we have been involved in a process of deciding how similar or dissimilar pieces of data are. We have formulated a basic understanding of data within specific categories and identified some common links between categories. As such, by this stage we have got a pretty good 'feel' for the data. However, the common links identified need to be explored further, their existence ratified and their nature determined. Further, as we have already noted, by breaking up the data and reorganising it in sorted categories we have taken the data out of context and lost information regarding the relationships between individual databits. In other words, we have lost information concerning how the data 'hang together'. Linking and connecting concern the process of trying to identify how the reorganised data 'hang together' within the context of the original transcripts. Rather than just being able to say how alike or similar the data are, we are trying to identify and understand the nature of relationships between data: how things are associated and how things interact. There is a difference between association and

interaction. In **association**, separate events occur together (when one thing happens so does another; A → B). In **interaction**, there is an engagement between the two events (A ← → B).

Clearly, things that are associated or interact with each other need not be similar (e.g., a person is very different from a pavement but there is clearly a link as people use pavements). Whereas categorisation and comparison concerned the identification of formal connections (similar/different), finding associations and links concerns the identification of **substantive connections**. There are two different types of substantive connections (see Sayer, 1992). **Internal relations** are necessary; one does not exist without the other. For example, without sight you are blind (conversely if you did have sight you would not be blind). There are no alternatives. **External relations** are contingent or non-necessary; a substantive relationship may exist but need not do so. For example, a blind person can, but does not need to, use a guide dog to navigate. Many of the approaches within the social sciences seek to identify substantive relations (laws) between social entities: those things that are necessary for certain things to happen. Of course, most social relations are extremely complex, consisting of a series of contingent measures. Here, many social scientists try to work out what happens when some contingent relations exist and others are absent. This is the same whether the researcher is from a quantitative or qualitative background, is seeking understanding or explanation, or is using deductive or inductive approaches: the researcher is trying to figure out how data are substantively linked (internally or externally).

So far, we have not attempted to identify **substantive links** between our data. We have not, for example, examined whether there is a link – internal or external – between a certain type of *spatial confusion* and a certain type of *coping strategy*. If a visually impaired person becomes disorientated by miscounting streets, do they necessarily undertake a retracing coping strategy (that is, all visually impaired people in this situation will retrace) or is the choice more contingent (that is, they undertake a coping strategy but this varies from person to person)? The next stage, then, is to try to identify the links between data and the nature of those links (internal/external). As we have stated, these links do not need to be causal (e.g., X causes Y) but can take any form (e.g., X dislikes Y, X scares Y, X supports Y, etc.).

The best place to start is by going back to our original transcripts and seeing what links we can identify within the text (Box 8.15). The links we find within the transcripts will provide the basis for exploring wider connections between the data now stored within the sorted categories. Some of these links will be intuitive and may have formed the basis of categorisation (e.g., A1 [reasons for spatial confusion] and A2 [coping strategies]), but it is best to go formally through each transcript so that some of the less obvious links are established.

It is clear from Box 8.15 that there are clear links between spatial and situational confusion, personal emotions and feelings (e.g., wariness, fear, disorientation), coping strategies and learning strategies. Having established this in an informal manner we now move on to explore the nature of these links more fully. At present, we do not know how these links operate or the relationships between links – just that they seem to exist.

To try to connect the bones of our skeleton together we can use a few different techniques. At a simple level we can use a **matrix** to compare associations between individual transcripts (see Table 8.1). This matrix reveals all the categories an individual

has databits assigned to (identified by an 'x' in the appropriate cell). By looking for similar patterns we can try to establish how the data relate to each other. Unfortunately, from our data, little clear pattern is emerging. This is in part because we have used only four data sets, but it may also be because the links between our data are contingent and therefore things like coping strategy can differ between individuals. What is clear is that coping strategies do seem to differ slightly for spatial and situational confusion. While three of the respondents said that they would ask for help when spatially confused, none would do so when situationally confused. There are also a wide range of coping strategies.

A second strategy would be to categorise the links identified within the transcript, placing the linked data into new sorted categories. For example, Box 8.16 contains all the link data between situational confusion and coping strategies. Nearly the same effect could be produced by splicing together all the coping category data (e.g., B21–B24). However, the data would not be as rich in nature, as not all the databits presented in Box 8.16 would be present (see next strategy). Looking at these data it is clear that there are very strong links, as might be expected, between situational confusion and the use of coping strategies. This substantive link is internal, in that without situational confusion there would be no need for coping strategies; however, the link is also contingent, as the particular coping strategy employed is not determined. If we were to look now at the links between situational confusion and people's feelings, we would see that there is also a strong link between them (Box 8.17 on p. 251). It would therefore be reasonable to suggest that coping strategies are employed in times of situational confusion because of fear, anxiety and wariness. However, people also worry about these situations because they feel that they make them look silly, and such situations make them feel embarrassed. The link be-tween coping strategies and anxiety/fear is rarely stated in the data. The link, although inferred, has been determined through **lateral thinking** by examining two related data categories. Although intuitive, it has taken extensive data analysis through description, categorisation and linking to provide sufficient evidence to support this claim.

A third strategy builds upon the last strategy but relies exclusively upon the sorted data. This approach, rather than searching for explicit links and identifying some implicit connections, focuses solely upon investigating possible **implicit links**. As such, this strategy involves lateral rather than literal thinking.

Box 8.15 Establishing links

Respondent A's transcript	Initial links

INT: *How do you approach and resolve problems encountered en route?*
RES: Such as?

INT: *Well, how about detours?*
RES: I'm OK with them. But I got lost once. It tends to knock you sideways a bit. I crossed one too many streets and I ended up going in the opposite direction. Well, I would first of all – if I didn't know the area – I would ask someone. Or I would go back on myself and retrace, and maybe get a taxi or some other transport. Or maybe I would go to a friend's house nearby.

Spatial confusion ←→ coping strategy

INT: *How about if you found yourself off-route?*
RES: I don't really like finding myself in this situation. If I was caught in a security scare or something like that, I would wait until someone came along, or I would go to the door of a house. Or I would retrace.

Spatial confusion → wariness → coping strategy
└ ← ← ← ← ← ┘

INT: *Do you come across any specific hazards?*
RES: Sometimes you have to be very wary, if there's a lorry parked on the pavement, or some awkward street furniture blocking the way. When two or three things happen at once, it can be really confusing. The only thing you can do is to backtrack.

Situational confusion ←→ coping strategy

INT: *Do you think environmental modifications might make navigating easier?*
RES: There is a conflict between dropped kerbs and flush pavements. I can't tell the difference. If you go from Castle Street, from Castle Court, the place where the pavement should be, it just isn't there. There is nothing to tell you where it is, it is all flush. Those bobbly tiles – I don't like those.

Situational confusion ←→ planning improvements

INT: *How do you find out about new routes?*
RES: By asking people. I haven't been down in the town for a while, but I knew there was work done on this part of the road. I didn't realise the Westlink went under the Falls Road. Nobody had ever told me that. I was walking across towards Divis Street, and suddenly I heard the roaring of a heavy thundering truck. 'Shit, this one's going to get me', I thought. Then it went underneath and I was OK.

Spatial confusion ←→ learning/feelings

INT: *How often do you learn a new route?*
RES: Very rarely, once or twice a year, most places I go, I go regular.

INT: *How do you prepare for learning a new route?*
RES: If I can get the information in advance I may go over it beforehand. If I can get detailed instructions, 'cross three roads, turn left, on the up kerb, turn right', it's a great help.

INT: *Have any route learning strategies proved to be particularly successful or disastrous?*
RES: No.

Table 8.1   Using a matrix to identify links.

RES	Reason for Sp. Con.				Coping strategies for Sp. Con.							Feelings				Sp. Beh	
	A11	A12	A13	A14	A21	A22	A23	A24	A25	A26	A27	A31	A32	A33	A34	A41	A42
A	x				x	x			x					x		x	
B		x	x		x		x			x		x	x	x	x	x	x
C				x	x		x	x		x	x						
D						x										x	

RES	Reasons for Sit. Con.					Coping strategies for Sit. Con.				Feelings			
	B11	B12	B13	B14	B15	B21	B22	B23	B24	B31	B32	B33	B34
A	x		x		x	x		x	x	x		x	x
B	x	x			x		x						
C	x		x								x		
D	x	x			x			x					

RES	Learning Strategy							Time	Train
	C11	C12	C13	C14	C15	C16	C17	C2	C3
A	x					x	x	x	
B		x	x	x				x	
C		x				x		x	x
D					x			x	

Box 8.16   Identifying links 1

Linking situational confusion (*italics*) and coping strategies (<u>underlined</u>)

RES A: *Sometimes you have to be very wary, if there's a lorry parked on the pavement, or some awkward street furniture blocking the way.* When two or three things happen at once, it can be really confusing. <u>The only thing you can do is to backtrack.</u>

RES B: *Say I arrived in town and I wanted to go to a department store, I would get off the bus at the appropriate stop to find that the road had been dug up. The noise of the men working lets me know that they are working, but it is then off-putting as you can't find other sound clues to get around that.* <u>I would, perhaps . . . if in town, I try to follow the flow of people – use them as guides even though they don't know it.</u> The same is true <u>for crossing roads, rather than asking each time, I would find somebody who I think is reasonably competent and then follow them.</u>

RES C: <u>I tend to cloak people, wait on somebody that's going across and then shadow them.</u>

RES D: *Well, most obstacles are small and in areas that I'm familiar with it certainly wouldn't be a problem – I would just take an automatic route to bypass them. For example, I was on the Antrim Road, lines of shops, traffic and there was CableTel digging holes. There were pneumatic drills going, it was really disorientating – I couldn't use the traffic noise to work out which way I was facing.* <u>So I turned, took the dog to the kerb to reorientate myself, turned back and walked through the parallel streets that I know.</u>

You start by asking whether a link could possibly exist between two categories (e.g., X1 and Y1) and what the form of this link might be. Then by looking through the data in X1 and Y1 you try to determine if there is any evidence to suggest that such a link might exist even if the data do not explicitly state such a link. Within the sorted categories, any explicit links have often been removed by categorising the

---

Box 8.17 Identifying links 2

Linking situational confusion (*italics*) and feelings (underlined)

RES A: <u>Sometimes you have to be very wary,</u> *if there's a lorry parked on the pavement, or some awkward street furniture blocking the way.* When two or three things happen at once, <u>it can be really confusing.</u> The only thing you can do is to backtrack.

RES A: By asking people. *I haven't been down in the town for a while, but I knew there was work done on this part of the road. I didn't realise the Westlink went under the Falls Road. Nobody had ever told me that. I was walking across towards Divis Street, and suddenly I* heard the roaring of a heavy thundering truck. <u>'Shit, this one's going to get me', I thought.</u> Then it went underneath and I was OK.

RES B: *I keep bumping into a lot of them [hazards] such as when I come across them outside shops and the like. Very often I'd be kicking things.* <u>And people tend to think, what's up with him, it is a bit early in the day to be drinking.</u>

RES C: *When I do go into shops everything just brights out for a moment.* <u>So you stand there and look like a wally, everybody looking at you. So it is a lot easier with someone to guide me.</u>

---

data into separate, distinctive databits. It is at this point that we really start to **interpret** qualitative data – to try to make inferences in the light of annotating, categorising and sorting the data. This stage is probably the most difficult as it relies on the ability of the researcher to think laterally and connect data together in meaningful ways, despite the lack of explicit links.

There are two ways of trying to negate the problem of missing links and make informed inferences. First, we can compare the data assigned to different categories, and look for evidence of possible connections between them. It may be the case that some of the data might have been assigned to both categories. If so then obviously there is a link between the two. If not, is there evidence to link the categories? Using our example, in Box 8.18 a spliced situational confusion category (all the reasons for situational confusion spliced together) is compared with a spliced coping strategies category (all the strategies for coping with situational confusion spliced together). Note that we do not need to use spliced categories – we have done so only to provide sufficient data for comparison given our small data set. It is clear from comparing these spliced categories that there is an integrative link between situational confusion and coping strategies: when a blind person encounters a confusing situation they employ coping strategies to try to minimise the confusion (situational confusion ←→ coping strategies). This becomes increasingly evident if we compare the sequence of line numbers between respondents – coping statements almost invariably follow statements of confusion. However, just from the difference between the number of entries, it is clear that many of the confusion statements were not accompanied by coping statements. We therefore have to infer that coping strategies did occur in all

these other situations – they were just not articulated in the interview. This might seem a reasonable conclusion to draw. You should, however, be careful of drawing definitive, universal conclusions from inferences. It is highly probable that coping strategies are used when confusing situations occur – from our data we do not know whether this happens on *all* occasions. There might be exceptions that we do know about.

A second approach to dealing with missing links might be to look through a single sorted category to try to infer any associations or links. If any connections are identified then we can look through other sorted categories for more evidence. For example, looking through the entries in the sorted category B11 (Box 8.13), it is clear that there are a number of issues which hinder visually impaired people's use of pavement areas. The implied tone within the statements is that something needs to be done to improve the planning and building of these areas. There is an implied link between situational confusion and better planning. This link is, however, only implied. To strengthen our claims we need either to find an explicit link or to corroborate our evidence. It is to corroboration we now turn.

8.7 Corroborating evidence

The **corroboration of** conclusions is an extremely important part of qualitative data analysis. The processes of qualitative analysis is much like being a detective. You have the accounts from a number of sources which you then need to piece together and try to determine what is going on. Since the qualitative analysis is interpretative and relies on your ability to

Box 8.18 Comparing sorted categories

Spliced situational confusion category	Spliced coping strategy category

Spliced situational confusion category

RES A (7) If I was caught in a security scare or something like that,

RES A (11) Sometimes you have to be very wary, if there's a lorry parked on the pavement, or some awkward street furniture blocking the way.

RES A (13) There is a conflict between dropped kerbs and flush pavements. I can't tell the difference. If you go from Castle Street, from Castle Court, the place where the pavement should be, it just isn't there. There is nothing to tell you where it is, it is all flush.

RES A (15) I haven't been down in the town for a while, but I knew there was work done on this part of the road.

RES A (16) I didn't realise the Westlink went under the Falls Road. Nobody had ever told me that.

RES B (1) Say I arrived in town and I wanted to go to a department store, I would get off the bus at the appropriate stop to find that the road had been dug up. The noise of the men working lets me know that they are working, but it is then off-putting as you can't find other sound clues to get around that.

RES B (12) A lot of shops now have huge amounts of pavement clutter. Sometimes I can see these, pick them out OK.

RES B (12) But if it is bright and there are lots of people then I will bump into things.

RES C (3) When I do go into shops everything just brights out for a moment.

RES C (11) I keep bumping into a lot of them [hazards] such as when I come across them outside shops and the like. Very often I'd be kicking things.

RES C (14) Those wiggly pavement bobbles are good for totals, but for me they are a menace – it throws you.

RES C (15) Some places in the shopping centres you have to go up and grope the shopfront looking for the door. You don't know if it is a window or a door. There are just so many things we have to put up with.

RES D (2) For example, I was on the Antrim Road, lines of shops, traffic and there was CableTel digging holes. There were pneumatic drills going, it was really disorientating – I couldn't use the traffic noise to work out which way I was facing.

RES D (5) One time another dog came up and we got entangled and spun around,

RES D (7) Tactile pavements need to be standardised and there need to be more of them.

RES D (8) Shopfronts with baskets and stalls are a nightmare, but you can't outlaw that.

Spliced coping strategy category

RES A (8) I would wait until someone came along,

RES A (9) or I would go to the door of a house.

RES A (10) Or I would retrace.

RES A (12) When two or three things happen at once, it can be really confusing. The only thing you can do is to backtrack.

RES B (2) I would, perhaps . . . if in town, I try to follow the flow of people – use them as guides even though they don't know it. The same is true for crossing roads, rather than asking each time, I would find somebody who I think is reasonably competent and then follow them.

RES D (3) So I turned, took the dog to the kerb to reorientate myself, turned back and walked through the parallel streets that I know.

RES D (6) there was no traffic to orientate myself so again I had to take my dog to the kerb and get my bearings.

Box 8.19 Corroborating conclusions

Thinking of alternatives

- Can you think of any possible alternative conclusions which can be drawn from your data?
- If alternative conclusions do exist:
  — which conclusion is more credible?
  — which conclusion is more internally coherent (hangs together best)?
  — which conclusion might have wider scope to other groups and places?
- If you cannot think of alternative explanations:
  — does your conclusion seem plausible? If not, why not?
  — is your conclusion internally coherent?
  — does your conclusion have universality?
  — are there data that do not fit your conclusions? If yes, why are there exceptions and misfits?

Checking the quality of data

- Are your conclusions the results of your observations, or a result of your fitting the data to what you want it to say?
- Have any other studies reported the same results or did they conclude differently? If their conclusions were different, why might that be?
- In what context were the data generated? Has the context unduly influenced the results?
- How reliable were the respondents in the study? Were they telling the 'truth'?
- Could any biases have been introduced at any stage of the data generation and analysis?

Source: Adapted from Dey 1993: 219–236.

---

make informed and impartial judgements, the process is open to abuse. Evidence can be fabricated, discounted and misinterpreted. The first two practices are generally undertaken maliciously in an attempt to get the data to 'say' what you want. Understandably, such practices are seriously frowned upon within academia. If your project is an assessed piece of work, the detection of such practices may well lead to a poor mark and depending upon the rules of the institution, disqualification. Remember that, unless your data are highly confidential, in general researchers are meant to share data upon request so that the results can be verified.

Misinterpreting evidence is not a deliberate attempt to fabricate the results of a study. It is quite easy to get the 'wrong end of the stick' with some data. Corroborating evidence concerns the **cross-checking** of conclusions to try to avoid '**genuine' errors** in analysis and interpretation. One of the main criticisms of qualitative data analysis is that it is subjective, relying on the ability of the researcher to make **subjective judgements** concerning categorisation, to place value on and interpret data, and to think laterally. Corroboration is aimed at avoiding some of these criticisms by strengthening the claims made from qualitative data; it is concerned with **integrity** and **validity**. There are two main ways to corroborate your conclusions (see Box 8.19). The first is to try to think of possible alternative conclusions and then check whether these are more likely or valid. The second is to check the quality of the data and to compare the conclusions to those drawn from other studies. Both

of these strategies *should* be undertaken before you write up your results. It is important that you are confident about the validity of your claims and that you are prepared to stand by your conclusions.

8.8 Analysing qualitative data quantitatively

So far we have concerned ourselves with analysing qualitative data through interpretative techniques. We can, however, also analyse some aspects of qualitative data using **quantitative techniques**. This necessarily involves the conversion of the qualitative data into a numeric format. This conversion can take the form of counts and tallies and will be nominal or ordinal in character (see Section 4.7). Once in a numeric format we could then apply statistical tests to describe and compare data (see Chapter 5). For example, one of us (Kitchin, 1997) has used this approach to help analyse 'talk-aloud protocol' data. Talk-aloud protocols are where respondents detail what they are doing as they do it. In our case, respondents described verbally what they were thinking about and doing while trying to complete various tests designed to measure their cognitive map knowledge of the Swansea area. The interviews were transcribed and analysed using the strategy detailed in this chapter. Eight different strategies of spatial thought (ways to think about geographical concepts) were identified. These results were then mapped into matrices (see Tables 8.2 and

Table 8.2 The strategy structural frames and the codes for Table 8.3.

Code Description

Identifier
F11   Female respondent 11
M11   Male respondent 11

Common strategies
C1   Imagining or constructing various types of maps.
C2   Referring to the coastline.
C3   Imagining the route or travelling between two locations.
C4   Using travel time to work out the separation between locations.
C5   Imagining standing at a location and 'looking' in the direction of another location.
C6   Imagining looking down vertically or obliquely.
C7   Working out where places are in relation to the current location.
C8   Just know – Propositional coding.

Common task strategies
CT1   Elimination.
CT2   Logical deduction.
CT3   Draw a map.
CT4   Where the sun sets.
CT5   Look back at former answers.

Task-specific strategies
Projective convergence
T4   Work out the direction between two locations by working out the direction from each place to a third place. e.g., for A to B work out direction from A to C, and then C to B.
T5   Imagine flying as a crow would between two locations.
T6   Instead of working out the route from A to B, work out B to A.
T7   Draw a route map across the projective convergence circle.
T8   Imagining a map directly in front of yourself and lining a pencil up between the two locations and moving the pencil down across the circle.

Orientation specification
T9   Work out how it should look like then scan all the squares for one that fits.
T10   Work out how it should look like then work systematically through the squares until one fits, choose that and ignore the rest of the squares.
T11   Draw the coastline onto the configurations.

Quantitative results
PR   Projective convergence bidimensional regression r^2 value.
OS   Orientation specification completion score.

Table codes
p   Projective convergence
o   Orientation specification
x   Category response best fits into.

8.3). An ANOVA test (see Chapter 5) was then used to determine if there was a relationship between the common strategies of spatial thought adopted and the quantitative results. The ANOVA tests indicated that the adoption of certain strategies of thought did not lead to more accurate results on any of the tests (projective convergence test, $F = 0.55$, $p = 0.791$; orientation specification test, $F = 0.40$, $p = 0.804$).

In addition, the number of strategies adopted by respondents does not differ across the tests, that is, certain tests do not increase the likelihood of adopting more strategies of thought (one-sample chi-square test, $\chi^2 = 2.97$, $p < 0.95$). It was clear from the ANOVA tests that respondents adopt a range of common strategies, but that these strategies are equally likely to provide a solution to the task set.

Table 8.3   Structural frame analysis for projective convergence and orientation specification tests in Kitchin (1997) study.

	C1	C2	C3	C4	C5	C6	C7	C8	CT1	CT2	CT3	CT4	CT5	T4	T5	T6	T7	T8	T9	T10	T11	PR	OS
F11	po	p	p					o	o	o				x								52.09	95
F12	po	p	p			po	p								x							78.90	85
F13	p	o	p												x							70.17	40
F14	po	o																	x			62.99	0†
F15	po					p		o	o	o									x			78.68	10‡
F16	po								o	o												68.83	55
F17	po	o	p		p	po			o							x						86.37	85
F18	po	o	p		p			p	o	o									x			88.21	95
F19			po		o				o	o	o						x					33.88	65
F20	po	o	p	p		o			o									x				63.66	85
M11	po	o							o													18.13	80
M12	p	o	p		p						p									x		45.36	80
M13	p	o	p											x								76.79	65
M14	po		p					o		po				x								21.24	20
M15	o		p		p			po						x								17.13	40
M16	po	p		p				po											x			62.20	90
M17							po															33.67	—
M18	po	p	p			o			o	o	o											63.62	40
M19	p	o			o	o																35.13	45
M20	po	po	p	p	p							po	po								x	66.30	95

† 95 when rotated 90 degrees.
‡ 65 when rotated 90 degrees.

Depending upon your point of view, analysing qualitative data quantitatively can be a useful tool in determining the similarity or difference between data. Of course, some qualitative researchers would suggest that such an approach reduces the richness of the data to just numbers. Others would argue that, although this richness is not lost, as it is still used to illustrate why data sets differ or are similar, quantitative data just add a certain degree of confidence to interpretative findings. While we ourselves are happy to analyse qualitative data quantitatively, it is really up to you to decide whether you agree with such an approach.

8.9   Summary

After reading this chapter you should:

- be able to describe, categorise and connect qualitative data;
- understand the reasons for splitting and splicing qualitative data;
- understand the importance of data corroboration.

In this chapter we have presented a prescriptive approach to analysing qualitative data. We divided our discussion into a number of related themes centred around description, classification and connection. By following the prescription of transcribing and annotating your data, categorising the data, splitting and splicing categories where necessary, linking and connecting data, and corroborating evidence, you should ensure that your data have been rigorously examined. Where appropriate, quantitative analysis of qualitative data should be considered. Each step within the pro-

cess needs to be carefully executed to ensure validity when interpreting the findings. In the following chapter we describe how to perform the same analysis using NUD-IST, a computer package specifically designed to aid qualitative data analysis.

8.10   Questions for reflection

- *Is qualitative data analysis best viewed as an art or a science?*
- *Can you think of an alternative way of categorising our sample data?*
- *Why is it important to carefully transcribe and annotate your data?*
- *Why might you want to split or splice data?*
- *What insights might the process of connection reveal?*
- *Why is corroboration important?*

Further reading

Dey, I. (1993) *Qualitative Data Analysis: A User Friendly Guide for Social Scientists*. Routledge, London.

Miles, M.B. and Huberman, A.M. (1992) *Qualitative Data Analysis: An Expanded Sourcebook*. Sage, London.

Patton, M. (1990) *Qualitative Evaluation and Research Methods*, 2nd edition. Sage, London.

Silverman, D. (1993) *Interpreting Qualitative Data: Methods for Analysing Talk, Text and Interaction*. Sage, London.

CHAPTER 9

Analysing qualitative data using a computer

This chapter covers

9.1 Introduction
9.2 Getting started and description
9.3 Classifying qualitative data using a computer
9.4 Connecting qualitative data using a computer
9.5 Summary
9.6 Questions for reflection

9.1 Introduction

Using a computer to help analyse your data can have several advantages. The main advantage is that it is **efficient** in terms of time and data management. By hand, the analysis of qualitative data can be messy, confusing and time consuming. By using a computer package you can input the transcripts, code the entries and then group, compare, contrast and link data. The computer provides **flexibility**, removes much of the tedium of handworked analysis, and allows tasks that would be almost impossible to calculate with large data sets to be processed effortlessly. In this section we explore how to use one particular package especially designed to help analyse qualitative data. However, before we continue you should remember that:

A computer can *help* us analyse our data, but it cannot analyse our data . . . we must do the analysis.

(Dey, 1993: 55; our italics).

That is, you are responsible for telling the computer what to do and you still have to interpret the results. It is not simply a case of inputting data and awaiting the results.

There are a number of different commercial computer packages which can be used to manage and help analyse qualitative data. All approach the task in slightly different ways, with varying degrees of sophistication. For an overview of 24 different packages,

see Weitzman and Miles (1995). For the purposes of this book we have used **NUD-IST** (Non-numerical Unstructured Data Indexing Searching and Theorising) (see Appendix C for details) to demonstrate how qualitative data can be effectively organised, analysed and interpreted through a computer package. As the online manual details, NUD-IST allows users to:

- manage, explore and search texts of documents;
- manage and explore ideas about the data;
- link ideas and construct theories about the data;
- test theories about the data;
- generate reports including statistical summaries.

We have chosen NUD-IST for three principal reasons. First, the package came recommended by colleagues. Second, we have familiarity with its use. Third, NUD-IST seems to be well known and used within the research community.

This chapter seeks to build upon the conceptual issues and the handworked examples outlined in the previous chapter by offering practical guidance in the computer-based analysis of qualitative data. We advise that you read this chapter in front of a computer, with the NUD-IST program running, and work through the examples at the appropriate places. It is only in this way that the true power of the package in aiding qualitative analysis can be appreciated. Where appropriate, you should refer back to Chapter 8 for guidance. The sample data for the examples can be downloaded from the Pearson Education web site (http://www.awl-he.com).

9.2 Getting started and description

Using NUD-IST, at first, can seem quite daunting and complicated. Just playing around with the pull-down menus does not get you very far – it is a computer package that has to be learnt. However, it can be learnt very quickly and once done provides a very powerful way to manage and analyse qualitative data. The following account is designed to allow you to run and manage a qualitative study using NUD-IST. We have only detailed how to use the basic functionality. However, this is sufficient in the majority of cases to allow detailed analysis. Our discussion mirrors the prescription of qualitative data analysis outlined in the last chapter: **description**, **classification** and **connection**, outlining how to undertake each of these phases using NUD-IST (version 4.0).

To aid explanation, we have used the same notation as we used to illustrate MINITAB in Chapter 4. The symbol ➡ is used to indicate either a selection from a menu or a button, which needs to be clicked on with the mouse. Where this is followed by a name and two square brackets (e.g., ➡ File Name [ ]), this means that you should click on a box/section with the name 'File Name'. Other information to be entered into the box is then explicitly requested (e.g., 'Type in . . .'; 'Select . . .') at the appropriate place in the sequence of instructions.

9.2.1 Preparing the data

Before you can start to use NUD-IST, you will need to transcribe your data and to type them into ASCII files (plain text – with no word-processing codes). Each individual or focus-group interview should be stored in a separate file. To aid file use in NUD-IST, you should carefully format your transcription data following the steps in Box 9.1). Typing in data and formatting files can be quite tedious and time consuming. Remember to leave sufficient time to undertake this task. Once you have completed these tasks you can now use NUD-IST to assist with your analysis.

Box 9.1 Formatting raw data files

Step 1   Store your data in ASCII text files. These are plain files which contain only the text, with no word-processing or other coding.

Step 2   Prefix header information and interview questions with an asterisk. This is to distinguish respondent answers from other information. For example, using Respondent A's data (see Box 8.1), the start of the file would look like this:

```
* Male
* Age 43
* Age since blind: 19
* Location of interview: Respondent's
home.
* Spatial context: Urban estate, pri-
vately owned (formerly council), semi-
detached.
* Date/Time of interview: */*/97; 4 pm.
* Who was present: DJ and Respondent
A.
* Interview was recorded on tape and
video.
* Notes: Interviewee was cautious but
reasonably forthcoming.

* INT: How do you approach and resolve
problems encountered en route? <HR>
RES: Such as? <HR>
```

```
* INT: Well, how about detours? <HR>
RES: I'm OK with them. But I got lost
once. It tends to knock you sideways
a bit. I crossed one too many streets
and I ended up going in the opposite
direction. Well, I would first of all
- if I didn't know the area - I would
ask someone. Or I would go back on
myself and retrace, and maybe get a
taxi or some other transport. Or maybe
I would go to a friend's house nearby.
<HR>
```

Notes:
- Each carriage return (starting a new line – <HR>) creates a new text unit (group of text) when loaded into NUD-IST. In the above example, 'Such as?' would be the first text unit and 'I'm OK with them ... nearby' would be the third.
- Sometimes you might wish each line of text to be a text unit; at other times you might assign whole paragraphs. In the example data, we have used paragraphs.
- Because each carriage return introduces a new text unit, include only those necessary (i.e., do not have blank lines in your document except between the header information and interview text).
- Make sure each question and each answer is a separate text unit.

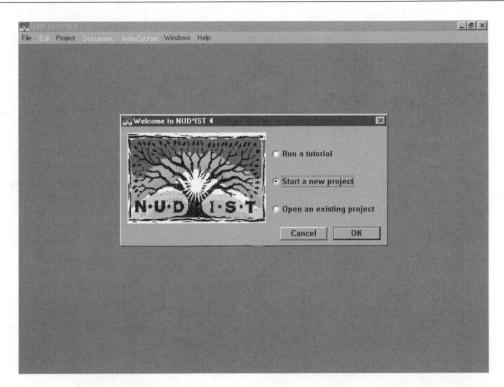

Figure 9.1   NUD-IST start screen.

9.2.2   Creating a new project

Upon starting NUD-IST you will be presented with a screen as displayed in Figure 9.1.

NUD-IST refers to each set of analyses upon a set of data as a **project**. The first task we need to undertake is to create a new project. Check 'start a new project', or if you have already started a project and are restarting NUD-IST check 'open an existing project', and select OK. We now need to name our project. In our case, we have chosen the name Blind to indicate that it refers to analysis of transcripts with blind individuals:

➥ File name [ ]
Type filename (e.g., blind.stp)
➥ Save
Type your name (e.g., Rob Kitchin)
➥ OK

Two new boxes will appear entitled '*Node Explorer*' and '*Document Explorer*' (see Figure 9.2). This process will automatically create a whole new series of sub-directories which NUD-IST will use to store your analyses. Minimise the NUD-IST window and

transfer your transcript files into the *Rawfiles* directory. You are now ready to start importing data.

9.2.3   Importing data

Once you have created a new project you need to import your data. Data sets within NUD-IST are referred to as '**documents**' and we will use the same terminology. To import sample data into data documents within the NUD-IST package:

➥ Documents
➥ Import
➥ File name [ ]
Select file (e.g., Resa.txt)
➥ OK
➥ OK

A modified copy of *Resa.txt* has now been put into the BLIND project as a data document. To introduce other data sets (e.g., for Respondents B, C and D) repeat the process.

To view individual data documents once loaded into NUD-IST use the *Document Explorer* window:

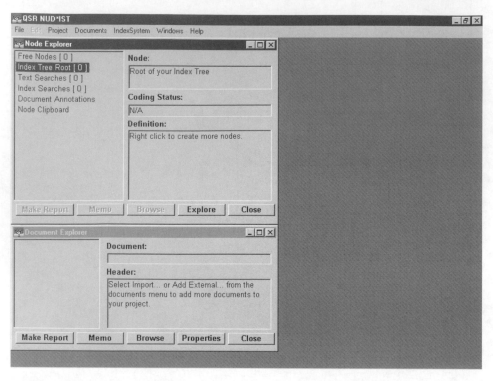

Figure 9.2   Basic NUD-IST environment.

Document Explorer [ ]
Highlight file (e.g., RESA)
➤ Make Report
➤ OK
➤ OK

You will now be presented with a window containing some data about the document, the header and the selected data (Figure 9.3). Note the text unit numbers are shown on the right. You can scroll up and down the document to read the text units.

To close a report (or any active window):

➤ File
➤ Close
Select 'No' if you do not want to save a report and 'Yes' if you do.
If you select 'Yes':
➤ File name [ ]
Type in file name.
➤ Save

9.2.4   Annotating data

NUD-IST allows you to annotate documents once you have started to use the package, saving the annotations as memos. Memos can be attached either to data documents or to index categories (see next section). You can create a Memo using the *Document Explorer* window by:

➤ Document Explorer [ ]
Highlight RESA
➤ Memo
➤ Yes
Type in your memo

9.3   Classifying qualitative data using a computer

9.3.1   Assigning category codes

As discussed in the previous chapter, qualitative data analysis requires data to be classified (sorted into relevant categories). This process is made easier by NUD-IST because it allows you to use the power of the computer to manage and manipulate data. In NUD-IST, category codes are called **index categories**.

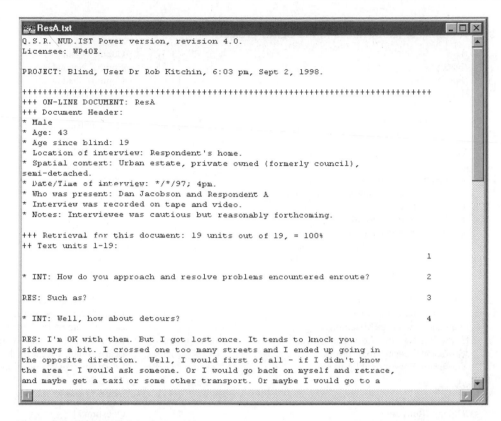

Figure 9.3   Report of data document.

NUD-IST holds the index categories you create in a hierarchical structure. The simplest way to create an index category is to follow the steps below:

➡ IndexSystem
➡ Create Nodes …
➡ Type a node address [ ]
Type the number of the index category to be created (e.g., 3)
➡ OK
➡ Please enter a title for node (n) [ ]
Type index category name (e.g., Coping)
➡ OK
➡ Cancel

An alternative is to use the '*Tree Display at <root>*' window, or the process of assigning text units to index categories, to create the index categories (see below). Whichever method is chosen, index categories are not set in stone and the data within them can be split and spliced easily (see below). Box 9.2

provides a complete list of index categories for our example project. Note how the hierarchical structure contrasts with the handworked example in the last chapter (there are now eight main categories as opposed to three in the handworked example). The rationale behind this new structure is the easier storage and organisation of text units assigned to multiple index categories, and is designed to aid analysis (see Section 9.4).

9.3.2   Creating a tree of index categories

As you create index categories, and assign your data to different categories (see below), it can be useful to **visualise** how the categories relate to each other (see Figure 9.5). The master category of the hierarchy of index categories is called the **Root** from which all index categories branch.

A visual record of your index categories is easy to create:

Box 9.2   A complete list of nodes in the BLIND project

(1)	/base	(3 4)	/Coping /O and M
(1 1)	/base/Sex	(3 5)	/Coping /Secondary
(1 1 1)	/base/Sex/Female	(3 6)	/Coping /Shadowing
(1 1 2)	/base/Sex/Male	(3 7)	/Coping /Other
(1 2)	/base/visual extent		
(1 2 1)	/base/visual extent/Total	(4)	/Feelings
(1 2 2)	/base/visual extent/Partial	(4 1)	/Feelings/Panic fear
		(4 2)	/Feelings/Confid Embarr
(2)	/confusion	(4 3)	/Feelings/Wariness
(2 1)	/confusion/types	(4 4)	/Feelings/Frustration
(2 1 1)	/confusion/types/Spatial	(4 5)	/Feelings/dislikes
(2 1 2)	/confusion/types/Situation		
(2 3)	/confusion/Reasons	(5)	/Sp Beh
(2 3 1)	/confusion/Reasons/Road	(5 1)	/Sp Beh/set pattern
(2 3 2)	/confusion/Reasons/Pavement	(5 2)	/Sp Beh/Explore
(2 3 2 1)	/confusion/Reasons/Pavement/vehicle		
(2 3 2 2)	/confusion/Reasons/Pavement/Street Furn	(6)	/Learn
(2 3 2 3)	/confusion/Reasons/Pavement/Pave Design	(6 1)	/Learn/Strategy
(2 3 3)	/confusion/Reasons/Never told	(6 1 1)	/Learn/Strategy/Asking
(2 3 4)	/confusion/Reasons/shops	(6 1 2)	/Learn/Strategy/Fam Friends
(2 3 5)	/confusion/Reasons/other	(6 1 3)	/Learn/Strategy/Retrace
(2 3 6)	/confusion/Reasons/Miscount	(6 1 4)	/Learn/Strategy/Secondary
(2 3 7)	/confusion/Reasons/Attempted sh cut	(6 1 5)	/Learn/Strategy/Trial error
(2 3 8)	/confusion/Reasons/road layout	(6 1 6)	/Learn/Strategy/O and M train
(2 3 9)	/confusion/Reasons/public trans	(6 1 7)	/Learn/Strategy/Rehearsal
		(6 1 8)	/Learn/Strategy/Propositions
(3)	/Coping	(6 2)	/Learn/No of times
(3 1)	/Coping /asking	(6 3)	/Learn/O and M training
(3 2)	/Coping /Retracing		
(3 3)	/Coping /Fam Friends	(8)	/Question
		(10)	/Analysis

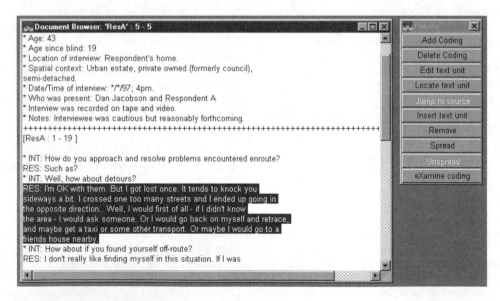

Figure 9.4   Adding text units to index categories.

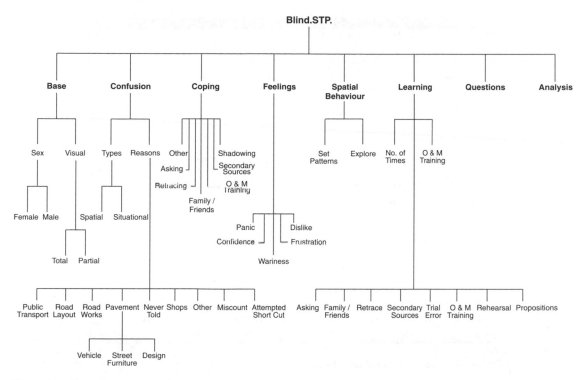

Figure 9.5   Tree structure of whole project.

➡ IndexSystem
➡ Display Tree
Move the mouse pointer to the pane (rectangle) in its upper left corner and double-click. A tree structure of the category indexing already created will automatically appear (see Figure 9.6).

To navigate up and down the hierarchy simply click on the desired arrows.

9.3.3 Adding text units to index categories

To assign text within data documents to an index category, a process is adopted that is much the same as performed manually in the last chapter:

➡ Documents
➡ Browse Document
➡ Select Document [ ]
Highlight data document required (e.g., RESA)
➡ OK

A *Document Browser* window will appear displaying the text of the data document selected, along with a palette window for controlling the categorisation process (Figure 9.4).

Scroll through the text until you come to the section you wish to code. Once you have identified the text for coding, select it by clicking on it (the text will now be highlighted). To add the text (e.g., 'I'm OK with them . . .': see Box 9.1) to an index category (e.g., 3 1 – coping/asking):

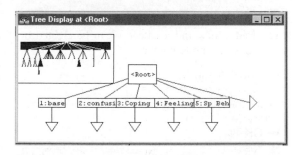

Figure 9.6   Examining the whole tree structure in NUD-IST.

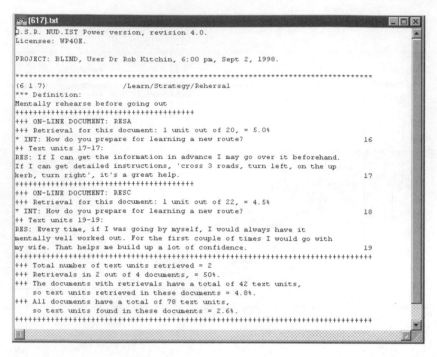

Figure 9.7   Report of Investigate Node.

Highlight text unit
➡ Add Coding
➡ Type a node address [ ]
Type the number of the index category (e.g., 3 1).
➡ OK

If you cannot remember the index category you wish to assign a category to:

➡ …Select
Double-click on Index Tree Root [ ]
Work your way down the hierarchy by clicking on the +/− signs until the index category you desire is highlighted.
➡ OK
➡ OK

If required, a new index category can be created by typing in a new node number:

➡ Add Coding
➡ Type a node address [ ]
Type the number (n) of the new index category (e.g., 3 1)
➡ OK
➡ Yes
➡ Please enter a title for node (n) [ ]
Type index category name (e.g., Asking)
➡ OK

By working your way through the data document all the text units can be assigned to an index category. Remember, the same text units can be assigned to more than one index category. For example, the example piece of text concerning being lost might be coded in categories concerning blindness level, gender, age, question number, spatial confusion, coping and feelings. This cross-referencing is crucial for the connection phase.

To close the *Document Browser* window simply:

➡ File
➡ Close

Note you can add Memos (annotations) to index categories. This can be done using the following commands in the *Node Explorer* window:

Double-click on Index Tree Route [ ]
Work your way down the hierarchy by clicking on the +/− signs until the index category you desire is highlighted.
➡ Memo
➡ Yes
Type in memo/annotation

At any point, all the data assigned to an index category can be examined using *Make Report* in the *Node Explorer* window. An editable report window will appear containing the requested information, plus some statistical summary data (Figure 9.7).

9.3.4   Searching data documents

An alternative to scrutinising each data set manually is to use NUD-IST to search the data documents for you. NUD-IST will allow you to search for individual words or phrases, for example:

➡ Documents
➡ Search Text
➡ String Search.

A window will appear (Figure 9.8).
Next:

➡ Search For [ ]
Type your search string (e.g., 'ask' – without quote marks)
➡ OK

All data documents are searched for your selected search string. A summary window and an edit window will appear displaying all instances of this string.

If you want to restrict the search in any way, you can use the *Exclude* option. This is useful if you want to filter out any text units (e.g., selecting node 1 1 2 in the BLIND project will exclude all interview data with men – see Box 9.2). Simply:

➡ Documents
➡ Search Text
➡ String Search.
➡ Search For [ ]
Type your search string (e.g., ask)
➡ Restrict
➡ Exclude
➡ Coded At [ ]
Type the number of the index category (e.g., 1 1 2) or use Select
➡ OK
➡ OK

Any instances of the text string are automatically placed onto the *Text Searches* clipboard. To move the data detected by the search to an index category we can use the *Merge* command:

➡ Tree Display at <Root>
Select the index category to merge the data with (e.g., 3 1)
➡ Merge
➡ Yes

If you wish to form a new category, instead of using *Merge*, use the *Attach* command:

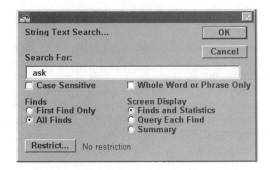

Figure 9.8   Search window.

➡ Tree Display at <Root>
Select the index category to append the data to (e.g., 3)
➡ Attach
➡ Yes

The text will then become a child of the node you append the data to.

We can also search for patterns within your data documents ('either/or' searches, e.g., search for 'lost' and/or 'disorientated'):

➡ Documents:
➡ Search Text
➡ Pattern Search.
➡ Search For [ ]
Type your search string within square brackets (e.g., [lost|disorientated] – the vertical line means 'or').
➡ OK

You can add context to the data stored on the clipboard by using the *Spread* command in the *Node Explorer* window. This will add to the text findings the preceding and succeeding text.

Select Node Clipboard
Click right-hand mouse button
➡ Spread
➡ Context [ ]
Select number of text units (e.g., 5)
➡ OK

In our example, five text units on either side of every search finding are added to the clipboard.

When searching your data documents, remember that not all relevant passages will contain keywords. You should therefore also be careful to search the text manually, using the method described above.

9.3.5   Reshaping index categories

Once all the text units have been assigned to appropriate categories, there is likely to be a need to split and splice the index categories. This can be achieved by using either the *Merge* and *Attach* commands (see above) or the '*Tree Display at <Root>*' window. Reshaping the tree, adding in new branches and reassigning others, is relatively straightforward. To add in new categories:

➡ Tree Display at <Root>
Select index category to add new index category to (e.g., <Root>)
➡ Create Node
➡ /Question/ [ ]
Type in a new index category name (e.g., Learning)
➡ Definition [ ]
Type in a definition (e.g., Branch concerning how a route is learnt)
➡ OK

The main tree display will show the new node, created as a child of <Root>. To *Delete* an index category:

➡ Tree Display at <Root>
Select index category to add new category to (e.g., Learning)
➡ Delete
➡ Yes

Sometimes you might want to move a category to a new position within the data tree:

➡ Tree Display at <Root>
Select index category to be moved
➡ Cut
Select index category the cut category is to be appended to
➡ Attach
➡ Yes

or *Merge* two categories together:

➡ Tree Display at <Root>
Select index category to be moved
➡ Cut
Select index category the cut category is to be merged with
➡ Merge
➡ Yes

Through careful construction, a complex, hierarchical structure describing the data can be developed. Figure 9.5 displays the tree structure for the BLIND project. To explore the tree structure you have created, click on the arrows to move up and down the hierarchy. The index category codes (see Box 9.2) mirror this structure (e.g., [2 1] means the second node of the root and the first child of this node). By clicking in the index category boxes within the '*Tree Display at <root>*' window and selecting a category name, details about the category can be examined. Opening more 'Tree Display' windows can help you to view more of the tree. To do this:

➡ IndexSystem
➡ Display Tree

9.4   Connecting qualitative data using a computer

Computer packages can aid the process of classification greatly because of their flexibility and speed of reassigning data within index categories. However, it is in the ability to allow you to make connections that they come into their own. NUD-IST is no exception, allowing you to query your data extensively. NUD-IST uses an Index System Search (ISS) to compare different categories and their coding. The ISS allows you to compare, contrast and link text relating to different issues. Our new data structure (as opposed to the handworked example – see Figure 8.4) is designed to take advantage of the querying capabilities of ISS. Instead of having to divide the data into 'spatial' and 'situational' categories, we can use the computer to make these distinctions at a later date. For example, if we wanted to determine the reasons for spatial confusion we could compare the text units stored in node 2 1 1 (confusion/types/spatial) with those in the sub-categories of node 2 3 (confusion/ Reasons). To merge all reason data (e.g., 2 3 1, 2 3 2, 2 3 3, etc.) into 2 3, use the copy and merge commands:

➡ Tree Display at <Root>
Select index category to be copied (e.g., 2 3 1)
➡ Copy
Select index category the copied category is to be merged with (e.g., 2 3)
➡ Merge
➡ Yes

Repeat for the other index categories to be copied. We suggest that to save confusion, at this point, you create a new node 'Analysis'. This node will be used to store all your query results.

The text units which appear in both (i.e., *Intersect*) 2 1 1 and 2 3 reveal the reasons for spatial confusion:

➥ IndexSystem
➥ Search Index System
➥ Intersect.
➥ Intersect [ ]
Type node number of index categories in brackets (e.g., (2 1 1) (3 1)) or use Select
➥ OK
➥ OK

The findings are saved in the *Index Search* in the Node Explorer, and can be viewed using *Make Report*.

We could perform the same analysis for 'situational confusion', or even work out which reasons are common to both 'spatial' and 'situational confusion'. This would have been difficult and time consuming to compute by hand (hence the different structure adopted for hand analysis). The computer provides much greater flexibility, ease of analysis, and speed.

The ISS allows many different sorts of queries to be processed beyond intersection (there are 18 different operations). For example, *If-inside* allows you to see if different groups thought/did the same things. For example, we might ask 'is it only totally blind people who panic when lost, or do people with residual vision also panic?'. This sort of question might yield interesting insights into the spatial behaviour of totally blind and visually impaired people:

➥ IndexSystem:
➥ Search Index System
➥ If-inside.
In the left-hand box add the search node (e.g., 4 1 (panic)).
In the right-hand box add the node which is to be searched (e.g., 1 2 1 (totally blind)).
➥ OK
➥ OK

The findings will be saved in the *Index Search* in the Node Explorer. We could now repeat the process for partially sighted respondents (e.g., 1 2 2). By making reports for new nodes we can compare the findings (see Box 9.3), but note that *you* still have to interpret the results! Box 9.3 reveals that one of the totally

blind respondents mentions panic once, whereas both partially sighted respondents mention panic several times. Whilst we cannot draw any conclusions from this because of our limited data set, it would be interesting to compare the results of the full data set. If the results were continued across the whole data set, we might then draw in more evidence to try to find out why this is the case. For example, we might compare the spatial behaviour of totally and partially blind people – do partially blind people experience more panic because they use their residual vision to explore more, whereas totally blind people stick to set patterns of spatial behaviour? Other questions might relate to the congenitally blind (from birth) versus the adventitiously blind (lost sight later in life), or guide dog users against non-users, or women versus men.

Similarly we could perform the *Near* analysis. *Near* lets you determine whether text units of different nodes appear close together in the text. In general, the closer together text units are, the more related they are. For example, we might expect feelings of fear to be related to coping strategies. While the text might not overlap (intersect), we might want to know if there is a general relationship.

➥ IndexSystem:
➥ Search Index System
➥ Near
In the left-hand box add the search node (e.g., 4 1 (panic))
In the right-hand box add the 'near to' node (e.g., 3 (merged coping category index))
➥ Context [ ]
Choose how near the two issues should be (e.g., section, text unit (number)).
➥ OK
➥ OK

Followed-by could be used to indicate whether coping strategies always succeeded statements of fear (same procedure as *Near* but select *Followed-by*). We could also determine whether coping strategies always succeeded statements of fear across groups (as with our *If-inside* example). In our example data, because the text units are paragraphs and not individual sentences or lines, the *Near* and *Followed-by* commands are relatively ineffective as related issues appear in the same paragraph (be aware of shortcomings in your data structure). However, other operators (such as *Intersect* and *If-inside*) can still be used to query our data successfully.

Box 9.3   The results of an if-inside query

Panic in totally blind

Panic in partially blind

Search for (IF-INSIDE (4 1) (1 2 1))
++++++++++++++++++++++++++++++++++++++
ON-LINE DOCUMENT: RESA
+++ Retrieval for this document: 1 unit out of 20, = 5.0%
* INT: How do you find out about new routes?
++ Text units 13–13:
RES: By asking people. I haven't been down in the town for a while, but I knew there was work done on this part of the road. I didn't realise the Westlink went under the Falls Road. Nobody had ever told me that. I was walking across towards Divis Street, and suddenly I heard the roaring of a heavy thundering truck. 'Shit, this one's going to get me' I thought. Then it went underneath and I was OK.
++++++++++++++++++++++++++++++++++++++
+++ Total number of text units retrieved = 1
+++ Retrievals in 1 out of 4 documents, = 25%.
+++ The documents with retrievals have a total of 20 text units, so text units retrieved in these documents = 5.0%.
+++ All documents have a total of 78 text units, so text units found in these documents = 1.3%.
++++++++++++++++++++++++++++++++++++++

Search for (IF-INSIDE (4 1) (1 2 1))
++++++++++++++++++++++++++++++++++++++
ON-LINE DOCUMENT: RESB
+++ Retrieval for this document: 3 units out of 18, = 17%
* INT: How about off route?
++ Text units 5–5:
RES: Generally I don't like this. I can get anxious. Sometimes it is OK though, but especially . . . if I'm not familiar with it I may panic a bit and wonder 'am I where I want to be?', 'Is this where I want to be?' This can happen. One day a few weeks ago I was in big hurry, I was taking my daughter to the opticians, and we had to come back in half an hour. We had a few other things to do so we left the shop to do those. They were in another part of town and then we had to get back to Castle Court shopping centre. Our normal route would be up Royal Avenue and Castle Court would then be on your left. But we found ourselves around Queen Victoria Street and I found myself thinking, 'I wonder if I can cut through these streets?', 'will it bring me out at Castle Court?' Very much a guessing game to head in the right direction. We did get there in the end, but I had my sighted child with me. If I had been there on my own I probably would have gone the long way around.
++ Text units 7–7:
RES: I don't like it, I hate being 'out of control'. It is not a good feeling. It can be an anxious feeling. It's also quite a confidence downer. All of a sudden you are not where you think you are and you feel a bit stupid. You think any other adult could manage this and maybe you can't so you feel inadequate in some way. How I would resolve it would be to ask somebody. This has, believe it or not, taken me nearly 42 years to reach the point where I can ask somebody. You get a whole range of answers that you need emotional strength for, like the 'are you blind?' question.
++ Text units 15–15:
RES: Friends, an A to Z with a magnifying glass, and generally try to get as much information as possible to be prepared before I go. Last year I was told I was going away for the weekend but I wasn't told where I was going, it was to be a surprise, and the person I was going with was completely blind. Then I panicked totally. 'How will we ever find our way around?', 'What will we do?', 'How will we manage?' Just total panic really. We flew to Edinburgh and by basically asking people where to get the bus etc. we made our way to the town. Once in town we asked for the street with the hotel on, where we were staying. The first thing we did was to get a map of Edinburgh, I was able to see some of this with my magnifier, although I had trouble translating that into my surroundings. The person I was with could translate this much easier, by not looking for visual cues as he can't see.

Box 9.3   (cont'd)

Panic in totally blind	Panic in partially blind
	++
	ON-LINE DOCUMENT: RESC
	+++ Retrieval for this document: 1 unit out of 22, = 4.5%
	++ Text units 9–9:
	RES: I would ask somebody. A few times I've been, not lost, but mislaid. Recently, when I was on a course I got the wrong bus one morning. There were two buses leaving Lisburn, one was an express to the back of the City Hall in Belfast and the other one went to Newry, somewhere. I soon knew the bus I was on wasn't going to the right place and I started sort of panicking, and then it was more really of a mental panic 'I'm in deep shit'. Although in the end I recovered easily enough
	++
	+++ Total number of text units retrieved = 4
	+++ Retrievals in 2 out of 4 documents, = 50%.
	+++ The documents with retrievals have a total of 40 text units, so text units retrieved in these documents = 10%.
	+++ All documents have a total of 78 text units, so text units found in these documents = 5.1%.
	++

9.5   Summary

After reading this chapter you should:

* be able to analyse data using NUD-IST;
* recognise the power of the computer to simplify qualitative data analysis and interpretation.

In this chapter we have extended the discussion of qualitative data analysis, illustrating how qualitative data can be analysed effectively using a dedicated computer package (in this case NUD-IST). While NUD-IST might not seem intuitive immediately, it does not take long to master its commands. As with quantitative analysis, care is needed at all stages of analysis from data entry (see Section 4.2.2) to interpreting the results. Of particular note is how you can use the NUD-IST package to help in the process of categorising and linking your data. You should take full advantage of the computing power to explore all the (appropriate) relationships within your data set. In the following chapter, we move the focus of our attention away from data analysis to concentrate upon how to write up your project and disseminate your findings.

9.6   Questions for reflection

* *Why is the computer a useful tool for analysing qualitative data?*
* *How do the handworked and NUD-IST tree structures for the BLIND project differ and why?*

Further reading

Crang, M.A., Hudson, A.C., Reimer, S.M. and Hinchliffe, S.J. (1997) Software for qualitative research. 1. Prospectus and overview. *Environment and Planning* **A29**: 771–787.

Hinchliffe, S.J., Crang, M.A., Reimer, S.M. and Hudson, A.C. (1997) Software for qualitative research. 2. Some thoughts on 'aiding' analysis. *Environment and Planning* **A29**: 1109–1124.

Weitzman, E.A. and Miles, M.B. (1995) *Computer Programs for Qualitative Data Analysis*, Sage, London.

CHAPTER 10

Writing-up and dissemination

This chapter covers	**10.1** Introduction
	10.2 Writing as part of the research process
	10.3 Writing a final report
	10.4 Presenting your research as a talk
	10.5 Before submitting your work
	10.6 After submitting your work
	10.7 Summary
	10.8 Questions for reflection

10.1 Introduction

In this chapter, we discuss the process of closing your research: writing a final report, presenting your findings and disseminating your thoughts. Conducting research is rarely for just personal pleasure and gain. The research is usually aimed at informing a wider audience (even if that is just your project marker!). The **finished product** (final report) of research, then, is just as important as the process of generating and analysing your data. Despite how well things may have gone during the research process, *a piece of work is usually judged by the standard of the finished product* (unless the reader was involved, there is nothing else to go on!). Whether we think it fair or not, the quality of the write-up is paramount. It is therefore essential that you can communicate in a clear and unambiguous manner the rationale for the project, the research process, the research findings and the main theoretical arguments to be drawn from your work. You will need to balance depth and detail with breadth to produce a well-rounded report, and to develop a coherent and consistent plotline that tells the 'story' of your research. Much of what we have to say in this chapter might seem like common sense, but we know from experience that the standard of writing up is extremely variable (even among top students) and often lets a good project down. While you may feel that the process of writing up is self-explanatory, we urge you to read through the hints and guidelines presented in this chapter and to take heed of our advice. Although also of importance, we do not consider English grammar or punctuation. Remember that spelling, grammar and punctuation errors will detract from your ability to communicate. If you feel that your writing ability does need to be improved, we recommend that you invest a little time and seek out appropriate advice. Access to a good dictionary and thesaurus are essential.

10.2 Writing as part of the research process

Many researchers consider the process of writing an important part of research. In other words, writing should not be viewed as just the finishing-off phase but rather as an important part of your analytical toolkit. Writing helps you to think through and reassess ideas and research undertaken. As such, rather than waiting until the whole research process is completed before starting the process of writing, we suggest that you try to write what you can in rough as each stage of the research is completed. This has two distinct advantages. First, the process of writing helps clarify your thoughts on a particular part of the research. As Wolcott (1990: 21) states, '*writing is thinking*'. Second, the early organisation of material on paper will aid the writing of a final report. There is nothing worse than finishing the research and progressing to the process of writing up only to have forgotten why a particular strategy was taken or what particular results indicate. Moreover, any drafts will provide a rough framework to guide the structure of a final report and help overcome the problem of where to start. We recommend that you **draft** out rough summaries of all aspects of the research, including any thoughts and interpretations, as you progress through the project. This can consist of as little as detailed notes or as much as detailed drafts of sections/chapters. The process of writing begins as soon as you start to think and read about your research topic, even if this is just a series of notes on the back of an envelope. Indeed it is quite common for some project supervisors to insist that a detailed rationale for the research, and appropriate literature review, be completed *before* any research commences. In this way the research will be properly grounded, the project will be more focused, and a relatively time-consuming part of the research process will be complete.

One of the difficulties many students experience is knowing **when to stop** their research. If the project has been particularly successful there may be a reluctance to let go. You may be a perfectionist or you may feel that you have only partially completed what you set out to achieve. In many cases, student researchers are unaware that they have succeeded in completing a project – especially since most projects throw up a whole new set of tempting questions to examine. A typical student concern is whether 'enough' work has been done. It is quite common for students to undervalue any research they have done so far, and assume that more and more work needs to be done. This is more likely to be the case for the postgraduate student who is often not as constrained by time. For many professional researchers, the process of research is continuous. The questions generated in one study form the basis for the next. However, even if they start a new project before the old one is finished, they are always careful to close a project fully – to write a final report of their findings. You *must* be able to make this break, to produce a final report. One of us recalls the pragmatic advice given to him when a PhD student, which was to spend your allotted number of months doing research, stop and immediately write up: simple! Of course, some people cannot wait until the project is finished and will rush to complete the process of writing up and submitting their work. These projects are invariably poor in quality, achieve low marks or are rejected by publishers. Probably even worse are those who never manage to make the

break to produce a final report, whose projects or dissertations never seem to get completely finished. The worst kind of final report is one that never gets written.

10.3 Writing a final report

There is more to writing a final report to the research project than just reporting the findings. In general, you will be required to reflect upon the whole research process from rationale to data generation and to concluding statements. It is important to remember that:

[A report of a research project] is far more than a passive record of your research and generally involves presenting an argument or point of view. In other words, it must 'say' something and be substantiated with reasoned argument and evidence.

(Barnes, 1995: 100).

The aim of the final report, then, is usually twofold. First, to report upon the research project – what was done, what was found. Second, to use your research findings to develop some conclusions – to explain what your results mean. The report should then be **accessible** to the reader with your conceptual ideas and empirical data clearly described and explained. In this section we detail the process of writing a final report: reporting the project and developing an explanation.

10.3.1 Writing for an audience

As Wolcott (1990) explains, writing up a project is informed by both the **research process** itself and the **intended audience**. In other words, the sort of study you undertook (e.g., qualitative or quantitative) and the position and competence of the people intended to read the final report should inform how you undertake the process of writing up and the dissemination of your research. We will address issues concerning the writing up and presentation of particular types of study later in the chapter. At this point, the most important question you need to ask is '*Who is the intended audience?*'; '*Who am I writing for?*'. If your research project was part of an undergraduate course then, in general, the intended audience is the project assessor. If your research was part of a postgraduate course then you might be writing several reports aimed at different audiences (e.g., the project marker, academic/non-academic journals, a summary report to your sponsors, a summary report to the people you generated data from). *Different audiences require different reports* which vary in level of detail and sophistication. Writing for the lay person who has no expertise in your area of study and writing for someone who is an expert in the field you have studied are very different tasks. The level of detail required to satisfy and inform an expert about your knowledge of the subject area might seem impenetrable to a lay person. This does not mean that expert reports should use over-complex language and seek to baffle lay people. **Appropriate language** should be used when addressing different audiences. Some things are complex and require specialised writing and notation (mathematical equations, for example). Similarly, a report to a lay person in plain English is no less valid than an expert report. The manner in which the report is written aims to explain the research process and findings in more simplified terms. It is a brave or brilliant person who believes that one report will satisfy all audiences.

Writing for your project assessor(s)

The first rule of writing for a project assessor is to know the **rules and regulations** of the institution you are writing for. If the institution requires your final report to adopt a certain presentation style, you must implement this style. When trying to secure the best mark possible, there is little point in losing marks for the incorrect formatting of your text. The discussion throughout this chapter relates mainly to writing for an assessor. The most important thing to remember is that the assessor wants you to demonstrate that you know the subject area, that the project was well thought out using appropriate methods of data generation and analysis, and that you have interpreted the findings appropriately, developing a coherent and consistent argument. Your assessor is not expecting to read a Nobel prize-winning piece of research. He or she is, however, hoping to read a well-written account of a well-planned and thought-out project. Do not worry if your data generation was a disaster. As long as you explain the circumstances and suggest alternative research strategies you should satisfy the assessor.

Writing for an academic journal

Academic journals provide a specialised place in which to disseminate your research. They are usually peer-reviewed (two or three people read your manuscript to judge whether it is suitable for publication)

Box 10.1   Writing for an academic journal

- **Select the journal carefully**, which would be the most suitable for publication of your article. Most journals are not generic but are specialised. Sending an article which does not concern gender issues to *Gender, Place and Culture* is unlikely to meet with success. If in doubt ask for advice from your supervisor.
- **Follow the advice given to authors**. Most journals provide specific guidelines to prospective authors concerning the length, format and contents of prospective manuscripts.
- **Remember you are writing for a specialised audience** and temper your language and style accordingly.
- **Your article should have at least one specific argument** and should not just be 'I did this . . . and I found this'. Remember that research is not atheoretical. While reporting the research process is important (fellow researchers want to know how you came to your conclusions), presenting a clear argument and conclusion is paramount. Your article must have a purpose.
- **Read recent editions of the journal to get a feel for the level and style.**
- **Don't be disheartened if your paper is rejected, or accepted subject to major corrections.**

---

and generally read by an expert group of scholars. It is extremely rare to see undergraduate projects published in a peer-reviewed academic journal, and then usually only with the help of the student's supervisor. However, it is now usually expected that coherent sections of postgraduate projects will be submitted for publication in academic journals. When writing for an academic journal there are a number of general rules to follow (see Box 10.1).

Writing for your research sponsors

Research sponsors generally require a **written report** detailing the research they have funded or supported. The type and style of the report varies from sponsor to sponsor and you should seek advice as to exactly what they require. The report can vary from a very detailed report of all stages of the research project from rationale to conclusion, to just an executive summary detailing the main findings and conclusions. Whatever the type and style of report your sponsors will expect a professional production which they can distribute to interested parties.

Writing a lay-person report

In addition to writing a fully detailed project report you might also need to produce a **lay-person report**. The agency through whom you have obtained participants might want a report to circulate to any interested parties, or you might want to submit your findings to a generic (non-academic) magazine/newspaper. Occasionally, participants themselves would like to know about the main findings of the project to which they contributed. Writing a lay-person report can be extremely difficult. Research is generally a complex process, and describing the theoretical underpinnings, critically appraising the literature, detailing the methods of data generation and analysis, and explaining the findings without using technical terms, in a lucid manner that is understandable to all, is no easy task. Even if you have been trained to write in a technical manner then writing for a lay person can be difficult. We suggest that when writing for a lay person you should avoid technical language completely. If you are trying to describe a complex procedure or theory, then break down the text into a number of short sentences, each of which details a specific aspect of what you are trying to convey. Use everyday examples to illustrate your descriptions. Before you submit the work get friends or family who are lay people themselves (i.e., they have no real knowledge of the project) to read through your report. If they have to ask you questions of clarification or tell you that something is unclear, then you will need to redraft and refine your text further. Never assume that people will understand what you mean!

10.3.2 The writing process

Once the audience has been identified, the process of writing the report can begin in earnest. However, this is rarely a simple, linear task. That is, most researchers will rarely start the writing process by drafting the introduction and progressing through each individual section until reaching the conclusion. Many start with **rough sketches**, building up drafts of each section – these sketches might differ substantially from any drafts written as part of the research process. Only when a clear picture emerges of what is going to be said, and where it is to be said, is each section written up fully. Even then, the writing order of each section might not be progressive. For example, a section on 'methodology' might be written before the 'introduction'. Generally speaking, the process of writing up, like the research itself, is *iterative*. There is a sequential

pattern to the writing, as each section is generally done in turn, but at any stage we can return to previous sections and update our thoughts and ideas, reshaping the text. With the advent of the word-processor writing up in an iterative fashion has become easier. It is now easy to move text around by cutting-and-pasting, to rewrite sections and rephrase sentences without the page becoming a jumbled assembly of crossed-out jottings and rewritten passages. However the writing task is undertaken, you *must* remember that your ultimate aim is a report which communicates in a **clear** and **unambiguous** manner, has a good structure, and is presented in a professional manner. You should write using **formal English** – no slang, no colloquialisms, and no contractions (e.g., I'm, don't, isn't, etc.).

Getting started

It is often difficult to know where to start. You have done so much work that the process of writing up your thoughts seems daunting. In other cases, you might feel that you have done so little, yet so many words are expected! You might simply lack motivation or are bored with the topic, or think that what you have done is not very good (most researchers have crises of confidence over the quality of their research). It is easy to procrastinate; to find reasons for putting off the process of writing. Regardless of how much you delay this process, you are going to have to start at some point if you are going to complete the research project and meet deadlines. If you are having problems getting started we suggest that you think carefully about what you want to say and why. Draft out a contents page, or play around constructing a chapter structure. Once you have a potential structure, start drafting what seems like the easiest section. This might be a methodology section where you just need to document what you did. Slowly build up a number of sections and then try to link them together. In this way your chapters and ideas will start to take shape. However much you procrastinate, it is important that you **start writing**. It is unlikely that you will produce a final report at the first attempt, but you will produce a draft from which you can construct a final report. Lofland and Lofland (1995: 205–212) provide some useful advice for getting your final report started and kept on track (see Box 10.2).

Remember that you should **never underestimate** the time it takes to write up a project. Writing can be extremely time consuming. For example, in many

Box 10.2   Getting the process of writing started and kept on track

- **Have something you want to say about something**. Writing is the communication of ideas and practices. Writers' block is usually an ideas block. Think carefully about what you want to say, to argue, and why, and jot down your thoughts. Use these notes to structure your report.
- **Write on any project aspect, but write**. If you are still having difficulty deciding upon what you want to argue, then write up another section. Concentrate upon detailing procedures and maybe you will gain inspiration as to their wider relevance.
- **Trust in 'discovery and surprise in writing'**. Clearly, before you start your project you do not know what you are going to say or argue. It is only once you have gained your results that the process of interpretation starts. Writing about these results can help you to think through their implications and provide a platform from which to develop your arguments.
- **Admit aversion and write regularly anyway**. It is easy to put off the process of writing, to find something else to do instead. This is particularly the case when the main line of argument is undecided. You need to be disciplined and force yourself to start writing and thinking about your research. Set a daily routine and stick to it. Approach the task as a job that needs to be done regardless of whether you feel like doing it.
- **Do not seek perfection or the right way**. It is common for writers to feel dissatisfied with their work. There is no such thing as a perfect report or the right way to do something. Instead seek to produce a well-structured and argued report.
- **Be prepared to omit cherished writing**. Once you have started to write not all the text will be relevant to your final report. Be prepared to leave out writings that detract from the main argument. Do not try to include everything. Your report should be a concise, coherent story.
- **Have a secluded and indulging place to write in**. Try to find a place to write that is free from distractions.
- **Reread and revise your drafts**. This will improve the text and help you think through your ideas.

Source: Summarised from Lofland and Lofland 1995: 207–212.

up. If nothing else, this will provide a buffer you can eat into if your research takes a little longer than you expected. Remember that time invested in writing up is time well spent – readers will judge the quality of your research project from your final report.

Developing an appropriate structure plan

In general, the structure of a final report of a research project takes the form detailed in Box 10.3. This traditional format provides a strong organisational structure which ensures that all aspects of the research project are reported. The structure plan of a final report is the key to a well-organised and coherent report. In many ways, a final report should be told like a **story** – there should be a beginning, a middle and an end with a strong, coherent plot-line running throughout, linking all the intervening sections. **Linkage** is an important aspect of structuring your final report. Sections should be **logically ordered** and should link to the previous and subsequent sections. It is often useful to number sections using a hierarchical system as this allows the reader to determine how each section relates to others:

Chapter 1   Introduction
    1.1   **The Thesis: Introduction**
            Para 1:   *what the thesis is concerned with*
            Para 2:   *central argument of the research*
            Para 3:   *description of the chapter contents*
    1.2   **Motivation**
            1.2.1   A New View of Nature
                    Para 1: general intro about '...'
                    etc. ....
            1.2.2   New Tools for the Geographer
            1.2.3   Caveat Actor
    1.3   **Manifesto**
            etc. ....

We would advise that you do not progress beyond a third or fourth level of numbering (i.e., 1.2.1.3). Although the structure of each chapter or section will be different, it is always advisable to start with an 'Introduction' (e.g., 1.1 above) and end with a 'Summary'. The introduction can be used to introduce the material which will be discussed, and the summary is a useful tool to recapitulate some of the main points made in the chapter/section. Although not numbered, each paragraph can be summarised and included within

three-year PhDs over one year may be spent writing up the research (we both know people who took a great deal longer!). We suggest that you should allow about 25–33% of your total project time for writing

Box 10.3   The general format for a research project report/thesis

- *Title page*: a short, 10–12 word, statement that informs the reader as to the general contents of the final report.
- *Copyright page* (if required)
- *Thesis Committee approval page* (if required)
- *Dedication page* (optional)
- *Acknowledgement page*
- *Frontispiece* (optional)
- *Abstract*: a 200–300 word summary of what was undertaken and the main findings.
- *Table of contents*: a list of the report's contents which usually includes chapter/section headings and page numbers.
- *List of tables*
- *List of figures*
- *List of abbreviations used* (optional)
- *List of symbols used* (optional)
- *Introduction*: introduces the reader to what is being studied and the rationale behind the project and approach adopted.
- *Literature review*: discusses and critically appraises the findings and theoretical arguments presented by other researchers working on the same or related subjects.
- *Methodology*: details the methods of data generation and analysis.
- *Results*: details the main findings of the research with summary explanation.
- *Discussion*: critically compares the results to the findings of others and develops an argument as to what the results mean.
- *Conclusions*: provides a summary of the main findings and arguments developed.
- *Appendices*: provides extra material to supplement that detailed in the final report.
- *Endnotes* (if these are used)
- *Reference list*: provides a full list of the literature referred to and quoted in the final report.
- *Viva* (if required)

Source: Adapted from Miller and Taylor 1987.

the structure plan as displayed above. Once a structure plan has been organised to the paragraph level, all that is required is to flesh out each paragraph with text. This is by far the easiest method to write a final report in the form of a report or thesis, although sometimes it is easier simply to start writing at the top left of the page and organise later! The latter approach can take considerably longer to complete,

and it is often very hard to scrap a section of text if you later deem it to be irrelevant to your argument. Final reports that are poorly organised are difficult to read, often repetitive and should be avoided. The decision on what to include in a structure plan is not always easy. However, in general, your report should be structured around answering the following questions (outlined by Parsons and Knight, 1995: 112):

- What was the aim of the work?
- Why is this aim important?
- How does the project fit into the context of other research?
- How did you set out to investigate your aim?
- What findings did you produce?
- What do these findings tell you?
- What do you conclude?

Notice how these questions are linked in a logical sequence that, if answered, will provide the story of your research. Remember, your report should start with the *general* (aims grounded in a wider context), move to the *specific* (what you did), and then back to the general (placing what you found into the wider context).

You do not have to stick to the structure outlined in Box 10.3. In recent years there has been a discernible difference in the structure and writing style of final reports produced by researchers who adopt different approaches to research. A traditional, academic report of a research project is often a very staid, conservative affair which adopts a prescriptive approach to writing up. These reports are written in the **third person** (e.g., *It was found . . .*) and seek to be objective and neutral. Recently there has been an acknowledgement that research is a social process and that knowledge is rarely neutral or objective. Within approaches which recognise the social production of knowledge there has been a softening of the writing style and report structure. Reports often shift to the first person (e.g., *I found that . . .*) and there is a recognition that the research process is subjective and 'messy' (note that we are not advocating an 'I did this, then I did that' style of writing). While the final report is still structured and not just a chaotic collection of thoughts and findings, it lacks the formal prescriptive approach of a traditional report. For example, rather than there being formalised, separate literature review and methodology sections, these sections might be merged into a more specific discussion. Some postmodernist writers reject highly structured reports of research altogether and try to find a better way to convey the message they wish to impart.

While we would encourage you to think about writing styles, we must remind you that you are writing for an audience. If you are undertaking a project that is to be assessed by a firm believer in the traditional writing structure then it may be prudent to adopt this structure. If your assessor is more relaxed in their views on how the final report should be structured, then an alternative structure might be acceptable. If in doubt, we would advise that you take the more traditional route. We are aware of undergraduate dissertations that have received widely varying marks from different assessors (e.g., a 3rd from a traditionalist and a 1st from a more liberal assessor). In all cases, the writing style, structure and format were cited as contributory factors for the poor mark by the traditionalist assessor. Similarly, if you are writing for a journal or producing a report for a sponsor, make sure you understand the type of article or report that is required. If they do require a report of your work in a certain style, then it is prudent to supply what is requested.

Developing your argument

As noted, writing a final report consists of more than just details about what you did and what you found. You also need to explain your results, to develop an **argument** as to what your research indicates, to demonstrate what conclusions can be drawn from your study. In other words, you have to **link theory with practice**. To link theory and practice effectively you need to be able to develop a coherent and sound argument as to why your conclusions are justified. For example, just because your statistical test says there is a positive correlation between two variables does not mean that they *are* related or tell us how or why they are related, just that there is a statistical association. Given the evidence gathered during your study, and from your reading of the literature, you need to try to deduce what the statistical association means – to either construct a theory or verify/falsify a theory.

The rationale for the study and arguments you develop provides the *context* for the final written report and provides the link between the various sections. Your argument should be the **strong, coherent plot-line** that links all the sections into a 'story'. Each section or chapter that you write should add support to your argument. However, developing a coherent argument is not always an easy task. Not only do you have to make sense of your own findings but you also need to compare and interweave your interpretations with those of other researchers. In

general, this means being able to make a critical comparison between your thoughts and those of other researchers. This does not mean that you set out to rubbish another person's work. Criticism is measured even-handed evaluation, not all-out attack. As such, you need to identify the quality of other researchers' data, the merits and limitations of other people's thoughts in comparison to your own, and why they might have come to different conclusions (e.g., their data were generated under different circumstances). Remember to keep *your* work on centre stage – it is often easy to shift the focus of your writing to other people's work, leaving your own in the wings. You are developing an argument using other researchers' conclusions for points of reference, not producing a literature review that also refers to your own conclusions. You need to control your sources with some care, moulding them to your purpose, rather than let their format and arguments guide your discussion. In general, the main thrust of your arguments is developed and explained in the discussion section of your final report.

Revising and editing your text

As already noted, a first draft is rarely the final report. The process of writing consists of revising and editing – reworking the text into a suitable format. **Revising** concerns the content and consists of redrafting sections of text, whereas **editing** refers to the style and grammatical changes. There are several reasons why you might need to revise a text (see Box 10.4). Revision and editing are aimed at making your report as good as possible. Time invested in redrafting and editing is time well spent. While redrafting is a necessary part of the writing process you should be aware of when to stop. Some people find it difficult to accept that their text is sufficient – they want to continue redrafting in the search for a perfect manuscript. You should be aiming to generate a manuscript that you are happy with but not one that is perfect. A text can always be improved. Therefore, you should stop the process of editing and revising when you are happy that the text will satisfy the audience it is intended for, i.e., it is *good enough*. There are no general guidelines as to when a text is finished. We suggest that as a general rule, you put down your pen (or turn off your computer) when all the sections have been written, there is a strong structure, and the text reads in a coherent and consistent manner much like a story with a beginning, a middle and an end.

Reaching the word limit

You should always write a final report with any word or length limit (number of pages) firmly in mind. Often, final reports will have word limits attached – whether they be student projects, academic articles or sponsor reports. If you exceed any such limit, it will be at your peril. Most universities will penalise excessive length in a thesis or dissertation. In the process of planning and writing your report you should aim to write to the word limit. It is harder to condense and edit down text if you have overshot your limit, especially if you have to literally throw away text which you have written with great labour and care.

The easiest way to write to the **correct length** is to assign a set number of words or pages to each section, and then to stick to your plan. If you are finding it difficult to stick to your limits, edit the text, taking out all the minor points to leave just the most important, or recalculate the number of words per section, reducing the length of other sections. If at the end of the writing process you are still way over the length limit then you will have to edit your text further.

One problem that we regularly encounter from students is the attitude that 'I generated these data and analysed them and I'm not going to waste them even though they are *not relevant* to my arguments'. Students often feel that they have to include everything that they did. This is not the case. Project assessors and other readers want a report written like a good story. Most good stories do not go off on wild tangents or include information which has no relevance to the plot. You should therefore remove any repetition and unnecessary text. Although it is hard to

specify any rules for determining whether material is relevant, if you are in doubt ask yourself the questions: '*Is this sentence/paragraph/section important to the development of my argument?*', and '*What would be lost if I left it out?*'.

Probably the most difficult type of editing is the reduction of the length of a completed manuscript. Imagine that you have finished the manuscript, that you are happy with the contents and structure, but in spite of the advice above, you have written double the word limit. How do you reduce the length of the final report? This is a difficult task – when you first look at the text every sentence seems important. Again, you need to decide what are the major points and remove all ancillary text. Remember, if you can summarise the whole project in just a 200–300 word abstract you *can* reduce the length of the manuscript and leave the reader with a clear idea of what you did, what you found, and your conclusions. Blaxter *et al.* (1996: 212) provide some useful tips for reducing the length of your manuscript (see Box 10.5).

Alternatively, you may be having trouble reaching your length limit. This can be an equally difficult problem. How do you expand your writing without the text becoming verbose and unclear? Usually if the research project has been a success, there should be enough material to reach any word limit. If so, the report is often short because the author has left out material that they feel is obvious or unnecessary, particularly if they assume that the reader has some knowledge of the topic. This might be a fair point if their supervisor is an expert in the subject. However, the purpose of the report is to *demonstrate* to your assessor that you *understand* the problem, your analysis, and the results. In other words, your understanding needs to be spelled out as clearly as possible, even if the reader is an expert. One way to extend your final report, then, is to extend your description of the methodology, mindful of the points we have raised about relevance above. A more **detailed description** of some of the methods and procedures you used will clarify your analysis to the reader. It may also reveal that you do not really understand the details of what you did, but the onus is on you to ensure that you do! If the report is still not long enough then do not worry. If the project is sound and you achieved what you set out to do then you will still obtain the mark you deserve. Bigger does not mean better and repetition only serves to frustrate the reader. Most assessors will prefer a short, succinct, tightly written piece of work where there is no spurious material, to a long rambling epic! In the main, word limits are the maximum length of a final report, not the minimum. We suggest that as long as you manage to write about two-thirds of the length limit, you will not be penalised.

10.3.3 Writing the components of the final report

Finding a title

Finding a title for your final report might seem like an easy task. However, finding an appropriate and informative title can be difficult. You need to decide upon a title that conveys immediately the precise concerns detailed in your final report. This is particularly the case when you are writing for an audience beyond your immediate project assessor, where people often judge whether to read an article on the basis of its title. Moreover, with the increase in library search software, if keywords are missing from the title then your article might not be found. You can either utilise just one main title (e.g., *Conducting Research in Human Geography*) or attach a subtitle (e.g., *Conducting Research in Human Geography: Theory, Methodology and Practice*). The general rules for composing a title are:

* Keep the main title short, snappy and to the point.
* Use a secondary title only for further explanation or to situate your work.
* Try to construct a title that is interesting or amusing (puns are quite commonly used) that will attract readers. A dry, plain title suggests a dry, plain read.

Writing an abstract

The abstract is a 200–300 word summary (**synopsis**) of your research project. It seeks to boil down every-thing you have done, found and concluded to just a few sentences – no easy task! In the abstract you are trying to convey the **essential essence** of your research. It should clearly and unambiguously tell the potential reader exactly what to expect if they read the rest of the report. It should not just detail the results and conclusions but answer the questions (O'Connor, 1991: 70):

* Why did you start?
* What did you do and why?
* What did you find?
* What do your findings mean?

The abstract should be written in **non-technical language** and be understandable to a lay reader. Remember that the abstract gives the reader their first impression of the research. Time taken to express what you want to say in a clear, succinct manner is time well spent. One way to try to construct an abstract is to summarise each section of the report into just one sentence. In this way you say a little about the whole project. However, there must be a flow and linkage, with the abstract giving a strong sense of purpose. Refer to Box 10.6 for an example abstract. You should also browse through a selection of academic journals relevant to your subject area for more varied examples.

Writing an introduction

The introduction is literally what it says – it introduces the study, sets the scene, and provides the reader with an insight into what will follow. The introduction should provide a strong rationale for the research, detail and justify the approach used, and set out the main arguments developed throughout the thesis. This should start off in general terms and progress to the specific. For example, the first paragraphs should outline the rationale for study and overall aims. Here, the **broad context** of the study should be described. The following paragraphs should become more specific, detailing the exact aims and placing the study into a theoretical framework. Kidder (1981) provides the following advice for writing a good introduction:

1 Write in plain English, not geographical jargon.
2 Do not plunge your reader into the middle of your problem or theory. Take the time and space to lead the general reader up to a more formal explanation.

Box 10.6   An example abstract

Below is an abstract for a project entitled '*Issues of safety in spatial decision making*'.

ABSTRACT

Form: Introducing statement, problem, aim, methodology, findings, conclusions.

Model: The aim of this project was to . . . This is important because . . . This follows from earlier research on this topic in that . . . The research was carried out by . . . It was observed/found that . . . This suggests . . . It is therefore concluded that . . . (Parsons and Knight, 1995: 121).

Example: Researchers of spatial decision making seek to understand how we decide where to go in the geographical environment and how to get there. Studies to date have mainly examined the effects of spatial layout upon spatial decision making. Other potential influencing factors such as safety and travel cost have largely been ignored. The aim of this study was to examine spatial decision making by comparing the influence of spatial variables such as distance, time and street layout with personal factors. Particular attention was paid to the influence of safety. One hundred residents of Swords, Dublin, estimated the relative importance of spatial and personal variables when choosing a route. Estimates were reported using a series of rating scales. Analysis revealed that personal factors were deemed to be more important in choosing a route than spatial variables. Safety was found to be of paramount concern. These findings indicate that research into the processes which underlie spatial decision making have neglected the role of personal factors, and issues of safety in particular. Given this concern, a new conceptual model of spatial decision making is presented. It is concluded that safety is the dominant, influencing factor in spatial decision making.

Tips:

- Don't refer to information in the abstract that is not in the final report.
- Write as a single paragraph.
- Write complete sentences and avoid acronyms or abbreviations.
- Key words for indexing/abstracting are often included at the end of the abstract – choose the most important specific terms (O'Connor 1991: 70).

3   Try to open the introduction with a sentence concerning the topic, not what other researchers have done or found.

4   Use (hypothetical) examples to introduce or illustrate theoretical points.

Writing a literature review

In general, the literature review is a discussion of what other researchers have found and concluded on the topic you are researching. Primarily, it provides the context in which your study is situated. Secondly, it demonstrates that you know of and understand other research on the same topic. Usually the review is a **critical appraisal** of all the main theories and findings relevant to your topic. A critical appraisal literally means to compare assessments, identifying the main merits of a position and the main criticisms that can be levelled at an interpretation. It should provide a detailed and **balanced picture** of the various thoughts relating to a particular topic. The literature review should not be a sterile report of the writings of others, which seems divorced from the rest of the final report. It must be linked to the rationale for the study and provide an **external context** in which your study is grounded. In other words, it *justifies* your project – it should be obvious why you have chosen a particular issue to study once someone has read your literature review. Your thoughts relating to this literature should be interwoven within the text. This should not take the form of 'I think that this . . .'; rather your thoughts should be in the third person. In the following example, the second sentence (which we have italicised for emphasis) represents the writer's thoughts on another researcher's claim:

Invented (1997) argues that the spatial decisions people make are based solely on issues of distance and spatial layout. *While there is truth in his assertion, it would seem intuitive that other factors such as safety also influence the spatial decision-making process.*

In the literature review, you should present only the results or substantive findings from other writers. The findings of *your* research are presented, compared and contrasted with other researchers' work in subsequent sections on results and discussion. Remember, you are not seeking to incorporate everything that has ever been written on the topic in your literature review, but just the points that are relevant to your study.

Detailing the methodology

The methodology section details the **methods** and **procedures** used, as well as the rationale behind the choice of methodology. Sufficient detail should be included to allow another researcher to replicate your study, if they so desired. You should include:

- The choice and rationale for the methodological approach adopted.
- The hypotheses, if the study is deductive in nature.
- A description of the study area.
- Full, relevant demographic details of the sample population (e.g., age, sex, employment, etc.).
- How the sample population were selected (sampling strategy).
- A full description of the method(s) and procedures of data generation.
- A full description of the method(s) and procedures of data analysis.
- A discussion of ethical issues if appropriate.

There are two ways to present the methodology. The first is to use a very prescriptive, scientific report detailing information in specific subsections. For example, section headings might be *Approach adopted*; *Hypotheses tested*; *Description of study area*; *Sample population*; *Sampling procedure*; *Method of data generation*; *Method of data analysis*. The second is just to present all the information in the form of a general methodological discussion. Studies that are more positivistic or quasi-scientific in nature usually employ the more prescriptive approach. Others are more relaxed in their presentation style. This section might also contain a section relating to the social relations of the research (how you were accepted, problems encountered, etc.). Remember that the quality of your data and the validity of your conclusions are judged from how your data were generated. Careful and accurate description of your methodology will engender confidence in your work, and allow the reader to assess in full the validity of your results and conclusions.

Writing a results section

The results section details the **main findings** of the research and provides a **summary explanation**. It should include all the relevant results from the data analysis with an explanation as to what they mean. If you have tested hypotheses, it is here that you detail whether they can be accepted or rejected and interpret the significance of the outcomes. Whether to include a results section is a matter of choice. It is quite common to see the results of data analysis in a discussion section. If you are using a scientific approach, we recommend that you keep the results and discussion sections separate. The results section is a more objective statement of what was found, the discussion section being a subjective interpretation of

what the results mean. In the results section, you should (O'Connor, 1991: 64–65):

- emphasise results which answer the questions you are examining by putting these first;
- not suppress valid results which contradict your hypothesis (failing to reject a hypothesis is an interesting result in itself);
- make it clear how any observations relate to your argument.

If you are reporting the results of statistical analysis, you will need to:

- report the data in such a way that readers can assess the degree of experimental variation;
- estimate the variability or precision of the findings.

Where your write-up is the result of several years' research (for example, towards a PhD), then you may find that your results section will have to be broken down into separate chapters. This is particularly so if you are using a variety of different methodologies to analyse your data.

Writing a discussion section

In the discussion section you seek to describe what your results mean, developing a logical argument (see the section on *Developing your argument*) based on the evidence of your results. In essence, the discussion section acts as a bridge between your results and conclusions in the context of the questions raised in the introduction, and literature surveyed in the literature review. The discussion turns your evidence (findings) into an argument which supports and illustrates your conclusions regarding these questions. In general, this will require you to compare and contrast your findings with those from other studies which were surveyed in the literature review. This will help to place your findings in a **broader context** and illustrate their importance. You should be careful to avoid the temptation to repeat earlier description from your results section, and distinguish between discussion based upon analysis and unsupported speculation. In addition, you should identify any potential errors in the work with suggestions concerning design improvements.

Writing the conclusion

The conclusion is a **summary statement** of the main findings of the research and the conclusions that have been drawn from their interpretation. In essence, it is a summary of the problems and aims set at the start of the project, in the light of the results of your analysis and subsequent discussion, i.e., X was the problem, Y was the aim, Z is the answer. Within the conclusion, any **shortcomings** of the research are usually acknowledged, and an agenda for future research identified. When writing the conclusion, rather than just restate what you did and why, concentrate upon describing the main findings and their significance, and expound the main conclusions which can be drawn from the research. Your concluding statements should not be too long. Try to keep the conclusion concise and avoid unnecessary repetition of material stated in previous chapters/sections.

10.3.4 Practical considerations

As a companion to the section above, we include some advice below on the more practical elements of writing a report. There are a variety of texts available which give advice on questions of style and format (e.g., Parsons and Knight, 1995; Miller and Taylor, 1987); however, beyond the basics, many details are determined by the guidelines and regulations supplied from the Faculty of Graduate Studies or the department of the university in which you are registered. In this light, the details given here can be used in the absence of any information, or as a general guide. The most important practical consideration, if you are writing your report yourself, is to keep **copies** of the text in a secure location, preferably in a separate building from your office (e.g., at home), and make regular backups as you progress through the writing process.

Compiling a Contents list

In the final report of a research project you should include a list of contents. This is to allow the reader to easily find information that may be of use. Generally, the list of contents consists of a standard set of components:

- Chapter headings and subheadings
- List of tables
- List of figures
- List of abbreviations (optional)
- List of symbols (optional)
- List of appendices

although strictly (as in Box 10.3) the list of contents refers to the chapter headings/subheadings. All

chapter headings and subheadings should be listed, indented on the left side of the page, along with the page numbers on which they can be found on the right. Lists of tables and figures should also include page numbers and be recorded on separate sheets. Lists of abbreviations and mathematical symbols should be used only if their inclusion will aid understanding and increase clarity. The amount of time required to compile a list of contents should not be underestimated. Ideally, each component list should be compiled as the report is written, to save time and guard against the loss of information.

Presenting the text

Your final report should be **typed** or **word-processed**. The presentation should be both **consistent** and appropriate, and conform to any thesis submission regulations in terms of line spacing, size and type of font used, and margin width. Each page should be neatly laid out, with chapter and section headings clearly distinguishable from the rest of the text. In this book we have chosen a standard form and size of headings. For example, from earlier in this chapter:

Writing-up and dissemination

10.3 Writing a final report

10.3.1 Writing for an audience

Writing for your project assessor(s)

The text should be divided into easily distinguishable paragraphs. All pages should be numbered. The contents pages and any preliminary pages are usually numbered using Roman numerals (i.e., i, ii, iii, iv etc.), with the numbering reverting to Arabic numerals starting from '1' at the beginning of Chapter 1. If your report is divided into chapters, start each chapter on a new page. A header may be placed at the top of each page, which may be used to indicate the chapter and/ or page number. Throughout the process of writing the text, remember that presentation is important. A poorly presented report suggests a poorly conducted research project; consequently you should make every effort to ensure your text looks professionally produced.

Including statistical formulae and results

Beyond simple mathematical operations such as addition, subtraction, etc., you should include all statistical formulae, ideally as numbered equations centred within the text. You should be careful to use the full form of mathematical expressions, employ **standard mathematical symbols**, and fully define all variables. For example, to calculate a mean:

$$\bar{x} = \frac{\sum_{i=1}^{N} x_i}{N} \qquad \text{is preferable to} \qquad v = \frac{\sum v_i}{G}$$

The correct use of the equation on the left – which uses more standard symbols – would be followed by a definition of the terms x_i, N and $\bar{x}$. Although, strictly speaking, new terms will need to be defined only on their first appearance in the text, subsequent inclusions may increase the clarity of a complicated equation. As with all components of the written report, sympathy for the reader, who might not be familiar with standard notation, is essential. It may be helpful to provide a list of symbols in the Contents page, particularly if you employ a great variety of symbols. Similarly if you have used a standard statistical test upon your data, you should adopt a standardised style of reporting the results. There are several conventions used to report the results of hypothesis tests (*cf.* Chapter 5). For example, if we had used a two-sample *t*-test to explore whether males and females differed in their ability to estimate distances, we might report the results: 'using a two-sample *t*-test ($t = 0.52$, $p = 0.61$) it was found that no significant difference existed between men's and women's ability to estimate distances' (remember that $p < 0.05$ means that a result is significant at the 95% level).

Using tables, diagrams and maps

Tables, diagrams and maps can be extremely useful **summary devices** for conveying information or illustrating a point. However, they should be used *only when relevant* and *must* be referred to in the text. We have both marked student projects containing tables that are never referred to or explained and have been included only to demonstrate that they have been constructed, or to pad out the report. This misuse of summary devices detracts from the final report and serves only to frustrate the reader. You should use a table or diagram only when communication by textual means is ineffective or inefficient, and then there is no need for the information displayed to be replicated in a textual form as well. Remember, summary devices play a **supportive role**, helping you to illustrate the main points raised in the text. Where possible,

you should produce any maps or diagrams using a computer. There are a variety of appropriate computer software packages, such as Adobe Photoshop and Adobe Illustrator, with which you can construct high-quality illustrations. Other spreadsheet (e.g., Excel), graphing (e.g., Cricketgraph) and mapping (e.g., Map-Viewer) software can be used to ensure professional results. If you do not have access to such software, it is possible to use traditional cartographic techniques, using drawing pens and/or Letraset/Letratone (or similar) products, but the effective use of such materials requires some prior familiarity and training. Try to be consistent in presentation style. If you wish to reproduce someone else's table/diagram/map, where possible (and time permitting), rather than just photocopying and pasting it in, you should redraw it so that it conforms to your presentation style. The widespread availability of flatbed scanners enables existing material to be scanned into a computer and modified accordingly. For the purposes of an undergraduate or postgraduate project, you will not need to obtain **copyright permission** to reproduce a table, figure, plate, map or section of text of up to 200 words. For journal articles or a final report available to the general public, permission will have to be obtained from the copyright holder (usually the publisher).

Place your summary devices at appropriate points in the text. Space permitting, this is usually within the body of the text. If there is insufficient space at this point then the map or table can be placed at the top of the next page or the end of the following section. Do not place all your summary devices at the end of the relevant chapter or at the end of the report as this is inconvenient for the reader.

Not only do you have to decide whether a summary device would be the best way to illustrate a point but also what type of device is most appropriate. In many cases, this will be a commonsense decision. For example, a map is the most appropriate way to detail the location of a study area, and a schematic diagram is the most appropriate way to detail the relationships in a conceptual framework (see Figure 2.2). However, data can be presented to the reader in a number of different ways. For example, the same data could be presented within a table, as a bar chart, as a pie chart, or as a map displaying proportional circles (see Figure 10.1). In these cases, you need to decide which method of communication is most effective. As a general rule, graphic displays are more effective than tables. Whichever method of communication you choose you must ensure that your summary device is presented correctly (see Box 10.7).

(a)

Shop 1	13,021
Shop 2	11,382
Shop 3	16,233
Shop 4	12,139
Shop 5	9,673

(b)
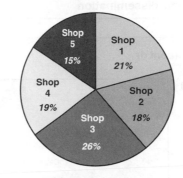

(c)
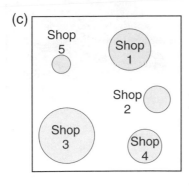

(d)
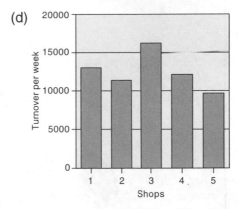

Figure 10.1 Displaying the same data using different methods: (a) table; (b) pie chart; (c) map made up of proportional circles; (d) bar chart.

Box 10.7 Presenting summary devices

For all summary devices
- **Provide a reference and a title** (e.g., Table 1.1: The percentage of workers in different sectors of the UK economy 1996–1997).
- **Display at an appropriate size** so that visual interpretation is easy and text readable.
- **Provide an acknowledgement** immediately underneath the reproduction if the summary device is copied from another source (e.g., Source: Kitchin and Tate (1999: 11)). If you are publishing the work you will need to obtain copyright permission to reproduce the summary device.

Specific guidelines

Tables
- Label every column and row with appropriate titles (including data units).
- Column headings should refer to independent variables and row headings to dependent variables.
- If the data are numeric then use appropriate decimal places and be consistent.
- For missing data insert n.a. (an abbreviation of not available).
- Format the table so that stronger lines demarcate specific rows and columns, and columns are appropriately spaced and interpretation is easy.

Graphs
- Label both axes with appropriate titles and provide units of measurement for each axis.
- The independent variable should be plotted on the horizontal axis and the dependent variable on the vertical axis.
- Use appropriate scales on each axis.
- Make sure different data sets are clearly distinguishable and are identified by a key.

Illustrations
- Make sure the message the diagram is trying to convey is clear and unambiguous.
- Make sure the diagram is uncluttered and is as simple as possible.
- Box text and use arrows where appropriate.

Maps
- Make sure the map is uncluttered and contains the necessary information at an appropriate scale.
- Include a legend, scale and north arrow.

Figure 10.2 displays examples of a draft and a final map of the same area.

Remember: You should list all figures, tables and maps in the contents pages at the beginning of the report.

Box 10.7   (cont'd)

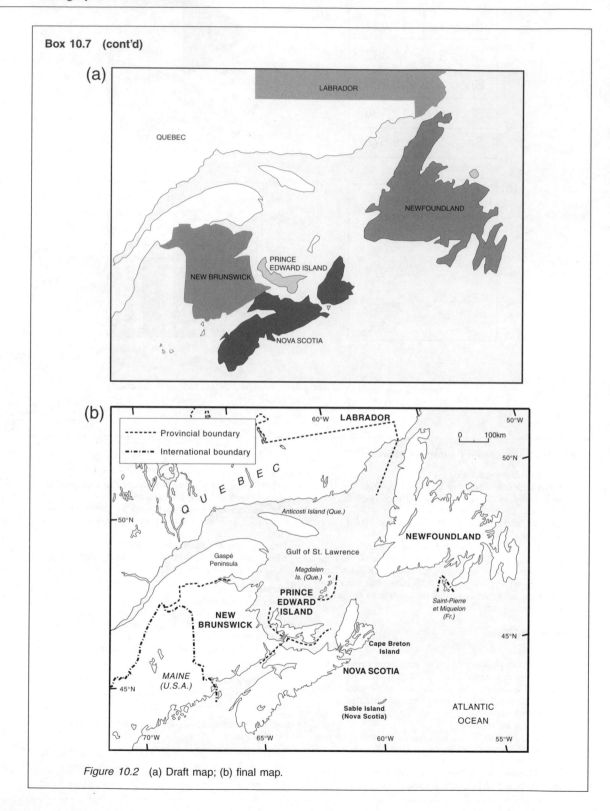

Figure 10.2   (a) Draft map; (b) final map.

Box 10.8 An example of using qualitative sample data

Spatial confusion
It is clear from the analysis of the interview transcripts that spatial confusion is of paramount concern. As illustrated by the following interview extract, getting lost or disorientated is a panic-inducing situation that all of our respondents feared:

RES B: I don't like it [being lost]. I hate being 'out of control'. It is not a good feeling. It can be an anxious feeling. It's also quite a confidence downer. All of a sudden you are not where you think you are and you feel a bit stupid. You think any other adult could manage this and maybe you can't so you feel inadequate in some way.

Presenting qualitative data

As discussed in Chapter 8, qualitative data can be summarised using tables. However, it is more common for researchers who have used a qualitative approach to use sample sections from their data to illustrate their arguments. Here, sample data is interwoven into the text to illustrate typical (or atypical) responses. As with tables, diagrams and maps, sample qualitative data should be used judiciously. Do not fall into the trap of allowing your text to become a series of linked sets of data with little substantive discussion of their meaning. Sample data should only be used as **supportive evidence**. As such, your discussion should dominate the data, not the other way around. Dyck (1995) provides a good example of using data from interview transcripts to illustrate her arguments. When using sample data you should indicate to the reader the source of the statement while protecting the respondents' anonymity. We always use a standardised coding system. Our respondents become RES A, RES B, RES C, etc. or M1 (male 1), M2, . . . , F1 (female 1), F2, . . . A summary table to accompany these codes reveals more details about the individuals and the place and context of data generation. Through this coding system, all sample data can be placed in context. To indicate the use of illustrative sample data we suggest that you indent the text and use a different font/size. An example of using sample data is illustrated in Box 10.8.

Referencing and using quotes

All the material which has been derived from other sources must be **referenced** within the text and fully recorded in the bibliography. Failure to reference material is **plagiarism** (using somebody else's work as your own). Plagiarism is taken very seriously in academia and in many universities it can lead to failure and disqualification. There are various different ways to acknowledge the ideas of others within the text. These are often referred to as **styles of documentation** (Miller and Taylor, 1987: 115). The rules and form of documentation vary from discipline to discipline, and between universities; however, four general forms are usually used (Miller and Taylor, 1987):

1 **Endnotes** – numbered citations in the text referring to notes at the end of chapters, and lists of references at the end of the text.
2 **Footnotes** – numbered citations in the text referring to notes at the bottom of each page, and lists of references at the end of the text.
3 **Author–date citations** – author's name, publication date, and page numbers in the text, with the full reference at the end of the text.
4 **Parenthetical citations** – author's name and page numbers in the text, with the full reference at the end of the text.

There are also a variety of style manuals produced which describe the specifics of documentation for different areas of academic study. For example, Miller and Taylor (1987) list *The Chicago Manual of Style* and *The MLA Handbook for Writers of Research Papers*, and note that these form the basis of many US university thesis format requirements. The style of documentation we have chosen for referencing is similar to the author–date citation above, and is commonly referred to as the **Harvard system** which gives as a minimum both a name and a year reference in the text (see Box 10.9) and a list of full references at the end of the text. By referring to this reference list the reader can then track down that article themselves. In some cases you might decide that a fellow researcher has made a pertinent statement that you wish to quote in full. If the quote is short it is generally retained in the main text. However, if the quote is longer or has particular relevance it may be separated from the text and indented to distinguish from the rest of the text. When using a quote it is standard also to provide the page number from which the quote was taken. Box 10.9 provides sample illustrations of referencing the text and using quotes.

Constructing a reference list

A **reference list** contains the full citations for the references and quotes referred to in the text. In contrast a

Box 10.9   Referencing and quoting

Referencing

Referencing is used to acknowledge the use of other people's ideas or to refer the reader to a similar study. Text can be referenced in different ways but the text always includes the author name(s) and the date of publication. For example:

Made-up (1997) concluded that safety is a key concern in spatial decision making.

Safety is of particular concern in spatial decision making (Made-up, 1997).

For a second author also include their name (e.g., Made-up and Invented, 1997). Where there are more than two authors, use *et al.* as an abbreviation (e.g., Made-up *et al.*, 1997). If the same idea or point has been made or illustrated in different sources use multiple references (e.g., Invented, 1995; Made-up, 1997). Order references either alphabetically or by year, but be consistent. If the same author published more than two articles in the same year and you want to reference both, indicate the difference (e.g., Made-up, 1997a; Made-up, 1997b).

Quoting

Quoting consists of reproducing a statement from another person's report. As well as reproducing the statement, the author name(s), date, and page number(s) must be supplied. Short quotes might appear in the text. For example:

It seems intuitive that 'safety is paramount' (Made-up, 1997: 1), yet there are other factors which strongly influence spatial decision making.

Alternatively a whole passage might be reproduced (notice the indenting and different size font). For example:

As Made-up (1997: 1) argues:

> Safety is of paramount importance in spatial decision making. Regardless of context and familiarity, all the adult participants in this study listed safety as their number one priority when selecting a route between destinations.

Quotes do not have to appear in full. For example, the middle of the previous quote could be omitted:

> Safety is of paramount importance . . . when selecting a route between destinations. (Made-up, 1997: 1)

date of publication, paper title, publication title and, in the case of a book, the place of publication, or for a journal article the volume number and page numbers. Find out which system you need to use and then implement it consistently. Use all the **correct syntax** (full stops, commas, colons, abbreviations, etc.) and **font style** (e.g., italics or bold). Include all author names, even if you used *et al.* in the text. The list should be presented alphabetically using the author's surname. Every university and publisher we know insists upon a properly constructed reference list. Box 10.10 illustrates example reference list entries using a variant on the Harvard system.

Using appendices?

An appendix contains material that the writer feels will supplement the final report in some useful way. A final report can have several appendices, each relating to a separate part of the study. For example, this book has three appendices. One contains look-up tables for different statistical tests, the second contains annotated and categorised data used when explaining qualitative data analysis relevant to Chapter 8, and the third contains the addresses of software suppliers. In general, appendices in student projects contain a copy of the method to generate data (e.g., the questionnaire used) and summary data (e.g., question returns). We suggest that where possible you should avoid the use of appendices. Where used they should contain useful information that does not sit well in the actual text. Appendices should *not* be used as an overspill section when the length limit in the final report has been reached, and they should not contain reams of example data.

10.4   Presenting your research as a talk

The final product of research might not be just a written report. You may also have to make an oral presentation about your project details and conclusions to others. Perhaps more than anything else, the thought of talking in front of an audience causes considerable anxiety. In such a situation, we are all nervous and apprehensive to some degree, particularly if the audience is comprised of our peer group. Fear of potential embarrassment is entirely natural! Few people have the natural confidence and oratory skill to captivate an audience, and very often those who do, have put in a lot of work and practice to do so.

bibliography is list of all references relevant to your research topic. There are many different systems for presenting a reference list which are variants on a theme. All contain the same information – the author,

---

Box 10.10   Constructing a reference list

An article:
Form: Author surname, Initial. (date). Paper title. Journal title, Volume (issue number): pages.
Example: Made-up, A. (1997). Issues of safety in spatial decision making. *Geography Journal*, 1(1): 1–10.

A book:
Form: Author surname, Initial. (date). Book title. Publisher, Where published.
Example: Made-up, A. (1997). *Issues of Safety in Spatial Decision Making*. Fake Publisher, London.

A chapter in an edited book:
Form: Author surname, Initial. (date). Chapter title. In, Editor's name and initial (Ed(s)) Book title. Publisher, Where published. Page numbers.
Example: Made-up, A. (1997). Issues of safety in spatial decision making. In, Invented, A. (Ed.) *Spatial Decisions, Spatial Choices*. Fake Publisher, London. pp. 21–30.

A paper on the Web:
Form: Author surname, Initial. (date). Paper title. Journal title (if appropriate), Web address.
Example: Made-up, A. (1997). Issues of safety in spatial decision making. *Wonderful Web Journal*, 1(1) <http://www.fakeuniversity.edu/>.

If the source has not been published:
Form: Author surname, Initial. (date). Title. Unpublished, Where written.
Example: Made-up, A. (1997). *Issues of safety in spatial decision making*. Unpublished manuscript, Department of Geography, National University of Ireland, Maynooth.

If the source is not dated:
Write 'n.d.' in date parentheses.

If the source has been submitted and accepted for publication but not yet been published:
Write 'In Press' in date parentheses.

Reference order:
References should be ordered alphabetically by author. If there is more than one reference to the same author, order by year, e.g., Made-up (1992) comes before Made-up (1994). If an author has written a paper with someone else, list after the single entries, e.g., Made-up (1994) comes before Made-up and Invented (1992). If an author has written several joint papers, list alphabetically by second author, e.g., Made-up and Fake (1994) comes before Made-up and Invented (1992).

Presenting data orally is a task that many researchers have to undertake. This is especially true for postgraduate students who will often have to present their

work at department seminars and are also encouraged to speak at regional, national and international conferences. Presenting a talk to other people is a skill which can be learnt and develops with experience. Box 10.11 provides some general guidelines. For further advice see O'Connor (1991). You should be aware that away from academia, a large proportion of skilled jobs require employees to present the findings of a small project or the thinking of a working group, so these skills are particularly useful.

10.5   Before submitting your work

Once you have produced a final draft before submitting your work we suggest you run through the following guidelines (see Box 10.12). These guidelines are designed to add those finishing touches so that silly errors or poor presentation do not detract from what you have written. Once you have checked the manuscript thoroughly, make the required number of copies plus two extra copies. These two extra copies are your insurance against the manuscript being mislaid or defaced. You can now submit the work confident that you have done your best.

10.6   After submitting your work

You may feel that once you have submitted your final report it is time to put your feet up. In many ways you are right. However, the final report still needs to be assessed and the assessment dealt with. In many student projects, the assessment and mark is often final. There is, however, often an **oral defence/viva** which must be negotiated, and a right to appeal. A viva is where you discuss your project with an external examiner. If you strongly disagree with a mark you can appeal to the institution to reassess the mark awarded. This path should only be taken if you feel especially aggrieved. In the vast majority of cases, the original mark is ratified and in some the marks are even lowered!

If you have submitted your work elsewhere you may need to **make alterations** and resubmit the work in the light of suggestions and criticism. This is very common when articles are submitted to refereed journals. While criticism may at first seem difficult to swallow, you should try to look at it in a positive light. Instead of rejecting the criticism outright, sit

---

Box 10.11   Presenting a talk

The main rule for presenting a talk is to **prepare thoroughly**.

Talk content:
- **Know your project thoroughly** – re-read all relevant sections.
- **Read any relevant literature** that may have been published since you finished your project.
- **Be aware of your audience** to pitch the level of your talk appropriately (see Section 10.3.1, Writing for an audience).

Talk design:
- **Make use of visual aids**. Overheads and slides help explain and keep the audience interested.
- **Do not swamp the listener with information**. Use a few overheads and/or a few slides. An overhead/slide every 5 seconds is too much!
- **Keep the information on the overheads/slides to a minimum**. Just use key bullet point statements. Too much information on the overheads and people will stop listening and start reading.
- **Use large print for text** (minimum 24 font). The audience does not want to strain their eyes trying to read the text.
- **Make sure all figures, tables and maps are presented professionally** (see Box 10.7).

Talk rehearsal:
- **Practice presenting your talk** using whatever aids you will use at the presentation, keeping to any time limit. If necessary get a friend or family member to listen to what you have to say. Then rehearse it again, and again. Practice is the key.
- **Speak clearly, loudly, face the audience and look up from your notes** to engage the listeners. Try not to speak in a monotone.

On the day:
- **Always arrive a little early at the talk** and make sure any equipment you require (overhead projector, slide projector) is present, that it works, and that you know how to use it. At a conference, you should introduce yourself to the chair of the paper session in advance (if possible), confirm the presence of appropriate equipment, and check that the time slot for your talk has not changed.
- **Be prepared to answer questions at the end of the talk**. Listen to each question carefully. If you do not understand the question or know the answer, do not panic, be truthful and ask the enquirer to repeat their question or say that you are sorry but that you do not know the answer.
- **Be prepared to defend your work from criticism**.

Box 10.12   Finishing off

- **Check the text for the following:**
  - Does the manuscript have a title, your name and affiliation and other required information?
  - Are all the pages there? Are they numbered consecutively?
  - Are all the section and chapter titles present?
  - Is the formatting correct (margins and line spacing)?
  - Are all the diagrams, tables and plates professionally produced, titled and listed in the contents list?
  - Are all the references present in the bibliography?
  - Are references in the bibliography complete and in the correct format?
  - Has the text been checked for spelling and grammar errors?
- **Friends and family permitting, ask somebody to read through the text for you**. Somebody who is detached from your work will quickly be able to point out inconsistencies and incoherencies.
- **Time permitting, leave the manuscript unread for at least a week**. Then re-read the text from start to finish. This 'cooling off' period means you return to the text afresh, ready to spot errors and tidy up inconsistencies. It will also allow you to assess the coherency of your arguments and text.

---

back and reflect upon what the assessor or critic has said and try to understand why such criticism has been made. In the main, criticism is usually constructive – it is aimed at improving the final report. Try to use criticism positively either to make alterations to the final report or to help shape future projects. In general, those making any criticisms do so from experience, and often draw upon an expert knowledge of the field. This of course does not automatically make them right, but their views should be respected.

10.7   Summary

After reading this chapter you should:

- appreciate the need to start writing from the beginning of the project;
- understand what is required in a final report;
- be aware of the different ways in which to disseminate your work;
- understand the concept of writing for different audiences;
- be able to produce a well-written, structured and appropriately illustrated final report.

In this chapter, we have explored the process of writing up your research into a coherent report and disseminating the work. Time and effort invested in writing up is well spent because it is often the case that a project is judged by others on the basis of reading the final report. Every effort then should be made to produce a report that is well written, coherently structured, balanced, and professionally presented. Central to your final report must be an argument that seeks to explain the rationale for the study and findings. This will provide the strong, coherent plot-line that links all the sections into a 'story'. The process of writing should not be viewed as merely part of the finishing-off phase, but rather as an important part of the research process. As such, we suggest that the process of writing up commences when you start the project. It is also important to remember that you are writing for an audience, to whom your report should be targeted. In the next, and final, chapter we offer some final words to conclude this book.

10.8   Questions for reflection

- *What is the purpose of the final report?*
- *Why is it best to start writing at the beginning of the project?*
- *Why should your report be well written, coherently structured, balanced, and professionally presented?*
- *How is writing a lay person's report different from writing for a group of experts?*
- *In what way is a final report more than just reporting the findings of a study?*

Further reading

Barnes, R. (1995) *Successful study for degrees*, 2nd edition. Routledge, London.

Lofland, J. and Lofland, L.H. (1995) *Analysing Social Settings: A Guide to Qualitative Observation and Analysis*. Wadsworth, Belmont, CA.

Miller, J.I. and Taylor, B.J. (1987) *The Thesis Writers Handbook*. Alcove Publishing Company, West Linn, Oregon.

O'Connor, M. (1991) *Writing Successfully in Science*. Chapman and Hall, London.

Parsons, T. and Knight, P.G. (1995) *How to Do Your Dissertation in Geography and Related Disciplines*. Chapman and Hall, London.

Robson, C. (1993) *Real World Research: A Resource for Social Scientists and Practitioner-Researchers*. Blackwell, Oxford.

Wolcott, H. (1990) *Writing Up Qualitative Research*. Sage, London.

CHAPTER 11

Final words

This chapter covers

11.1 An overview of conducting research in human geography
11.2 Coping with problems
11.3 Independent research
11.4 Epilogue
11.5 Summary

11.1 An overview of conducting research in human geography

In this book we have sought to provide you with an insight into how to conduct successful research in human geography. We hope we have managed to convey the sentiment that whilst research is challenging, it is also stimulating and exciting. Through conducting research you can explore and discover things that no-one else yet knows. Conducting research is also a learning experience. Besides learning about your chosen topic, you get to experience at first hand the trials and tribulations of generating, analysing and interpreting data. In the course of doing so you may gain an insight into the processes which underlie other studies. Conducting research thus provides a practical benchmark by which to understand the findings and conclusions of other researchers. It also provides you with skills that will be of use in the workplace.

As we have illustrated, there is much to consider when undertaking research, from deciding upon a topic, to grounding your ideas conceptually, to choosing methods of data generation and analysis, and finally to interpreting the findings and writing up. There are many opportunities to make mistakes and to wander from your intended path. The advice proffered throughout the book is intended to minimise any mistakes. Box 11.1 stresses the most important things to do to ensure a sound project. Remember that your enthusiasm, creativity and energy are vital ingredients to a successful project. By following our advice, it is hoped that the stresses and strains of

Box 11.1 Ensuring a successful project

- Plan the whole project from start to finish before generating any data.
- Construct a timetable and stick to it.
- Read the relevant literature.
- Make sure your project is conceptually grounded: philosophically and theoretically anchored.
- Make sure you do actually get down to the business of generating, analysing and interpreting data.
- Be diligent in your data generation and analysis.
- Write up the research for the intended audience and invest time in presenting your research professionally.

conducting research are minimised, allowing you to enjoy the process of researching a particular topic.

Above all else, the most important thing to do when conducting research is to plan, plan, plan. Carefully think through the whole project, stage by stage, working out what you are going to do and why. Then, at every stage throughout the project, re-evaluate your original thinking and update your plans in the light of how the project has gone so far. Reading relevant literature will give you ideas, show you how other people have approached the same or similar issues, and provide a context for your research. Too many student projects suffer as a result of not knowing what others have done. Students tend to shy away from theoretical and conceptual issues but, as we argued in Chapters 1 and 2, they are vital in providing the ballast for your studies. Thinking theoretically about your study will help you in planning data generation and analysis, and aid you in the process of interpreting your results. You should not, however, over-theorise your study. You need to balance planning with the practicalities of researching your topic of interest. You need to generate good quality data and to analyse it thoroughly. There is a tendency among students to think that if the project is planned well at the start, the other stages can be rushed through. This is not the case and you should work through each stage carefully. Similarly, be careful not to overdo data generation and analysis. There is no point in undertaking unnecessary data generation and analysis. Time invested in the process of writing up is time well spent as people most often judge the quality and validity of a study by its write-up.

11.2 Coping with problems

All your planning and pilot-study work (see Chapter 2) is designed to try to prevent things going wrong. However, sometimes, despite all the planning in the world, things do start to go wrong. If this happens with your project, it is important that you do not become demoralised or disheartened. There are few experienced researchers who have not suffered a major setback whilst undertaking a project. The important thing is to try to **salvage** what work you have done and to fully **explain** what went wrong and why. If you have time, **change direction** and complete the project using a different strategy. Here, it may be possible to turn the problem to your advantage so that it becomes a focus of your study. If you do change direction, then be sure to explain in your write-up why you did. If it is too late in the day, then when writing up, be sure to explain fully what went wrong and why, and to give advice for those thinking of undertaking a similar project. Remember, many projects at an undergraduate level are as much about learning *how to* conduct research as about examining a substantive issue. If things do go wrong, remember that all is not lost!

There are also a whole series of problem scenarios which are not usually covered by textbooks – encounters that are difficult to anticipate and to deal with. These are usually one-off events that can dent your confidence and make the rest of the project daunting to complete. What we are referring to here are 'nightmare' encounters usually experienced while generating data. Generating primary data usually means engaging with members of the public. Not all members of the public will react to your research in the same way (recall from Chapter 3) A minority of

subjects may be hostile or abusive. When generating data concerning sensitive issues you need to be especially careful not to antagonise respondents while still probing relevant issues. We are aware of a number of researchers who have found themselves in situations that they would rather have avoided. For example, we know of one researcher who has been propositioned by one of her respondents; one who has conducted an interview with a respondent whilst a pair of Dobermans sat staring at her, with the husband constantly asking when she was going to leave; another who has had a respondent break down and pour their heart out; and yet another who was held at gunpoint by armed police while photographing a route that a blind respondent was walking (he also came back from one interview to find the bomb disposal squad deciding whether to do a controlled detonation on his car!). These sorts of situations are never easy to deal with. It is difficult to keep calm, collected and professional and also take a route which compromises neither the research, your personal safety, nor the feelings of the respondent. Our advice, if you find yourself in such a situation, is to put safety first:

1 Always make sure that somebody knows where you are going and at what time you expect to return.
2 If the situation is particularly stressful just abandon that encounter and chalk it up to experience.

There is a tendency among students to feel that their research has 'failed' if they do not achieve or discover the results they anticipated, for example if their null hypothesis cannot be rejected. No study ever 'fails'. Each study provides a little more knowledge about a particular phenomenon or situation. As long as the study has been well planned and executed, then we can accept the negative result(s) as valid reflection of the information contained within the data. Negative results tell us just as much as positive results. The important thing is to try to decide why you have produced such results and to explain this clearly to the reader of the final account, with suggestions for future studies.

11.3 Independent research

This book has been written with undergraduate students predominately in mind. Most student projects are undertaken as a requirement of their course. Although independent in nature, these projects are supported by a supervisor whose role is to offer support and advice. Upon completing a course and venturing forth into the world, many former students *want* to, or more commonly *need* to, undertake a truly independent study. In this situation there is often nobody to turn to when advice is needed. As such, undertaking an independent study can seem particularly daunting. This book, and the countless others concerning research methods, are there to provide a helping hand – to guide you through some of the more thorny theoretical and methodological problems. If you require more specific help concerning the research topic there a number of strategies you can adopt. For example, you can try to discuss the research through with your peer group, or if appropriate with work colleagues. Alternatively, you can visit the library and seek inspiration through the writings of others. As a last resort you can contact a former supervisor or lecturer, or get in touch with a leading expert in the field, to ask advice. In general, while many academics are very busy, most will find the time to have a quick chat with you, especially if the topic is interesting. Before you contact an expert make sure you have a clear idea about what you want to discuss and be able to state clearly what your project concerns and why you are undertaking it. It might also be prudent to have a knowledge of that person's work. It is our experience that people will take an interest in your work if you display enthusiasm and genuine interest in it yourself: if you demonstrate that you have invested time and effort into trying to solve a problem yourself. Few people want to help somebody who will not help themselves.

11.4 Epilogue

Human geography is a vast and diverse subject which provides a rich context in which to explore the many different aspects of the world. We encourage you to use this context and to grasp the opportunity to explore the geographical world of humanity. As well as helping you to pass your course, research can be tremendously interesting and exciting, broadening your skills and knowledge. We are sure that for the time and effort you invest in research you will get back ample rewards at both an academic and a personal level. We therefore hope that you will approach the task of conducting research with enthusiasm and verve, aiming to undertake any project to the best of your abilities. Last, and not least, we wish you **good luck** in conducting your research.

11.5 Summary

Having read through the book you should:

- understand the importance of linking theory and practice;
- know how to generate, analyse and interpret qualitative and quantitative data;
- be able to plan and conduct a research project in human geography.

Tables

Table A.1   Critical values of the *F*-distribution.

DF2	DF1															
	1	2	3	4	5	6	7	8	9	10	20	30	40	50	100	1000
α = 0.05																
1	161.45	199.50	215.71	224.58	230.16	233.99	236.77	238.88	240.54	241.88	248.02	250.10	251.14	251.77	253.04	254.19
2	18.51	19.00	19.16	19.25	19.30	19.33	19.35	19.37	19.38	19.40	19.45	19.46	19.47	19.48	19.49	19.49
3	10.13	9.55	9.28	9.12	9.01	8.94	8.89	8.85	8.81	8.79	8.66	8.62	8.59	8.58	8.55	8.53
4	7.71	6.94	6.59	6.39	6.26	6.16	6.09	6.04	6.00	5.96	5.80	5.75	5.72	5.70	5.66	5.63
5	6.61	5.79	5.41	5.19	5.05	4.95	4.88	4.82	4.77	4.74	4.56	4.50	4.46	4.44	4.41	4.37
6	5.99	5.14	4.76	4.53	4.39	4.28	4.21	4.15	4.10	4.06	3.87	3.81	3.77	3.75	3.71	3.67
7	5.59	4.74	4.35	4.12	3.97	3.87	3.79	3.73	3.68	3.64	3.44	3.38	3.34	3.32	3.27	3.23
8	5.32	4.46	4.07	3.84	3.69	3.58	3.50	3.44	3.39	3.35	3.15	3.08	3.04	3.02	2.97	2.93
9	5.12	4.26	3.86	3.63	3.48	3.37	3.29	3.23	3.18	3.14	2.94	2.86	2.83	2.80	2.76	2.71
10	4.96	4.10	3.71	3.48	3.33	3.22	3.14	3.07	3.02	2.98	2.77	2.70	2.66	2.64	2.59	2.54
11	4.84	3.98	3.59	3.36	3.20	3.09	3.01	2.95	2.90	2.85	2.65	2.57	2.53	2.51	2.46	2.41
12	4.75	3.89	3.49	3.26	3.11	3.00	2.91	2.85	2.80	2.75	2.54	2.47	2.43	2.40	2.35	2.30
13	4.67	3.81	3.41	3.18	3.03	2.92	2.83	2.77	2.71	2.67	2.46	2.38	2.34	2.31	2.26	2.21
14	4.60	3.74	3.34	3.11	2.96	2.85	2.76	2.70	2.65	2.60	2.39	2.31	2.27	2.24	2.19	2.14
15	4.54	3.68	3.29	3.06	2.90	2.79	2.71	2.64	2.59	2.54	2.33	2.25	2.20	2.18	2.12	2.07
16	4.49	3.63	3.24	3.01	2.85	2.74	2.66	2.59	2.54	2.49	2.28	2.19	2.15	2.12	2.07	2.02
17	4.45	3.59	3.20	2.96	2.81	2.70	2.61	2.55	2.49	2.45	2.23	2.15	2.10	2.08	2.02	1.97
18	4.41	3.55	3.16	2.93	2.77	2.66	2.58	2.51	2.46	2.41	2.19	2.11	2.06	2.04	1.98	1.92
19	4.38	3.52	3.13	2.90	2.74	2.63	2.54	2.48	2.42	2.38	2.16	2.07	2.03	2.00	1.94	1.88
20	4.35	3.49	3.10	2.87	2.71	2.60	2.51	2.45	2.39	2.35	2.12	2.04	1.99	1.97	1.91	1.85
21	4.32	3.47	3.07	2.84	2.68	2.57	2.49	2.42	2.37	2.32	2.10	2.01	1.96	1.94	1.88	1.82
22	4.30	3.44	3.05	2.82	2.66	2.55	2.46	2.40	2.34	2.30	2.07	1.98	1.94	1.91	1.85	1.79
23	4.28	3.42	3.03	2.80	2.64	2.53	2.44	2.37	2.32	2.27	2.05	1.96	1.91	1.88	1.82	1.76
24	4.26	3.40	3.01	2.78	2.62	2.51	2.42	2.36	2.30	2.25	2.03	1.94	1.89	1.86	1.80	1.74
25	4.24	3.39	2.99	2.76	2.60	2.49	2.40	2.34	2.28	2.24	2.01	1.92	1.87	1.84	1.78	1.72
26	4.23	3.37	2.98	2.74	2.59	2.47	2.39	2.32	2.27	2.22	1.99	1.90	1.85	1.82	1.76	1.70
27	4.21	3.35	2.96	2.73	2.57	2.46	2.37	2.31	2.25	2.20	1.97	1.88	1.84	1.81	1.74	1.68
28	4.20	3.34	2.95	2.71	2.56	2.45	2.36	2.29	2.24	2.19	1.96	1.87	1.82	1.79	1.73	1.66
29	4.18	3.33	2.93	2.70	2.55	2.43	2.35	2.28	2.22	2.18	1.94	1.85	1.81	1.77	1.71	1.65
30	4.17	3.32	2.92	2.69	2.53	2.42	2.33	2.27	2.21	2.16	1.93	1.84	1.79	1.76	1.70	1.63
40	4.08	3.23	2.84	2.61	2.45	2.34	2.25	2.18	2.12	2.08	1.84	1.74	1.69	1.66	1.59	1.52
50	4.03	3.18	2.79	2.56	2.40	2.29	2.20	2.13	2.07	2.03	1.78	1.69	1.63	1.60	1.52	1.45
60	4.00	3.15	2.76	2.53	2.37	2.25	2.17	2.10	2.04	1.99	1.75	1.65	1.59	1.56	1.48	1.40
70	3.98	3.13	2.74	2.50	2.35	2.23	2.14	2.07	2.02	1.97	1.72	1.62	1.57	1.53	1.45	1.36
80	3.96	3.11	2.72	2.49	2.33	2.21	2.13	2.06	2.00	1.95	1.70	1.60	1.54	1.51	1.43	1.34
90	3.95	3.10	2.71	2.47	2.32	2.20	2.11	2.04	1.99	1.94	1.69	1.59	1.53	1.49	1.41	1.31
100	3.94	3.09	2.70	2.46	2.31	2.19	2.10	2.03	1.97	1.93	1.68	1.57	1.52	1.48	1.39	1.30

α = 0.025

1	647.79	799.48	864.15	899.60	921.83	937.11	948.20	956.64	963.28	968.63	993.08	1001.40	1005.60	1008.10	1013.16	1017.76
2	38.51	39.00	39.17	39.25	39.30	39.33	39.36	39.37	39.39	39.40	39.45	39.46	39.47	39.48	39.49	39.50
3	17.44	16.04	15.44	15.10	14.88	14.73	14.62	14.54	14.47	14.42	14.17	14.08	14.04	14.01	13.96	13.91
4	12.22	10.65	9.98	9.60	9.36	9.20	9.07	8.98	8.90	8.84	8.56	8.46	8.41	8.38	8.32	8.26
5	10.01	8.43	7.76	7.39	7.15	6.98	6.85	6.76	6.68	6.62	6.33	6.23	6.18	6.14	6.08	6.02
6	8.81	7.26	6.60	6.23	5.99	5.82	5.70	5.60	5.52	5.46	5.17	5.07	5.01	4.98	4.92	4.86
7	8.07	6.54	5.89	5.52	5.29	5.12	4.99	4.90	4.82	4.76	4.47	4.36	4.31	4.28	4.21	4.15
8	7.57	6.06	5.42	5.05	4.82	4.65	4.53	4.43	4.36	4.30	4.00	3.89	3.84	3.81	3.74	3.68
9	7.21	5.71	5.08	4.72	4.48	4.32	4.20	4.10	4.03	3.96	3.67	3.56	3.51	3.47	3.40	3.34
10	6.94	5.46	4.83	4.47	4.24	4.07	3.95	3.85	3.78	3.72	3.42	3.31	3.26	3.22	3.15	3.09
11	6.72	5.26	4.63	4.28	4.04	3.88	3.76	3.66	3.59	3.53	3.23	3.12	3.06	3.03	2.96	2.89
12	6.55	5.10	4.47	4.12	3.89	3.73	3.61	3.51	3.44	3.37	3.07	2.96	2.91	2.87	2.80	2.73
13	6.41	4.97	4.35	4.00	3.77	3.60	3.48	3.39	3.31	3.25	2.95	2.84	2.78	2.74	2.67	2.60
14	6.30	4.86	4.24	3.89	3.66	3.50	3.38	3.29	3.21	3.15	2.84	2.73	2.67	2.64	2.56	2.50
15	6.20	4.77	4.15	3.80	3.58	3.41	3.29	3.20	3.12	3.06	2.76	2.64	2.59	2.55	2.47	2.40
16	6.12	4.69	4.08	3.73	3.50	3.34	3.22	3.12	3.05	2.99	2.68	2.57	2.51	2.47	2.40	2.32
17	6.04	4.62	4.01	3.66	3.44	3.28	3.16	3.06	2.98	2.92	2.62	2.50	2.44	2.41	2.33	2.26
18	5.98	4.56	3.95	3.61	3.38	3.22	3.10	3.01	2.93	2.87	2.56	2.44	2.38	2.35	2.27	2.20
19	5.92	4.51	3.90	3.56	3.33	3.17	3.05	2.96	2.88	2.82	2.51	2.39	2.33	2.30	2.22	2.14
20	5.87	4.46	3.86	3.51	3.29	3.13	3.01	2.91	2.84	2.77	2.46	2.35	2.29	2.25	2.17	2.09
21	5.83	4.42	3.82	3.48	3.25	3.09	2.97	2.87	2.80	2.73	2.42	2.31	2.25	2.21	2.13	2.05
22	5.79	4.38	3.78	3.44	3.22	3.05	2.93	2.84	2.76	2.70	2.39	2.27	2.21	2.17	2.09	2.01
23	5.75	4.35	3.75	3.41	3.18	3.02	2.90	2.81	2.73	2.67	2.36	2.24	2.18	2.14	2.06	1.98
24	5.72	4.32	3.72	3.38	3.15	2.99	2.87	2.78	2.70	2.64	2.33	2.21	2.15	2.11	2.02	1.94
25	5.69	4.29	3.69	3.35	3.13	2.97	2.85	2.75	2.68	2.61	2.30	2.18	2.12	2.08	2.00	1.91
26	5.66	4.27	3.67	3.33	3.10	2.94	2.82	2.73	2.65	2.59	2.28	2.16	2.09	2.05	1.97	1.89
27	5.63	4.24	3.65	3.31	3.08	2.92	2.80	2.71	2.63	2.57	2.25	2.13	2.07	2.03	1.94	1.86
28	5.61	4.22	3.63	3.29	3.06	2.90	2.78	2.69	2.61	2.55	2.23	2.11	2.05	2.01	1.92	1.84
29	5.59	4.20	3.61	3.27	3.04	2.88	2.76	2.67	2.59	2.53	2.21	2.09	2.03	1.99	1.90	1.82
30	5.57	4.18	3.59	3.25	3.03	2.87	2.75	2.65	2.57	2.51	2.20	2.07	2.01	1.97	1.88	1.80
40	5.42	4.05	3.46	3.13	2.90	2.74	2.62	2.53	2.45	2.39	2.07	1.94	1.88	1.83	1.74	1.65
50	5.34	3.97	3.39	3.05	2.83	2.67	2.55	2.46	2.38	2.32	1.99	1.87	1.80	1.75	1.66	1.56
60	5.29	3.93	3.34	3.01	2.79	2.63	2.51	2.41	2.33	2.27	1.94	1.82	1.74	1.70	1.60	1.49
70	5.25	3.89	3.31	2.97	2.75	2.59	2.47	2.38	2.30	2.24	1.91	1.78	1.71	1.66	1.56	1.45
80	5.22	3.86	3.28	2.95	2.73	2.57	2.45	2.35	2.28	2.21	1.88	1.75	1.68	1.63	1.53	1.41
90	5.20	3.84	3.26	2.93	2.71	2.55	2.43	2.34	2.26	2.19	1.86	1.73	1.66	1.61	1.50	1.39
100	5.18	3.83	3.25	2.92	2.70	2.54	2.42	2.32	2.24	2.18	1.85	1.71	1.64	1.59	1.48	1.36

Table A.1 (cont'd)

α = 0.01

DF2									DF1							
	1	2	3	4	5	6	7	8	9	10	20	30	40	50	100	1000
1	4052	4999	5404	5624	5764	5859	5928	5981	6022	6056	6209	6260	6286	6302	6334	6363
2	98.50	99.00	99.16	99.25	99.30	99.33	99.36	99.38	99.39	99.40	99.45	99.47	99.48	99.48	99.49	99.50
3	34.12	30.82	29.46	28.71	28.24	27.91	27.67	27.49	27.34	27.23	26.69	26.50	26.41	26.35	26.24	26.14
4	21.20	18.00	16.69	15.98	15.52	15.21	14.98	14.80	14.66	14.55	14.02	13.84	13.75	13.69	13.58	13.47
5	16.26	13.27	12.06	11.39	10.97	10.67	10.46	10.29	10.16	10.05	9.55	9.38	9.29	9.24	9.13	9.03
6	13.75	10.92	9.78	9.15	8.75	8.47	8.26	8.10	7.98	7.87	7.40	7.23	7.14	7.09	6.99	6.89
7	12.25	9.55	8.45	7.85	7.46	7.19	6.99	6.84	6.72	6.62	6.16	5.99	5.91	5.86	5.75	5.66
8	11.26	8.65	7.59	7.01	6.63	6.37	6.18	6.03	5.91	5.81	5.36	5.20	5.12	5.07	4.96	4.87
9	10.56	8.02	6.99	6.42	6.06	5.80	5.61	5.47	5.35	5.26	4.81	4.65	4.57	4.52	4.41	4.32
10	10.04	7.56	6.55	5.99	5.64	5.39	5.20	5.06	4.94	4.85	4.41	4.25	4.17	4.12	4.01	3.92
11	9.65	7.21	6.22	5.67	5.32	5.07	4.89	4.74	4.63	4.54	4.10	3.94	3.86	3.81	3.71	3.61
12	9.33	6.93	5.95	5.41	5.06	4.82	4.64	4.50	4.39	4.30	3.86	3.70	3.62	3.57	3.47	3.37
13	9.07	6.70	5.74	5.21	4.86	4.62	4.44	4.30	4.19	4.10	3.66	3.51	3.43	3.38	3.27	3.18
14	8.86	6.51	5.56	5.04	4.69	4.46	4.28	4.14	4.03	3.94	3.51	3.35	3.27	3.22	3.11	3.02
15	8.68	6.36	5.42	4.89	4.56	4.32	4.14	4.00	3.89	3.80	3.37	3.21	3.13	3.08	2.98	2.88
16	8.53	6.23	5.29	4.77	4.44	4.20	4.03	3.89	3.78	3.69	3.26	3.10	3.02	2.97	2.86	2.76
17	8.40	6.11	5.19	4.67	4.34	4.10	3.93	3.79	3.68	3.59	3.16	3.00	2.92	2.87	2.76	2.66
18	8.29	6.01	5.09	4.58	4.25	4.01	3.84	3.71	3.60	3.51	3.08	2.92	2.84	2.78	2.68	2.58
19	8.18	5.93	5.01	4.50	4.17	3.94	3.77	3.63	3.52	3.43	3.00	2.84	2.76	2.71	2.60	2.50
20	8.10	5.85	4.94	4.43	4.10	3.87	3.70	3.56	3.46	3.37	2.94	2.78	2.69	2.64	2.54	2.43
21	8.02	5.78	4.87	4.37	4.04	3.81	3.64	3.51	3.40	3.31	2.88	2.72	2.64	2.58	2.48	2.37
22	7.95	5.72	4.82	4.31	3.99	3.76	3.59	3.45	3.35	3.26	2.83	2.67	2.58	2.53	2.42	2.32
23	7.88	5.66	4.76	4.26	3.94	3.71	3.54	3.41	3.30	3.21	2.78	2.62	2.54	2.48	2.37	2.27
24	7.82	5.61	4.72	4.22	3.90	3.67	3.50	3.36	3.26	3.17	2.74	2.58	2.49	2.44	2.33	2.22
25	7.77	5.57	4.68	4.18	3.85	3.63	3.46	3.32	3.22	3.13	2.70	2.54	2.45	2.40	2.29	2.18
26	7.72	5.53	4.64	4.14	3.82	3.59	3.42	3.29	3.18	3.09	2.66	2.50	2.42	2.36	2.25	2.14
27	7.68	5.49	4.60	4.11	3.78	3.56	3.39	3.26	3.15	3.06	2.63	2.47	2.38	2.33	2.22	2.11
28	7.64	5.45	4.57	4.07	3.75	3.53	3.36	3.23	3.12	3.03	2.60	2.44	2.35	2.30	2.19	2.08
29	7.60	5.42	4.54	4.04	3.73	3.50	3.33	3.20	3.09	3.00	2.57	2.41	2.33	2.27	2.16	2.05
30	7.56	5.39	4.51	4.02	3.70	3.47	3.30	3.17	3.07	2.98	2.55	2.39	2.30	2.25	2.13	2.02
40	7.31	5.18	4.31	3.83	3.51	3.29	3.12	2.99	2.89	2.80	2.37	2.20	2.11	2.06	1.94	1.82
50	7.17	5.06	4.20	3.72	3.41	3.19	3.02	2.89	2.78	2.70	2.27	2.10	2.01	1.95	1.82	1.70
60	7.08	4.98	4.13	3.65	3.34	3.12	2.95	2.82	2.72	2.63	2.20	2.03	1.94	1.88	1.75	1.62
70	7.01	4.92	4.07	3.60	3.29	3.07	2.91	2.78	2.67	2.59	2.15	1.98	1.89	1.83	1.70	1.56
80	6.96	4.88	4.04	3.56	3.26	3.04	2.87	2.74	2.64	2.55	2.12	1.94	1.85	1.79	1.65	1.51
90	6.93	4.85	4.01	3.53	3.23	3.01	2.84	2.72	2.61	2.52	2.09	1.92	1.82	1.76	1.62	1.48
100	6.90	4.82	3.98	3.51	3.21	2.99	2.82	2.69	2.59	2.50	2.07	1.89	1.80	1.74	1.60	1.45

$\alpha = 0.005$

ν_2																
1	16212	19997	21614	22501	23056	23440	23715	23924	24091	24222	24837	25041	25146	25213	25339	25451
2	198.50	199.01	199.16	199.24	199.30	199.33	199.36	199.38	199.39	199.39	199.45	199.48	199.48	199.48	199.48	199.51
3	55.55	49.80	47.47	46.20	45.39	44.84	44.43	44.13	43.88	43.68	42.78	42.47	42.31	42.21	42.02	41.85
4	31.33	26.28	24.26	23.15	22.46	21.98	21.62	21.35	21.14	20.97	20.17	19.89	19.75	19.67	19.50	19.34
5	22.78	18.31	16.53	15.56	14.94	14.51	14.20	13.96	13.77	13.62	12.90	12.66	12.53	12.45	12.30	12.16
6	18.63	14.54	12.92	12.03	11.46	11.07	10.79	10.57	10.39	10.25	9.59	9.36	9.24	9.17	9.03	8.89
7	16.24	12.40	10.88	10.05	9.52	9.16	8.89	8.68	8.51	8.38	7.75	7.53	7.42	7.35	7.22	7.09
8	14.69	11.04	9.60	8.81	8.30	7.95	7.69	7.50	7.34	7.21	6.61	6.40	6.29	6.22	6.09	5.96
9	13.61	10.11	8.72	7.96	7.47	7.13	6.88	6.69	6.54	6.42	5.83	5.62	5.52	5.45	5.32	5.20
10	12.83	9.43	8.08	7.34	6.87	6.54	6.30	6.12	5.97	5.85	5.27	5.07	4.97	4.90	4.77	4.65
11	12.23	8.91	7.60	6.88	6.42	6.10	5.86	5.68	5.54	5.42	4.86	4.65	4.55	4.49	4.36	4.24
12	11.75	8.51	7.23	6.52	6.07	5.76	5.52	5.35	5.20	5.09	4.53	4.33	4.23	4.17	4.04	3.92
13	11.37	8.19	6.93	6.23	5.79	5.48	5.25	5.08	4.94	4.82	4.27	4.07	3.97	3.91	3.78	3.66
14	11.06	7.92	6.68	6.00	5.56	5.26	5.03	4.86	4.72	4.60	4.06	3.86	3.76	3.70	3.57	3.45
15	10.80	7.70	6.48	5.80	5.37	5.07	4.85	4.67	4.54	4.42	3.88	3.69	3.59	3.52	3.39	3.27
16	10.58	7.51	6.30	5.64	5.21	4.91	4.69	4.52	4.38	4.27	3.73	3.54	3.44	3.37	3.25	3.13
17	10.38	7.35	6.16	5.50	5.07	4.78	4.56	4.39	4.25	4.14	3.61	3.41	3.31	3.25	3.12	3.00
18	10.22	7.21	6.03	5.37	4.96	4.66	4.44	4.28	4.14	4.03	3.50	3.30	3.20	3.14	3.01	2.89
19	10.07	7.09	5.92	5.27	4.85	4.56	4.34	4.18	4.04	3.93	3.40	3.21	3.11	3.04	2.91	2.79
20	9.94	6.99	5.82	5.17	4.76	4.47	4.26	4.09	3.96	3.85	3.32	3.12	3.02	2.96	2.83	2.70
21	9.83	6.89	5.73	5.09	4.68	4.39	4.18	4.01	3.88	3.77	3.24	3.05	2.95	2.88	2.75	2.63
22	9.73	6.81	5.65	5.02	4.61	4.32	4.11	3.94	3.81	3.70	3.18	2.98	2.88	2.82	2.69	2.56
23	9.63	6.73	5.58	4.95	4.54	4.26	4.05	3.88	3.75	3.64	3.12	2.92	2.82	2.76	2.62	2.50
24	9.55	6.66	5.52	4.89	4.49	4.20	3.99	3.83	3.69	3.59	3.06	2.87	2.77	2.70	2.57	2.44
25	9.48	6.60	5.46	4.84	4.43	4.15	3.94	3.78	3.64	3.54	3.01	2.82	2.72	2.65	2.52	2.39
26	9.41	6.54	5.41	4.79	4.38	4.10	3.89	3.73	3.60	3.49	2.97	2.77	2.67	2.61	2.47	2.34
27	9.34	6.49	5.36	4.74	4.34	4.06	3.85	3.69	3.56	3.45	2.93	2.73	2.63	2.57	2.43	2.30
28	9.28	6.44	5.32	4.70	4.30	4.02	3.81	3.65	3.52	3.41	2.89	2.69	2.59	2.53	2.39	2.26
29	9.23	6.40	5.28	4.66	4.26	3.98	3.77	3.61	3.48	3.38	2.86	2.66	2.56	2.49	2.36	2.23
30	9.18	6.35	5.24	4.62	4.23	3.95	3.74	3.58	3.45	3.34	2.82	2.63	2.52	2.46	2.32	2.19
40	8.83	6.07	4.98	4.37	3.99	3.71	3.51	3.35	3.22	3.12	2.60	2.40	2.30	2.23	2.09	1.95
50	8.63	5.90	4.83	4.23	3.85	3.58	3.38	3.22	3.09	2.99	2.47	2.27	2.16	2.10	1.95	1.80
60	8.49	5.79	4.73	4.14	3.76	3.49	3.29	3.13	3.01	2.90	2.39	2.19	2.08	2.01	1.86	1.71
70	8.40	5.72	4.66	4.08	3.70	3.43	3.23	3.08	2.95	2.85	2.33	2.13	2.02	1.95	1.80	1.64
80	8.33	5.67	4.61	4.03	3.65	3.39	3.19	3.03	2.91	2.80	2.29	2.08	1.97	1.90	1.75	1.58
90	8.28	5.62	4.57	3.99	3.62	3.35	3.15	3.00	2.87	2.77	2.25	2.05	1.94	1.87	1.71	1.54
100	8.24	5.59	4.54	3.96	3.59	3.33	3.13	2.97	2.85	2.74	2.23	2.02	1.91	1.84	1.68	1.51

Table A.1 (cont'd)

DF2 | | | | | | | | | DF1 | | | | | | | |

α = 0.001

DF2	1	2	3	4	5	6	7	8	9	10	20	30	40	50	100	1000
1	405312	499725	540257	562668	576496	586033	593185	597954	602245	605583	620842	626087	628471	630379	633240	636101
2	998.38	998.84	999.31	999.31	999.31	999.31	999.31	999.31	999.31	999.31	999.31	999.31	999.31	999.31	999.31	999.31
3	167.06	148.49	141.10	137.08	134.58	132.83	131.61	130.62	129.86	129.22	126.43	125.44	124.97	124.68	124.07	123.52
4	74.13	61.25	56.17	53.43	51.72	50.52	49.65	49.00	48.47	48.05	46.10	45.43	45.08	44.88	44.47	44.09
5	47.18	37.12	33.20	31.08	29.75	28.83	28.17	27.65	27.24	26.91	25.39	24.87	24.60	24.44	24.11	23.82
6	35.51	27.00	23.71	21.92	20.80	20.03	19.46	19.03	18.69	18.41	17.12	16.67	16.44	16.31	16.03	15.77
7	29.25	21.69	18.77	17.20	16.21	15.52	15.02	14.63	14.33	14.08	12.93	12.53	12.33	12.20	11.95	11.72
8	25.41	18.49	15.83	14.39	13.48	12.86	12.40	12.05	11.77	11.54	10.48	10.11	9.92	9.80	9.57	9.36
9	22.86	16.39	13.90	12.56	11.71	11.13	10.70	10.37	10.11	9.89	8.90	8.55	8.37	8.26	8.04	7.84
10	21.04	14.90	12.55	11.28	10.48	9.93	9.52	9.20	8.96	8.75	7.80	7.47	7.30	7.19	6.98	6.78
11	19.69	13.81	11.56	10.35	9.58	9.05	8.65	8.35	8.12	7.92	7.01	6.68	6.52	6.42	6.21	6.02
12	18.64	12.97	10.80	9.63	8.89	8.38	8.00	7.71	7.48	7.29	6.40	6.09	5.93	5.83	5.63	5.44
13	17.82	12.31	10.21	9.07	8.35	7.86	7.49	7.21	6.98	6.80	5.93	5.63	5.47	5.37	5.17	4.99
14	17.14	11.78	9.73	8.62	7.92	7.44	7.08	6.80	6.58	6.40	5.56	5.25	5.10	5.00	4.81	4.62
15	16.59	11.34	9.34	8.25	7.57	7.09	6.74	6.47	6.26	6.08	5.25	4.95	4.80	4.70	4.51	4.33
16	16.12	10.97	9.01	7.94	7.27	6.80	6.46	6.20	5.98	5.81	4.99	4.70	4.54	4.45	4.26	4.08
17	15.72	10.66	8.73	7.68	7.02	6.56	6.22	5.96	5.75	5.58	4.78	4.48	4.33	4.24	4.05	3.87
18	15.38	10.39	8.49	7.46	6.81	6.35	6.02	5.76	5.56	5.39	4.59	4.30	4.15	4.06	3.87	3.69
19	15.08	10.16	8.28	7.27	6.62	6.18	5.85	5.59	5.39	5.22	4.43	4.14	3.99	3.90	3.71	3.53
20	14.82	9.95	8.10	7.10	6.46	6.02	5.69	5.44	5.24	5.08	4.29	4.00	3.86	3.77	3.58	3.40
21	14.59	9.77	7.94	6.95	6.32	5.88	5.56	5.31	5.11	4.95	4.17	3.88	3.74	3.64	3.46	3.28
22	14.38	9.61	7.80	6.81	6.19	5.76	5.44	5.19	4.99	4.83	4.06	3.78	3.63	3.54	3.35	3.17
23	14.20	9.47	7.67	6.70	6.08	5.65	5.33	5.09	4.89	4.73	3.96	3.68	3.53	3.44	3.25	3.08
24	14.03	9.34	7.55	6.59	5.98	5.55	5.24	4.99	4.80	4.64	3.87	3.59	3.45	3.36	3.17	2.99
25	13.88	9.22	7.45	6.49	5.89	5.46	5.15	4.91	4.71	4.56	3.79	3.52	3.37	3.28	3.09	2.91
26	13.74	9.12	7.36	6.41	5.80	5.38	5.07	4.83	4.64	4.48	3.72	3.44	3.30	3.21	3.02	2.84
27	13.61	9.02	7.27	6.33	5.73	5.31	5.00	4.76	4.57	4.41	3.66	3.38	3.23	3.14	2.96	2.78
28	13.50	8.93	7.19	6.25	5.66	5.24	4.93	4.69	4.50	4.35	3.60	3.32	3.18	3.09	2.90	2.72
29	13.39	8.85	7.12	6.19	5.59	5.18	4.87	4.64	4.45	4.29	3.54	3.27	3.12	3.03	2.84	2.66
30	13.29	8.77	7.05	6.12	5.53	5.12	4.82	4.58	4.39	4.24	3.49	3.22	3.07	2.98	2.79	2.61
40	12.61	8.25	6.59	5.70	5.13	4.73	4.44	4.21	4.02	3.87	3.15	2.87	2.73	2.64	2.44	2.25
50	12.22	7.96	6.34	5.46	4.90	4.51	4.22	4.00	3.82	3.67	2.95	2.68	2.53	2.44	2.25	2.05
60	11.97	7.77	6.17	5.31	4.76	4.37	4.09	3.86	3.69	3.54	2.83	2.55	2.41	2.32	2.12	1.92
70	11.80	7.64	6.06	5.20	4.66	4.28	3.99	3.77	3.60	3.45	2.74	2.47	2.32	2.23	2.03	1.82
80	11.67	7.54	5.97	5.12	4.58	4.20	3.92	3.70	3.53	3.39	2.68	2.41	2.26	2.16	1.96	1.75
90	11.57	7.47	5.91	5.06	4.53	4.15	3.87	3.65	3.48	3.34	2.63	2.36	2.21	2.11	1.91	1.69
100	11.50	7.41	5.86	5.02	4.48	4.11	3.83	3.61	3.44	3.30	2.59	2.32	2.17	2.08	1.87	1.64

Table A.2   Critical values for Student's *t*-distribution.

α one-tailed	0.05	0.025	0.01	0.005	0.001	0.0005
α two-tailed	0.1	0.05	0.02	0.01	0.002	0.001
DF						
1	6.314	12.706	31.821	63.656	318.289	636.578
2	2.920	4.303	6.965	9.925	22.328	31.600
3	2.353	3.182	4.541	5.841	10.214	12.924
4	2.132	2.776	3.747	4.604	7.173	8.610
5	2.015	2.571	3.365	4.032	5.894	6.869
6	1.943	2.447	3.143	3.707	5.208	5.959
7	1.895	2.365	2.998	3.499	4.785	5.408
8	1.860	2.306	2.896	3.355	4.501	5.041
9	1.833	2.262	2.821	3.250	4.297	4.781
10	1.812	2.228	2.764	3.169	4.144	4.587
11	1.796	2.201	2.718	3.106	4.025	4.437
12	1.782	2.179	2.681	3.055	3.930	4.318
13	1.771	2.160	2.650	3.012	3.852	4.221
14	1.761	2.145	2.624	2.977	3.787	4.140
15	1.753	2.131	2.602	2.947	3.733	4.073
16	1.746	2.120	2.583	2.921	3.686	4.015
17	1.740	2.110	2.567	2.898	3.646	3.965
18	1.734	2.101	2.552	2.878	3.610	3.922
19	1.729	2.093	2.539	2.861	3.579	3.883
20	1.725	2.086	2.528	2.845	3.552	3.850
21	1.721	2.080	2.518	2.831	3.527	3.819
22	1.717	2.074	2.508	2.819	3.505	3.792
23	1.714	2.069	2.500	2.807	3.485	3.768
24	1.711	2.064	2.492	2.797	3.467	3.745
25	1.708	2.060	2.485	2.787	3.450	3.725
26	1.706	2.056	2.479	2.779	3.435	3.707
27	1.703	2.052	2.473	2.771	3.421	3.689
28	1.701	2.048	2.467	2.763	3.408	3.674
29	1.699	2.045	2.462	2.756	3.396	3.660
30	1.697	2.042	2.457	2.750	3.385	3.646
40	1.684	2.021	2.423	2.704	3.307	3.551
50	1.676	2.009	2.403	2.678	3.261	3.496
60	1.671	2.000	2.390	2.660	3.232	3.460
70	1.667	1.994	2.381	2.648	3.211	3.435
80	1.664	1.990	2.374	2.639	3.195	3.416
90	1.662	1.987	2.368	2.632	3.183	3.402
100	1.660	1.984	2.364	2.626	3.174	3.390

Table A.3 Tukey's *T*-distribution.
Tukey Test: percentage points (q) of the Studentised range
($p = 0.05$)
k = the total number of means being compared
DF = degrees of freedom of denominator of F test

DF	k								
	2	3	4	5	6	7	8	9	10
1	17.97	26.98	32.82	37.08	40.41	43.12	45.40	47.36	49.07
2	6.08	8.33	9.80	10.88	11.74	12.44	13.03	13.54	13.99
3	4.50	5.91	6.82	7.50	8.04	8.48	8.85	9.18	9.46
4	3.93	5.04	5.76	6.29	6.71	7.05	7.35	7.60	7.83
5	3.64	4.60	5.22	5.67	6.03	6.33	6.58	6.80	6.99
6	3.46	4.34	4.90	5.30	5.63	5.90	6.12	6.32	6.49
7	3.34	4.16	4.68	5.06	5.36	5.61	5.82	6.00	6.19
8	3.26	4.04	4.53	4.89	5.17	5.40	5.60	5.77	5.92
9	3.20	3.95	4.41	4.76	5.02	5.24	5.43	5.59	5.74
10	3.15	3.88	4.33	4.65	4.91	5.12	5.30	5.46	5.60
11	3.11	3.82	4.26	4.57	4.82	5.03	5.20	5.35	5.49
12	3.08	3.77	4.20	4.51	4.75	4.95	5.12	5.27	5.39
13	3.06	3.73	4.15	4.45	4.69	4.88	5.05	5.19	5.32
14	3.03	3.70	4.11	4.41	4.64	4.83	4.99	5.13	5.25
15	3.01	3.67	4.08	4.37	4.59	4.78	4.94	5.08	5.20
16	3.00	3.65	4.05	4.33	4.56	4.74	4.90	5.03	5.15
17	2.98	3.63	4.02	4.30	4.52	4.70	4.86	4.99	5.11
18	2.97	3.61	4.00	4.28	4.49	4.67	4.82	4.96	5.07
19	2.96	3.59	3.98	4.25	4.47	4.65	4.79	4.92	5.04
20	2.95	3.58	3.96	4.23	4.45	4.62	4.77	4.90	5.01
24	2.92	3.53	3.90	4.17	4.37	4.54	4.68	4.81	4.92
30	2.89	3.49	3.85	4.10	4.30	4.46	4.60	4.72	4.82
40	2.86	3.44	3.79	4.04	4.23	4.39	4.52	4.63	4.73
60	2.83	3.40	3.74	3.98	4.16	4.31	4.44	4.55	4.65
120	2.80	3.36	3.68	3.92	4.10	4.24	4.36	4.47	4.56
∞	2.77	3.31	3.63	3.86	4.03	4.17	4.29	4.39	4.47

Source: From Pearson, E.S. and Hartley, H.O., editors; *Biometrika Tables for Statisticians*, edn 3, vol. 1. Cambridge: Cambridge University Press, 1966, with the kind permission of the trustees and publishers.

Table A.4   Critical values for the χ^2 (Chi-squared) distribution.

DF	α one-tailed					
	0.1	0.05	0.025	0.01	0.005	0.001
1	2.71	3.84	5.02	6.63	7.88	10.83
2	4.61	5.99	7.38	9.21	10.60	13.82
3	6.25	7.81	9.35	11.34	12.84	16.27
4	7.78	9.49	11.14	13.28	14.86	18.47
5	9.24	11.07	12.83	15.09	16.75	20.51
6	10.64	12.59	14.45	16.81	18.55	22.46
7	12.02	14.07	16.01	18.48	20.28	24.32
8	13.36	15.51	17.53	20.09	21.95	26.12
9	14.68	16.92	19.02	21.67	23.59	27.88
10	15.99	18.31	20.48	23.21	25.19	29.59
11	17.28	19.68	21.92	24.73	26.76	31.26
12	18.55	21.03	23.34	26.22	28.30	32.91
13	19.81	22.36	24.74	27.69	29.82	34.53
14	21.06	23.68	26.12	29.14	31.32	36.12
15	22.31	25.00	27.49	30.58	32.80	37.70
16	23.54	26.30	28.85	32.00	34.27	39.25
17	24.77	27.59	30.19	33.41	35.72	40.79
18	25.99	28.87	31.53	34.81	37.16	42.31
19	27.20	30.14	32.85	36.19	38.58	43.82
20	28.41	31.41	34.17	37.57	40.00	45.31
21	29.62	32.67	35.48	38.93	41.40	46.80
22	30.81	33.92	36.78	40.29	42.80	48.27
23	32.01	35.17	38.08	41.64	44.18	49.73
24	33.20	36.42	39.36	42.98	45.56	51.18
25	34.38	37.65	40.65	44.31	46.93	52.62
26	35.56	38.89	41.92	45.64	48.29	54.05
27	36.74	40.11	43.19	46.96	49.65	55.48
28	37.92	41.34	44.46	48.28	50.99	56.89
29	39.09	42.56	45.72	49.59	52.34	58.30
30	40.26	43.77	46.98	50.89	53.67	59.70
40	51.81	55.76	59.34	63.69	66.77	73.40
50	63.17	67.50	71.42	76.15	79.49	86.66
60	74.40	79.08	83.30	88.38	91.95	99.61
70	85.53	90.53	95.02	100.43	104.21	112.32
80	96.58	101.88	106.63	112.33	116.32	124.84
90	107.57	113.15	118.14	124.12	128.30	137.21
100	118.50	124.34	129.56	135.81	140.17	149.45

Table A.5a One-tailed 0.05 critical values of the Kolmogorov–Smirnov test for small samples.

n_1

n_2	3	4	5	6	7	8	9	10	11	12	13	14	15	16	17	18	19	20	21	22	23	24	25
3	9	10	13	15	16	19	21	22	25	27	28	31	33	34	35	39	40	41	45	46	47	51	52
4	10	16	16	18	21	24	25	28	29	36	33	38	38	44	44	46	49	52	52	56	57	60	61
5	13	16	20	21	24	26	28	35	35	36	40	42	50	46	49	51	56	60	60	62	65	67	75
6	15	18	21	30	25	30	33	36	38	48	43	48	51	54	56	66	61	66	69	70	73	78	78
7	16	21	24	25	35	34	36	40	43	45	50	56	56	58	61	64	68	72	77	77	79	83	85
8	19	24	26	30	34	40	40	44	48	52	53	58	60	72	65	72	73	80	81	84	89	96	95
9	21	25	28	33	36	44	54	46	51	57	57	63	69	68	74	81	80	83	90	91	94	99	101
10	22	28	35	36	40	44	46	60	57	60	62	68	75	76	77	82	85	100	91	98	101	106	110
11	25	29	35	38	43	48	51	57	66	64	67	72	76	80	83	87	92	95	101	110	108	111	116
12	27	36	36	48	45	52	57	60	64	72	71	78	84	88	89	96	98	104	108	110	113	132	120
13	28	33	40	43	50	53	57	62	67	71	91	78	86	90	94	98	102	108	112	117	120	124	131
14	31	38	42	48	56	58	63	68	72	78	78	98	92	96	99	104	108	114	126	124	127	132	136
15	33	38	50	51	56	60	69	75	76	84	86	92	105	101	105	111	113	125	126	130	134	141	145
16	34	44	46	54	58	65	68	76	80	88	90	96	101	112	109	116	120	128	130	136	140	152	148
17	35	44	49	56	61	72	74	77	83	89	84	99	105	109	136	118	125	130	135	141	146	150	156
18	39	46	51	66	64	73	81	82	87	96	98	104	111	116	118	144	127	136	144	148	151	162	161
19	40	49	56	61	68	80	80	85	92	98	102	108	113	120	125	127	152	144	147	151	159	162	168
20	41	52	60	66	72	81	83	100	95	104	108	114	125	128	130	136	144	160	154	160	163	172	180
21	45	52	60	69	77	84	90	91	101	108	112	126	126	130	135	144	147	154	168	163	170	177	182
22	46	56	62	70	77	89	91	98	110	110	117	124	130	136	141	148	151	160	163	198	173	182	188
23	47	57	65	73	79	96	94	101	108	113	120	127	134	140	146	151	159	163	170	173	207	183	194
24	51	60	67	78	83	95	99	106	111	132	124	132	141	152	150	162	162	172	177	182	183	216	204
25	52	61	75	78	85	95	101	110	116	120	131	136	145	148	156	161	168	180	182	188	194	204	225

Source: Adapted from Gail, M.H. and Green, S.B. (1976) Critical values for the one-sided two-sample Kolmogorov–Smirnov statistic, *Journal of the American Statistical Association,* **71**: 757–760.

Table A.5b Two-tailed 0.05 critical values of the Kolmogorov–Smirnov test for small samples.

n_2	n_1																							
	2	3	4	5	6	7	8	9	10	11	12	13	14	15	16	17	18	19	20	21	22	23	24	25
2							16	18	20	22	24	26	26	28	30	32	34	35	38	38	40	42	44	46
3				15	18	21	21	24	27	30	30	33	36	36	39	42	45	45	48	51	51	54	57	60
4			16	20	20	24	28	28	30	33	36	39	42	44	48	48	50	53	60	59	62	64	68	68
5		15	20	25	24	28	30	35	40	39	43	45	46	55	54	55	60	61	65	69	70	72	76	80
6		18	20	24	30	30	34	39	40	43	48	52	54	57	60	62	72	70	72	75	78	80	90	88
7		21	24	28	30	42	40	42	46	48	53	56	63	62	64	68	72	76	79	91	84	89	92	97
8	16	21	28	30	34	40	48	46	48	53	60	62	64	67	80	77	80	82	88	89	94	98	104	104
9	18	24	28	35	39	42	46	54	53	59	63	65	70	75	78	82	90	89	93	99	101	106	111	114
10	20	27	30	40	40	46	48	53	70	60	66	70	74	80	84	89	92	94	110	105	108	114	118	125
11	22	30	33	39	43	48	53	59	60	77	72	75	82	84	89	93	97	102	107	112	121	119	124	129
12	24	30	36	43	48	53	60	63	66	72	84	81	86	93	96	100	108	108	116	120	124	125	144	138
13	26	33	39	45	52	56	62	65	70	75	81	91	89	96	101	105	110	114	120	126	130	135	140	145
14	26	36	42	46	54	63	64	70	74	82	86	89	112	98	106	111	116	121	126	140	138	142	146	150
15	28	36	44	55	57	62	67	75	80	84	93	96	98	120	114	116	123	127	135	138	144	149	156	160
16	30	39	48	54	60	64	80	78	84	89	96	101	106	114	128	124	128	133	140	145	150	157	168	167
17	32	42	48	55	62	68	77	82	89	93	100	105	111	116	124	136	133	141	146	151	157	163	168	173
18	34	45	50	60	72	72	80	90	92	97	108	110	116	123	128	133	162	142	152	159	164	170	180	180
19	35	45	53	61	70	76	82	89	94	102	108	114	121	127	133	141	142	171	160	163	169	177	183	187
20	38	48	60	65	72	79	88	93	110	107	116	120	126	135	140	146	152	160	180	173	176	184	192	200
21	38	51	59	69	75	91	89	99	105	112	120	126	140	138	145	151	159	163	173	189	183	189	198	202
22	40	51	62	70	78	84	94	101	108	121	124	130	138	144	150	157	164	169	176	183	198	194	204	209
23	42	54	64	72	80	89	98	106	114	119	125	135	142	149	157	163	170	177	184	189	194	230	205	216
24	44	57	68	76	90	92	104	111	118	124	144	140	146	156	168	168	180	183	192	198	204	205	240	225
25	46	60	68	80	88	97	104	114	125	129	138	145	150	160	167	173	180	187	200	202	209	216	225	250

Source: Adapted from Table 51 of Pearson, E.S. and Hartley, H.O. (1972) *Biometrika Tables for Statisticians*, vol. 2, Cambridge University Press, New York, by permission of the publishers.

Table A.5c   Two-tailed critical values of the
Kolmogorov–Smirnov test for large samples.

Significance level	D
0.10	$1.22\sqrt{\dfrac{n_1+n_2}{n_1\times n_2}}$
0.05	$1.36\sqrt{\dfrac{n_1+n_2}{n_1\times n_2}}$
0.025	$1.48\sqrt{\dfrac{n_1+n_2}{n_1\times n_2}}$
0.01	$1.63\sqrt{\dfrac{n_1+n_2}{n_1\times n_2}}$
0.005	$1.73\sqrt{\dfrac{n_1+n_2}{n_1\times n_2}}$
0.001	$1.95\sqrt{\dfrac{n_1+n_2}{n_1\times n_2}}$

Source: Adapted from Smirnov, N. (1948) Tables for
estimating the goodness of fit of empirical distributions,
Annals of Mathematical Statistics, **19**: 280–281.

Table A.6   Two-tailed critical values of T for the Wilcoxon test.

N	Significance level 0.10	0.05	0.02	N	Significance level 0.10	0.05	0.02
5	0	—	—	28	130	116	101
6	2	0	—	29	140	126	110
7	3	2	0	30	151	137	120
8	5	3	1	31	163	147	130
9	8	5	3	32	175	159	140
10	10	8	5	33	187	170	151
11	13	10	7	34	200	182	162
12	17	13	9	35	213	195	173
13	21	17	12	36	227	208	185
14	25	21	15	37	241	221	198
15	30	25	19	38	256	235	211
16	35	29	23	39	271	249	224
17	41	34	27	40	286	264	238
18	47	40	32	41	302	279	252
19	53	46	37	42	319	294	266
20	60	52	43	43	336	310	281
21	67	58	49	44	353	327	296
22	75	65	55	45	371	343	312
23	83	73	62	46	389	361	328
24	91	81	69	47	407	378	345
25	100	89	76	78	426	396	362
26	110	98	84	49	446	415	379
27	119	107	92	50	466	434	397

Source: Adapted from Table I of Wilcoxon, F. (1949) *Some Rapid Approximate Statistical Procedures*, American
Cyanamid Company, New York, by permission of the publishers.

Table A.7 Cumulative probabilities for the normal distribution (Z-scores).

Z	0.00	0.01	0.02	0.03	0.04	0.05	0.06	0.07	0.08	0.09
0.0	0.50000	0.50399	0.50798	0.51197	0.51595	0.51994	0.52392	0.52790	0.53188	0.53586
0.1	0.53983	0.54380	0.54776	0.55172	0.55567	0.55962	0.56356	0.56749	0.57142	0.57535
0.2	0.57926	0.58317	0.58706	0.59095	0.59483	0.59871	0.60257	0.60642	0.61026	0.61409
0.3	0.61791	0.62172	0.62552	0.62930	0.63307	0.63683	0.64058	0.64431	0.64803	0.65173
0.4	0.65542	0.65910	0.66276	0.66640	0.67003	0.67364	0.67724	0.68082	0.68439	0.68793
0.5	0.69146	0.69497	0.69847	0.70194	0.70540	0.70884	0.71226	0.71566	0.71904	0.72240
0.6	0.72575	0.72907	0.73237	0.73565	0.73891	0.74215	0.74537	0.74857	0.75175	0.75490
0.7	0.75804	0.76115	0.76424	0.76730	0.77035	0.77337	0.77637	0.77935	0.78230	0.78524
0.8	0.78814	0.79103	0.79389	0.79673	0.79955	0.80234	0.80511	0.80785	0.81057	0.81327
0.9	0.81594	0.81859	0.82121	0.82381	0.82639	0.82894	0.83147	0.83398	0.83646	0.83891
1.0	0.84134	0.84375	0.84614	0.84849	0.85083	0.85314	0.85543	0.85769	0.85993	0.86214
1.1	0.86433	0.86650	0.86864	0.87076	0.87286	0.87493	0.87698	0.87900	0.88100	0.88298
1.2	0.88493	0.88686	0.88877	0.89065	0.89251	0.89435	0.89617	0.89796	0.89973	0.90147
1.3	0.90320	0.90490	0.90658	0.90824	0.90988	0.91149	0.91308	0.91466	0.91621	0.91774
1.4	0.91924	0.92073	0.92220	0.92364	0.92507	0.92647	0.92785	0.92922	0.93056	0.93189
1.5	0.93319	0.93448	0.93574	0.93699	0.93822	0.93943	0.94062	0.94179	0.94295	0.94408
1.6	0.94520	0.94630	0.94738	0.94845	0.94950	0.95053	0.95154	0.95254	0.95352	0.95449
1.7	0.95543	0.95637	0.95728	0.95818	0.95907	0.95994	0.96080	0.96164	0.96246	0.96327
1.8	0.96407	0.96485	0.96562	0.96638	0.96712	0.96784	0.96856	0.96926	0.96995	0.97062
1.9	0.97128	0.97193	0.97257	0.97320	0.97381	0.97441	0.97500	0.97558	0.97615	0.97670
2.0	0.97725	0.97778	0.97831	0.97882	0.97932	0.97982	0.98030	0.98077	0.98124	0.98169
2.1	0.98214	0.98257	0.98300	0.98341	0.98382	0.98422	0.98461	0.98500	0.98537	0.98574
2.2	0.98610	0.98645	0.98679	0.98713	0.98745	0.98778	0.98809	0.98840	0.98870	0.98899
2.3	0.98928	0.98956	0.98983	0.99010	0.99036	0.99061	0.99086	0.99111	0.99134	0.99158
2.4	0.99180	0.99202	0.99224	0.99245	0.99266	0.99286	0.99305	0.99324	0.99343	0.99361
2.5	0.99379	0.99396	0.99413	0.99430	0.99446	0.99461	0.99477	0.99492	0.99506	0.99520
2.6	0.99534	0.99547	0.99560	0.99573	0.99585	0.99598	0.99609	0.99621	0.99632	0.99643
2.7	0.99653	0.99664	0.99674	0.99683	0.99693	0.99702	0.99711	0.99720	0.99728	0.99736
2.8	0.99744	0.99752	0.99760	0.99767	0.99774	0.99781	0.99788	0.99795	0.99801	0.99807
2.9	0.99813	0.99819	0.99825	0.99831	0.99836	0.99841	0.99846	0.99851	0.99856	0.99861
3.0	0.99865	0.99869	0.99874	0.99878	0.99882	0.99886	0.99889	0.99893	0.99896	0.99900
3.1	0.99903	0.99906	0.99910	0.99913	0.99916	0.99918	0.99921	0.99924	0.99926	0.99929
3.2	0.99931	0.99934	0.99936	0.99938	0.99940	0.99942	0.99944	0.99946	0.99948	0.99950
3.3	0.99952	0.99953	0.99955	0.99957	0.99958	0.99960	0.99961	0.99962	0.99964	0.99965
3.4	0.99966	0.99968	0.99969	0.99970	0.99971	0.99972	0.99973	0.99974	0.99975	0.99976
3.5	0.99977	0.99978	0.99978	0.99979	0.99980	0.99981	0.99981	0.99982	0.99983	0.99983
3.6	0.99984	0.99985	0.99985	0.99986	0.99986	0.99987	0.99987	0.99988	0.99988	0.99989
3.7	0.99989	0.99990	0.99990	0.99990	0.99991	0.99991	0.99992	0.99992	0.99992	0.99992
3.8	0.99993	0.99993	0.99993	0.99994	0.99994	0.99994	0.99994	0.99995	0.99995	0.99995
3.9	0.99995	0.99995	0.99996	0.99996	0.99996	0.99996	0.99996	0.99996	0.99997	0.99997
4.0	0.99997	0.99997	0.99997	0.99997	0.99997	0.99997	0.99998	0.99998	0.99998	0.99998

Note: Some uses of z-scores (e.g., Mann–Whitney, Wilcoxon tests) require the 'tail-probability' which is 1 – number from the table above.

Table A.8a   Critical values of Mann–Whitney U at 0.025 one-tailed level and 0.05 two-tailed level.

n_2 \ n_1	1	2	3	4	5	6	7	8	9	10	11	12	13	14	15	16	17	18	19	20
1	—	—	—	—	—	—	—	—	—	—	—	—	—	—	—	—	—	—	—	—
2	—	—	—	—	—	—	—	0	0	0	0	1	1	1	1	1	2	2	2	2
3	—	—	—	—	0	1	1	2	2	3	3	4	4	5	5	6	6	7	7	8
4	—	—	—	0	1	2	3	4	4	5	6	7	8	9	10	11	11	12	13	13
5	—	—	0	1	2	3	5	6	7	8	9	11	12	13	14	15	17	18	19	20
6	—	—	1	2	3	5	6	8	10	11	13	14	16	17	19	21	22	24	25	27
7	—	—	1	3	5	6	8	10	12	14	16	18	20	22	24	26	28	30	32	34
8	—	0	2	4	6	8	10	13	15	17	19	22	24	26	29	31	34	36	38	41
9	—	0	2	4	7	10	12	15	17	20	23	26	28	31	34	37	39	42	45	48
10	—	0	3	5	8	11	14	17	20	23	26	29	33	36	39	42	45	48	52	55
11	—	0	3	6	9	13	16	19	23	26	30	33	37	40	44	47	51	55	58	62
12	—	1	4	7	11	14	18	22	26	29	33	37	41	45	49	53	57	61	65	69
13	—	1	4	8	12	16	20	24	28	33	37	41	45	50	54	59	63	67	72	76
14	—	1	5	9	13	17	22	26	31	36	40	45	50	55	59	64	67	74	78	83
15	—	1	5	10	14	19	24	29	34	39	44	49	54	59	64	70	75	80	85	90
16	—	1	6	11	15	21	26	31	37	42	47	53	59	64	70	75	81	86	92	98
17	—	2	6	11	17	22	28	34	39	45	51	57	63	67	75	81	87	93	99	105
18	—	2	7	12	18	24	30	36	42	48	55	61	67	74	80	86	93	99	106	112
19	—	2	7	13	19	25	32	38	45	52	58	65	72	78	85	92	99	106	113	119
20	—	2	8	13	20	27	34	41	48	55	62	69	76	83	90	98	105	112	119	127

Source: Adapted from Table I of Runyon, R.P. and Haber, A. (1991) *Fundamentals of Behavioral Statistics* (7th edn), McGraw-Hill, New York, by permission of the publishers.

Table A.8b Critical values of Mann–Whitney U at 0.05 one-tailed level and 0.10 two-tailed level.

n_2 \\ n_1	1	2	3	4	5	6	7	8	9	10	11	12	13	14	15	16	17	18	19	20
1	—	—	—	—	—	—	—	—	—	—	—	—	—	—	—	—	—	—	0	0
2	—	—	—	—	0	0	0	1	1	1	1	2	2	2	3	3	3	4	4	4
3	—	—	0	0	1	2	2	3	3	4	5	5	6	7	7	8	9	9	10	11
4	—	—	0	1	2	3	4	5	6	7	8	9	10	11	12	14	15	16	17	18
5	—	0	1	2	4	5	6	8	9	11	12	13	15	16	18	19	20	22	23	25
6	—	0	2	3	5	7	8	10	12	14	16	17	19	21	23	25	26	28	30	32
7	—	0	2	4	6	8	11	13	15	17	19	21	24	26	28	30	33	35	37	39
8	—	1	3	5	8	10	13	15	18	20	23	26	28	31	33	36	39	41	44	47
9	—	1	3	6	9	12	15	18	21	24	27	30	33	36	39	42	45	48	51	54
10	—	1	4	7	11	14	17	20	24	27	31	34	37	41	44	48	51	55	58	62
11	—	1	5	8	12	16	19	23	27	31	34	38	42	46	50	54	57	61	65	69
12	—	2	5	9	13	17	21	26	30	34	38	42	47	51	55	60	64	68	72	77
13	—	2	6	10	15	19	24	28	33	37	42	47	51	56	61	65	70	75	80	84
14	—	2	7	11	16	21	26	31	36	41	46	51	56	61	66	71	77	82	87	92
15	—	3	7	12	18	23	28	33	39	44	50	55	61	66	72	77	83	88	94	100
16	—	3	8	14	19	25	30	36	42	48	54	60	65	71	77	83	89	95	101	107
17	—	3	9	15	20	26	33	39	45	51	57	64	70	77	83	89	96	102	109	115
18	—	4	9	16	22	28	35	41	48	55	61	68	75	82	88	95	102	109	116	123
19	0	4	10	17	23	30	37	44	51	58	65	72	80	87	94	101	109	116	123	130
20	0	4	11	18	25	32	39	47	54	62	69	77	84	92	100	107	115	123	130	138

Source: Adapted from Table I of Runyon, R.P. and Haber, A. (1991) *Fundamentals of Behavioral Statistics* (7th edn), McGraw-Hill, New York, by permission of the publishers.

Annotation for the interviews with respondents B, C and D

RESPONDENT B	Full Annotation
INT: *How do you approach and resolve problems encountered enroute?* **RES:** Say I arrived in town and I wanted to go to a department store, I would get off the bus at the appropriate stop to find that the road had been dug up. The noise of the men working lets me know that they are working, but it is then off-putting as you can't find other sound clues to get around that. I would, perhaps . . . if in town, I try to follow the flow of people – use them as guides even though they don't know it. The same is true for crossing roads, rather than asking each time, I would find somebody who I think is reasonably competent and then follow them.	(m) Situational confusion: road works. (m) coping strategy (using relevant noises) – still confusing though. (i) anyway around this? What can road builders do to help? (m) coping strategy: shadowing people – to avoid asking. (i) How is competence assessed?
INT: *How about off route?* **RES:** Generally I don't like this. I can get anxious. Sometimes it is OK though, but especially . . . if I'm not familiar with it I may panic a bit and wonder 'am I where I want to be?', 'Is this where I want to be?' This can happen. One day a few weeks ago I was in a big hurry, I was taking my daughter to the opticians, and we had to come back in half and hour. We had a few other things to do so we left the shop to do those. They were in another part of town and then we had to get back to Castle Court shopping centre. Our normal route would be up Royal Avenue and Castle Court would then be on your left. But we found ourselves around Queen Victoria Street and I found myself thinking, 'I wonder if I can cut through these streets?', 'will it bring me out at Castle Court?' Very much a guessing game to head in the right direction. We did get there in the end, but I had my sighted child with me. If I had been there on my own I probably would have gone the long way around.	(m) fear of disorientation or being lost. (m) copying. (i) interrogating cognitive map knowledge? (m) spatial confusion: shortcutting. (i) Is this common? Risky to shortcut? (m) Normal navigation: Usual route. (m) Recognition of problems of navigation. (m) coping: family. Without help no shortcuts.
INT: *How do you cope with being lost?* **RES:** I don't like it, I hate being "out of control". It is not a good feeling. It can be an anxious feeling. It's also quite a confidence downer. All of a sudden you are not where you think you are and you feel a bit stupid. You think any other adult could manage this and maybe you can't so you feel inadequate in some way. How I would resolve it would be to ask somebody. This has, believe it or not, taken me nearly 42 years to reach the point where I can ask somebody. You get a whole range of answers that you need emotional strength for, like the "are you blind?" question.	(m) fear of being lost. Fear major factor in navigation? (m) fear/inadequacy. (i) confidence an issue in navigation? (m) coping: asking strategy. (m) coping: problems of asking. Confidence.
INT: *Are there specific hazards that you encounter?* **RES:** A lot of shops now have huge amounts of pavement clutter. Sometimes I can see these, pick them out OK. But if it is bright and there are lots of people then I will bump into things.	(m) Situational hazards: pavement clutter. (m) coping: problems – VI related. (i) what are others? How important are these? Does brightness affect spatial behaviour significantly – i.e., not going out at all.
INT: *How do you learn a new route?* **RES:** By trial and error and by talking to friends. If I am going alone I try to remember the way I have been and to follow it back.	(m) learning strategies: trial and error; friends; retracing. (m) learning: little and selective.
INT: *Do you learn many new routes?* **RES:** One a month. I would be open to trying out new routes.	(m) learning strategies: friends; using maps. (i) maps only available to VI not completely blind people?
INT: *How do you prepare for a new route?* **RES:** Friends, an A to Z with a magnifying glass, and generally try to get as much information as possible to be prepared before I go. Last year I was told I was going away for the weekend but I wasn't told where I was going, it was to be a surprise, and the person I was going with was completely blind. Then I panicked totally. 'How will we ever find our way around?', 'What will we do?', 'How will we manage?' Just total panic	(i) Is this common for VI people – unplanned trips. (m) fear/confidence. (m) coping strategy: asking. (m) coping strategy: maps.

really. We flew to Edinburgh and by basically asking people where to get the bus etc. we made our way to the town. Once in town we asked for the street with the hotel on, where we were staying. The first thing we did was to get a map of Edinburgh, I was able to see some of this with my magnifier, although I had trouble translating that into my surroundings. The person I was with could translate this much easier, by not looking for visual cues as he can't see.	
INT: *Have any route learning strategies proved to be particularly successful or disastrous?* **RES:** The trip to Edinburgh, it was just one weekend, at the time we came back I felt fairly confident about finding my way. This was news to me, as if somebody beforehand had told me you would go to Edinburgh and be able to find your way about and that you'll manage to do all the things that we did, I would not have believed them. It was a confidence booster and it did prove successful. Another one was unsuccessful. I went to Westport for the day. It was such a small place compared to Edinburgh, but I didn't manage too well. I was built on a square, but I never knew what section of the square I was at. I could have been three quarters of the way around, when I needed to only go back a quarter, but I would go around. There was a river and a bridge and everyone seemed to go by these, but they didn't mean anything to me. It was frustrating and I was quite disorientated.	(m) confidence boost. (i) successful adventures – confidence – more adventures – increased spatial behaviour. (m) confidence boost. (m) spatial confusion: replication. (m) no coping strategies? Lost and disorientated.

RESPONDENT C	Full Annotation
INT: *How do you approach and resolve problems encountered enroute?* **RES:** I do have a cane, and I suppose I would use it at night, but not during the day. "Watch that puddle, watch that kerb, watch that car". If I could watch the kerb I wouldn't need you to tell me to watch the kerb. I've had 3 months of O&M training with the long cane about 2 years ago for 2 hours a week. If ***** is there, it is a lot easier, picking out the gaps in the wall, or going into shops. When I do go into shops everything just brights out for a moment. So you stand there and look like a wally, everybody looking at you. So it is a lot easier with someone to guide me.	(m) coping strategies: Cane (i) other VI people rely on using specific aids? (m) learning strategies: O&M training. (m) coping strategies: benefits of using cane. (m) situational context: lighting. (m) situational context: embarrassment. (i) does this affect spatial behaviour a lot? (m) coping strategies: people helping.
INT: *How do you cope with more specific problems, such as crossing roads?* **RES:** I tend to cloak people, wait on somebody that's going across and then shadow them.	(m) coping strategies: shadowing.
INT: *How about attempting detours?* **RES:** They can be awkward. Generally *****, my partner, would be with me. If it was left to me I would be in for a lot of trouble, so I would probably ask.	(m) Coping: partner. (m) coping: asking.
INT: *How about if you were off-route or lost?* **RES:** I would ask somebody. A few times I've been, not lost, but mislaid. Recently, when I was on a course I got the wrong bus one morning. There were two buses leaving ******, one was an express to the back of the city hall in Belfast and the other one went to ******, somewhere. I soon knew the bus I was on wasn't going to the right place and I started sort of panicking, and then it was more really of a mental panic "I'm in deep shit". Although in the end I recovered easily enough.	(m) coping: ask. (m) disorientation but not lost. (i) distinct difference? More similar if VI? (m) situational confusion: wrong bus. (m) panic/fear. (i) what do you do in this situation? Just ask?
INT: *Do you come across many hazards?* **RES:** I keep bumping into a lot of them such as when I come across them outside shops and the like. Very often I'd be kicking things. And people tend to think. What's up with him, it is a bit early in the day to be drinking.	(m) situational confusion: street obstacles. (m) conscious of what people think.
INT: *Do you think improved training would help?* **RES:** Really it's not the training, the training I had was fine, but for me it was a bit of a let down. I was expecting something a little different, like fail-safe, or a "guide dog". What you think it is about and what it actually is are two different things. My expectations and reality didn't match.	(m) coping strategies: possible training. (m) training did not fulfil expectations.

INT: *Do you think environmental modifications might make navigating easier?* **RES:** Some changes are definitely needed. Those wiggly pavement bobbles are good for totals, but for me they are a menace – it throws you. Some places in the shopping centres you have to go up and grope the shop front looking for the door. You don't know if it is a window or a door. There are just so many things we have to put up with. But for me definitely the likes of kerbstones being painted helps.	(m) Planning needs. (i) planning design for VI people poor? (m) situational hazard: glass fronted shops. (m) suggestions: painted kerbs. (i) think of others?
INT: *How do you find out about new routes?* **RES:** I don't think I would get maps or anything like that. I would just learn by walking around with a friend. I only need to do something a couple of times and then I've got it. Probably a couple of times a year.	(m) learning: do not use secondary sources. (m) Learning strategy: friends. (m) quick learner (i) is this common?
INT: *How do you prepare for learning a new route?* **RES:** Every time, if I was going by myself, I would always have it mentally well worked out. For the first couple of times I would go with my wife. That helps me build up a lot of confidence.	(m) Learning strategy: rehearsal; with partner (confidence building). (i) need for confidence is important.
INT: *Have any route learning strategies proved to be particularly successful or disastrous?* **RES:** No, not really.	(m) no details.

RESPONDENT D	**Full Annotation**
INT: *How do you approach and resolve problems encountered enroute?* **RES:** Such as? **INT:** *Detours, for example?* **RES:** Well, most obstacles are small and in areas that I'm familiar with it certainly wouldn't be a problem – I would just take an automatic route to bypass them. For example, I was on the Antrim Road, lines of shops, traffic and there was CableTel digging holes. There were pneumatic drills going, it was really disorientating – I couldn't use the traffic noise to work out which way I was facing. So I turned, took the dog to the kerb to reorientate myself, turned back and walked through the parallel streets that I know.	(m) coping: in familiar areas not a problem – automatic avoidance. (i) how long before automatic? (m) situational confusion: building work. (m) disorientation. (m) coping strategy: traffic noise, kerb – O&M, guide dog training. (i) same strategy used by all guide dog users?
INT: *How about if you were off-route?* **RES:** It doesn't usually happen, but if I was unsure I would retrace my steps. I've never been totally lost, but disorientated, for sure. One time another dog came up and we got entangled and spun around, there was no traffic to orientate myself so again I had to take my dog to the kerb and get my bearings.	(m) coping strategy: retrace. (m) disorientated not lost. (m) situational confusion: dogs entangling. (m) coping strategy: O&M training.
INT: *Do you think environmental modifications might make navigating easier?* **RES:** Yes. Tactile pavements need to be standardised and there needs to be more of them. Shop fronts with baskets and stalls are a nightmare, but you can't outlaw that.	(m) planning: needs improvement and standardisation. Tactile pavements inadequate. (i) need new laws to aid disabled people's interaction with environments?
INT: *How do you find out about new routes?* **RES:** With some difficulty, but mainly when I came across them	(m) no details.
INT: *How often do you learn a new route?* **RES:** Not very often, but for example I've learned a new route for a course I am attending	(m) little learning. No details.
INT: *How do you prepare for a new route?* **RES:** I'd walk over it with someone from Guide Dogs – 'down kerb to up kerb, keeping the traffic on my right'.	(m) Learning: O&M training only. (i) is this common for guide dog users – only learn new routes with O&M trainer?
INT: *Have any route learning strategies proved to be particularly successful or disastrous?* **RES:** No.	(m) no details!

Category codes for interviews with respondents B, C and D

Respondent B	Category code
1. Say I arrived in town and I wanted to go to a department store, I would get off the bus at the appropriate stop to find that the road had been dug up. The noise of the men working lets me know that they are working, but it is then off-putting as you can't find other sound clues to get around that.	B12 [Sit Con – reason – road works]
2. I would, perhaps . . . if in town, I try to follow the flow of people – use them as guides even though they don't know it. The same is true for crossing roads, rather than asking each time, I would find somebody who I think is reasonably competent and then follow them.	B22 [Sit Con – coping – shadowing]
3. Generally I don't like this. I can get anxious. Sometimes it is OK though, but especially . . .	A33 [Sp Con – feel – wariness] A31 [Sp Con – feel – panic]
4. if I'm not familiar with it I may panic a bit and wonder 'am I where I want to be?', 'Is this where I want to be?' This can happen.	
5. One day a few weeks ago I was in a big hurry, I was taking my daughter to the opticians, and we had to come back in half an hour. We had a few other things to do so we left the shop to do those. They were in another part of town and then we had to get back to Castle Court shopping centre.	A12 [Sp Con – reason – att. sh. cut]
6. Our normal route would be up Royal Avenue and Castle Court would then be on your left.	A41 [Sp Con – Sp Beh – set pattern]
7. But we found ourselves around Queen Victoria Street and I found myself thinking, 'I wonder if I can cut through these streets?', 'will it bring me out at Castle Court?' Very much a guessing game to head in the right direction.	A12 [Sp Con – reason – att. sh. cut]
8. We did get there in the end, but I had my sighted child with me.	A23 [Sp Con – coping – fam/friends]
9. If I had been there on my own I probably would have gone the long way around.	A41 [Sp Con – Sp Beh – set pattern]
10. I don't like it, I hate being "out of control". It is not a good feeling. It can be an anxious feeling.	A31 [Sp Con – feel – panic]
11. It's also quite a confidence downer. All of a sudden you are not where you think you are and you feel a bit stupid. You think any other adult could manage this and maybe you can't so you feel inadequate in some way. How I would resolve it would be to ask somebody. This has, believe it or not, taken me nearly 42 years to reach the point where I can ask somebody. You get a whole range of answers that you need emotional strength for, like the "are you blind?" question.	A32 [Sp Con – feel – confidence] A21 [Sp Con – coping – asking]
12. A lot of shops now have huge amounts of pavement clutter. Sometimes I can see these, pick them out OK.	B11 [Sit Con – reason – pave]
13. But if it is bright and there are lots of people then I will bump into things.	B15 [Sit Con – reason – other]
14. By trial and error and	C14 [Learn – strategy – trial/error]
15. by talking to friends.	C12 [Learn – strategy – fam/friends]
16. If I am going alone I try to remember the way I have been and to follow it back.	C13 [Learn – strategy – retrace]
17. One a month. I would be open to trying out new routes.	C2 [Learn – No. of times]
18. Friends,	
19. an A to Z with a magnifying glass,	
20. and generally try to get as much information as possible to be prepared before I go.	
21. Last year I was told I was going away for the weekend but I wasn't told where I was going, it was to be a surprise, and the person I was going with was completely blind.	A42 [Sp Con – Sp Beh – Explore]
22. Then I panicked totally. 'How will we ever find our way around?', 'What will we do?', 'How will we manage?' Just total panic really.	A31 [Sp Con – feel – panic]
23. We flew to Edinburgh and by basically asking people where to get the bus etc. we made our way to the town. Once in town we asked for the street with the hotel on, where we were staying. The first thing we did was to get a map of Edinburgh, I was able to see some of this with my magnifier, although I had trouble translating that into my surroundings. The person I was with could translate this much easier, by not looking for visual clues as he can't see.	A21 [Sp Con – coping – asking] A26 [Sp Con – coping – secondary sources]

24. The trip of Edinburgh, it was just one weekend, at the time we came back I felt fairly confident about finding my way. This was news to me, as if somebody beforehand had told me you would go to Edinburgh and be able to find your way about and that you'll manage to do all the things that we did, I would not have believed them. It was a confidence booster and it did prove successful.	**A32** [Sp Con – feel – confidence]
25. Another one was unsuccessful. I went to Westport for the day. It was such a small place compared to Edinburgh, but I didn't manage too well. It was built on a square, but I never knew what section of the square I was at. I could have been three quarters of the way around, when I needed to only go back a quarter, but I would go around. There was a river and a bridge and everyone seemed to go by these, but they didn't mean anything to me.	**A13** [Sp Con – reason – layout]
26. It was frustrating and I was quite disorientated.	**A34** [Sp Con – feel – frustration]

Respondent C	**Category code**
1. I do have a cane, and I suppose I would use it at night, but not during the day. "Watch that puddle, watch that kerb, watch that car". If I could watch the kerb I wouldn't need you to tell me to watch the kerb. I've had 3 months of O&M training with the long cane about 2 years ago for 2 hours a week.	**A24** [Sp Con – coping – O&M training]
2. If ***** is there, it is a lot easier, picking out the gaps in the wall, or going into shops.	**A23** [Sp Con – coping – fam/friends]
3. When I do go into shops everything just brights out for a moment.	**B14** [Sit Con – reasons – shops]
4. So you stand there and look like a wally, everybody looking at you.	**B32** [Sit Con – feel – confidence]
5. So it is a lot easier with someone to guide me.	**A23** [Sp Con – coping fam/friends]
6. I tend to cloak people, wait on somebody that's going across and then shadow them.	**A27** [Sp Con – coping – shadowing]
7. They can be awkward. Generally *****, my partner, would be with me.	**A23** [Sp Con – coping – fam/friends]
8. If it was left to me I would be in for a lot of trouble, so I would probably ask.	**A21** [Sp Con – coping – asking]
9. I would ask somebody.	**A21** [Sp Con – coping – asking]
10. A few times I've been, not lost, but mislaid. Recently, when I was on a course I got the wrong bus one morning. There were two buses leaving ******, one was an express to the back of the city hall in Belfast and the other one went to ******, somewhere.	**A14** [Sp Con – reason – pub. trans]
11. I soon knew the bus I was on wasn't going to the right place and I started sort of panicking, and then it was more really of a mental panic "I'm in deep shit". Although in the end I recovered easily enough.	**A31** [Sp Con – feel – panic]
12. I keep bumping into a lot of them [hazards] such as when I come across them outside shops and the like. Very often I'd be kicking things.	**B11** [Sit Con – reasons – pave]
13. And people tend to think. What's up with him, it is a bit early in the day to be drinking.	**B32** [Sit Con – feel – confidence]
14. Really it's not the training, the training I had was fine, but for me it was a bit of a let down. I was expecting something a little different, like fail-safe, or a "guide dog". What you think it is about and what it actually is are two different things. My expectations and reality didn't match.	**C3** [Learn – training]
15. Those wiggly pavement bobbles are good for totals, but for me they are a menace – it throws you.	**B11** [Sit Con – reasons – pave]
16. Some places in the shopping centres you have to go up and grope the shop front looking for the door. You don't know if it is a window or a door. There are just so many things we have to put up with.	**B14** [Sit Con – reasons – shops]
17. I would just learn by walking around with a friend. I only need to do something a couple of times and then I've got it.	**C12** [Learn – strategy – fam/friends]
18. Probably a couple of times a year.	**C2** [Learn – No. of times]
19. Every time, if I was going by myself, I would always have it mentally well worked out.	**C16** [Learn – strategy – rehearsal]
20. For the first couple of times I would go with my wife. That helps me build up a lot of confidence.	**C12** [Learn – strategy – fam/friends]

Respondent D	Category codes
1. Well, most obstacles are small and in areas that I'm familiar with it certainly wouldn't be a problem – I would just take an automatic route to bypass them.	**A41** [Sp Con – Sp Beh – set pattern]
2. For example, I was on the Antrim Road, lines of shops, traffic and there was CableTel digging holes. There were pneumatic drills going, it was really disorientating – I couldn't use the traffic noise to work out which way I was facing.	**B12** [Sit Con – reasons – roads]
3. So I turned, took the dog to the kerb to reorientate myself, turned back and walked through the parallel streets that I know.	**B23** [Sit Con – coping – O&M training]
4. It doesn't usually happen, but if I was unsure I would retrace my steps. I've never been totally lost, but disorientated, for sure.	**A22** [Sp Con – coping – retrace]
5. One time another dog came up and we got entangled and spun around,	**B15** [Sit Con – reason – other]
6. there was no traffic to orientate myself so again I had to take my dog to the kerb and get my bearings.	**B23** [Sit Con – coping – O&M training]
7. Tactile pavements need to be standardised and there needs to be more of them.	**B11** [Sit Con – reason – pave]
8. Shop fronts with baskets and stalls are a nightmare, but you can't outlaw that.	**B14** [Sit Con – reason – shops]
9. Not very often, but for example I've learned a new route for a course I am attending.	**C2** [Learn – No. of times]
10. I'd walk over it with someone from Guide Dogs – 'down kerb to up kerb, keeping the traffic on my right'.	**C15** [Learn – strategy – O&M training]

To order any of the software used or referred to in this book, or to obtain demonstration software, contact:

MINITAB

In USA and Canada:
Minitab Inc., 3081 Enterprise Drive, State College, PA 16801-3008, USA.
Tel: (800) 448 3555   Fax: (814) 238 4383
E-mail: acadsales@minitab.com

In UK:
Minitab Ltd, 3 Mercia Business Village, Torwood Close, Westwood Business Park, Coventry CV4 8HX.
Tel: 01203 695730   Fax: 01203 695731
E-mail: uk@minitab.com

In the rest of the world:
Look at Minitab web-site for nearest vendor – http://www.minitab.com/

ARC/INFO and ARC/VIEW

In USA:
ESRI, 380 New York Street, Redlands, CA 92373-8100, USA.
Tel: (909) 793 2853   Fax: (909) 793 5953
E-mail: info@esri.com

In UK:
Peter Paisley/Paul Greenhalgh, 23 Woodford Road, Watford, Hertfordshire WD1 1PB.
Tel: 01923 210450   Fax: 01923 210739
E-mail: info@esriuk.com

In the rest of the world:
Look at ESRI web-site for nearest vendor – http://www.esri.com/

IDRISI

In USA:
IDRISI Project, The Clark Labs, Clark University, 950 Main Street, Worcester, MA 01610, USA.
Tel: (508) 793 7526   Fax: (508) 793 8842
E-mail: idrisi@clarku.edu

In the rest of the world:
Look at IDRISI web-site for nearest vendor – http://www.idrisi.clarku.edu/

MAP-INFO

In USA:
MapInfo Corporate Headquarters, One Global View, Troy, NY 12180, USA.
Tel: (800) 327 8627   Fax: (518) 285 6000
E-mail: sales@mapinfo.com

In UK:
MapInfo Ltd, Minton Place, Victoria Street, Windsor, Berkshire SL4 1EG.
Tel: 01753 848200   Fax: 01753 621140
E-mail: uk@mapinfo.com

In the rest of the world:
Look at IDRISI web-site for nearest vendor – http://www.mapinfo.com/

NUD-IST

In Americas:
Scolari, Sage Publications Software, 2455 Teller Road, Thousand Oaks, CA 91320, USA.
Tel: (805) 499 1325   Fax: (805) 499 0871
E-mail: nudist@sagepub.com

In Australia and New Zealand:
Qualitative Solutions and Research P/L, Box 171, La Trobe University PO, Victoria, Australia 3083.
Tel: +61(0)3 9459 1699   Fax: +61(0)3 9459 0435
E-mail: nudist@qsr.com.au

In the rest of the world:
Scolari, Sage Publications Software, 6 Bonhill St, London EC2A 4PU, United Kingdom.
Tel: +44(0)171 330 1222   Fax: +44(0)171 374 8741
E-mail: nudist@sagepub.co.uk

SASPAC

See MIDAS Web Page for information: http://midas.ac.uk/software.html

SURPOP

See Census Web Page for information and to run the program: http://census.ac.uk/cdu/surpop/

GASP

See Pearson Education's web site – http://www.awl-he.com (you will need to search for this book from that page, in order to access the link to the companion web site for this book).

References

Aitken, P.G. and Jones, B. (1997) *Teach Yourself C in 21 Days*. Sams Publishing, Indianapolis, IN.

Aitken, S.C. (1990) A transactional geography of the image-event: the films of Scottish director, Bill Forsyth. *Transactions of the Institute of British Geographers*, **16**: 105–118.

Aitken, S.C. (1991) Person–Environment theories in contemporary perceptual and behavioral geography 1: personality, attitudinal and spatial choice theories. *Progress in Human Geography*, **15**: 179–193.

Aitken, S.C. and Bjorklund, E.M. (1988) Transactional and transformational theories in behavioral geography. *Professional Geographer*, **40**: 54–64.

Anscombe, F.J. (1973) Graphs in statistical analysis. *American Statistician*, **27**: 17–21.

Aronoff, S. (1989) *Geographic Information Systems: a management perspective*. WDL Publications, Canada.

Ayer, A.J. (1969) *The Foundations of Empirical Knowledge*. Macmillan, London.

Bailey, T. and Gatrell, A.C. (1995) *Interactive Spatial Data Analysis*. Longman Scientific and Technical, Harlow.

Barlow, M. and Button, J. (eds) (1995) *ECO Directory of Environmental Databases in the UK 1995/6*. Environmental Information Trust, Bristol.

Barnes, C. and Mercer, G. (1997) Breaking the mould? An introduction to doing disability research. In Barnes, C. and Mercer, G. (eds), *Doing Disability Research*. Disability Press, University of Leeds, Leeds, pp. 1–14.

Barnes, R. (1995) *Successful Study for Degrees*, 2nd edition. Routledge, London.

Barrat, D. and Cole, T. (1991) *Sociology Projects: A Student's Guide*. Routledge, London.

Bates, D.M. and Watts, D.G. (1988) *Non-linear Regression and its Applications*. Wiley, New York.

Batty, M. and Longley, P. (1994) *Fractal Cities: A Geometry of Form*. Academic Press, London.

Beasley, J.D. (1988) *Practical Computing for Experimental Scientists*. Oxford University Press, Oxford.

Berman, M. (1992) Why modernism still matters. In Lash, S. and Friedman, J. (eds), *Modernity and Identity*. Blackwell, Oxford.

Berry, J.K. (1987) Fundamental operations in computer assisted map analysis. *International Journal of Geographical Information Systems*, **1**: 199–236.

Bird, J. (1993) *The Changing Worlds of Geography: A Critical Guide to Concepts and Methods*, 2nd edition. Clarendon Press, Oxford.

Blakemore, M. (1984) *Generalisation and error in spatial databases*, Cartographica 21: 131–139.

Blaxter, L., Hughes, C. and Tight, M. (1996) *How to Research*. Open University Press, Buckingham.

Bourque, L.B. and Clark, V.A. (1992) *Processing Data: The Survey Example*. Sage University Papers Series on Quantitative Applications in the Social Sciences 07-085. Sage, Newbury Park, CA.

British Sociological Association (1989a) *BSA Guidelines on Anti-Sexist Language*. BSA, London.

British Sociological Association (1989b) *Anti-Racist Language: Guidance for Good Practice*. BSA, London.

Bronfenbrenner, U. (1979) *The Ecology of Human Development: Experiments by Nature and Design*. Harvard University Press, Cambridge, MA.

Brown, S. (1995) Postmodernism, the wheel of retailing and the will to power. *International Review of Retail, Distribution and Consumer Research*, **5**: 387–414.

Bunting, T. and Guelke, L. (1979) Behavioral and perception geography: a critical appraisal. *Annals of the Association of American Geographers*, **69**: 448–462.

Burrough, P.A. (1986) *Principles of Geographical Information Systems and Resources Assessment*. Oxford University Press, Oxford.

Burrough, P.A. and McDonnell, R. (1998) *Principles of Geographical Information Systems*. Oxford University Press, Oxford.

Buttimer, A. (1974) *Values in Geography*. Resource Paper 24, Association of American Geographers, Washington DC.

Buttimer, A. (1976) The dynamism of lifeworld. *Annals of the Association of American Geographers*, **66**: 277–292.

Calkins, H. (1990) Unit 8 Socioeconomic data. In Goodchild, M.F. and Kemp, K. (eds), *NCGIA Core Curriculum*. NCGIA, Santa Barbara, CA.

Callender, J.T. and Jackson, R. (1995) *Exploring Probability and Statistics with Spreadsheets*. Prentice Hall, London.

Carter, J.R. (1989) On defining the Geographic Information System. In Ripple, W.J. (ed.), *Fundamentals of Geographic Information Systems: A Compendium*. ASPRS/ACSM, Falls Church, VA, pp. 3–7.

CESSDA (1998) Council of European Social Science Data Archives. <http://www.nsd.uib.no/cessda/>

Chalmers, A.F. (1982) *What is This Thing Called Science?* University of Queensland Press, St Lucia, Queensland, Australia.

Chatfield, C. (1995) *Problem Solving: A Statistician's Guide*. Chapman and Hall, London.

Chrisman, N.R. (1997) *Exploring Geographical Information Systems*. John Wiley, Chichester.

Clark, G. (1997) Secondary data sources. In Flowerdew, R. and Martin, D. (eds), *Methods in Human Geography: A Guide for Students Doing a Research Project*. Longman, Harlow, pp. 57–69.

Clark, W.A.V. and Hosking, P.L. (1986) *Statistical Methods for Geographers*. John Wiley, New York.

Cliff, A.D. and Ord, J.K. (1973) *Spatial Autocorrelation*. Pion, London.

Cloke, P., Philo, C. and Sadler, D. (1992) *Approaching Human Geography: An Introduction to Contemporary Theoretical Debates*. Paul Chapman, London.

Cochran, W.G. (1977) *Sampling Techniques*. John Wiley, New York.

Code, L. (1991) *What Can She Know? Feminist Theory and the Construction of Knowledge*. Cornell University Press, Ithaca, NY.

Cohen, J. (1960) A coefficient of agreement for nominal scales. *Educational and Psychological Measurement*, **20**: 37–46.

Cohen, L. and Holliday, M. (1982) *Statistics for Social Scientists*. Harper & Row, London.

Converse, J.M. and Presser, S. (1986) *Survey Questions: Handcrafting the Standardized Questionnaire*. Sage University Papers Series on Quantitative Applications in the Social Sciences 07-063, Sage, Newbury Park, CA.

Cooke, D., Craven, A.H. and Clarke, G.M. (1992) *Basic Statistical Computing*. Edward Arnold, London.

Coolican, H. (1990) *Research Methods and Statistics in Psychology*. Hodder and Stoughton, London.

Correll, S. (1995) The ethnography of an electronic bar: the lesbian café. *Journal of Contemporary Ethnography*, **24**: 270–298.

Coshall, J. (1988) The non-parametric analysis of variance and multiple comparison procedures in geography. *Professional Geographer*, **40**: 85–95.

Coshall, J. (1989) *The Application of Non-Parametric Statistical Tests in Geography*. Concepts and Techniques in Modern Geography 50, Geo Books, Norwich.

Couclelis, H. and Golledge, R.G. (1983) Analytical research, positivism and behavioural geography. *Annals of the Association of American Geographers*, **73**: 331–339.

Cox, K.R. and Golledge, R.G. (1981) Introduction. In Cox, K.R. and Golledge, R.G. (eds), *Behavioural Problems in Geography Revisited*. Northmerton University Press, Chicago.

Cramer, D. (1997) *Basic Statistics for Social Research: Step by Step Calculations and Computer Techniques using Minitab*. Routledge, London.

Crang, M.A., Hudson, A.C., Reimer, S.M. and Hinchliffe, S.J. (1997) Software for qualitative research. 1. Prospectus and overview. *Environment and Planning*, **A29**: 771–787.

Cresswell, J.W. (1998) *Qualitative Inquiry and Research Design: Choosing Among Five Traditions*. Sage, London.

Dale, A. and Marsh, C. (1993) *The 1991 Census User's Guide*. HMSO, London.

Dangermond, J. (1992) What is a Geographical Information System? In Johnson, A.I., Pettersson, C.B. and Fulton, J.L. (eds), *Geographic Information Systems (GIS) and Mapping: Practices and Standards*.

Daniels, S. (1988) In Cosgrove, D. and Daniels, S. (eds), *The Iconography of Landscape: Essays on the Symbolic Representation, Design, and Use of Past Environments*. Cambridge University Press, Cambridge.

Dear, M. (1988) The postmodern challenge: reconstructing human geography. *Transactions of the Institute of British Geographers*, **13**: 262–274.

Department of Education and Science and the Welsh Office (1990) *Geography Working Group's Interim Report*. HMSO, London.

Department of the Environment (1987) *Handling Geographic Information. The Report of the Committee of Enquiry, chaired by Lord Chorley.* HMSO, London.

Dewdney, J.C. (1985) *The UK Census of Population 1981.* Concepts and Techniques in Modern Geography 43, Geo Books, Norwich.

Dey, I. (1993) *Qualitative Data Analysis: A User Friendly Guide for Social Scientists.* Routledge, London.

Dixon, C. and Leach, B. (1978) *Sampling Methods for Geographical Research.* Concepts and Techniques in Modern Geography 17, Invicta Press, London.

Dixon, C. and Leach, B. (1984) *Survey Research in Underdeveloped Countries.* Concepts and Techniques in Modern Geography 39, Geo Books, Norwich.

Domosh, M. (1991a) Towards a feminist historiography of geography. *Transactions of the Institute of British Geographers*, **16**: 95–104.

Domosh, M. (1991b) Beyond the frontiers of geographical knowledge. *Transactions of the Institute of British Geographers*, **16**: 488–490.

Dorling, D. (1996) *Area Cartograms: Their Use and Creation.* Concepts and Techniques in Modern Geography 59, Geo Books, Norwich.

Dorling, D. and Fairbairn, D. (1997) *Mapping: Ways of Representing the World.* Longman, Harlow.

Downs, R.M. (1970) Geographic space perception: past approaches and future prospects. *Progress in Geography*, **2**: 65–108.

Dunford, M. (1981) *Historical materialism and geography.* Research paper in Geography, No. 4, University of Sussex, Falmer, Brighton.

Dyck, I. (1993) Ethnography: a feminist methodology. *Canadian Geographer*, **37**: 52–57.

Dyck, I. (1995) Hidden geographies: the changing lifeworlds of women with disabilities. *Social Science and Medicine*, **40**(3): 307–320.

Ebdon, D. (1985) *Statistics in Geography.* Blackwell, Oxford.

Ehrenberg, A.S.C. (1975) *Data Reduction: Analysing and Interpreting Statistical Data.* Wiley, Chichester.

Eichler, M. (1988) *Nonsexist Research Methods: A Practical Guide.* Unwin Hyman, London.

Erickson, B.H. and Nosanchuk, T.A. (1992) *Understanding Data.* University of Toronto Press, Toronto.

ESRI (1995) *Understanding GIS, The ARC/INFO Method: Self Study Workbook, Version 7 for UNIX and OpenVMS.* Environmental Systems Research Institute, GeoInformation International, Cambridge.

EUROSTAT (1993) *European Statistics: Official Sources.* EUROSTAT, Luxembourg.

Everitt, B.S. (1992) *The Analysis of Contingency Tables.* Chapman and Hall, London.

Eyles, J. (1989) The geography of everyday life. In Gregory, D. and Walford, R. (eds), *Horizons in Human Geography.* Macmillian, London, pp. 102–117.

FGDC (1994) *Content Standards for Digital Spatial Metadata* (June 8 draft). Federal Geographic Data Committee, Washington DC.

Flowerdew, R. (1997) Finding previous work on the topic. In Flowerdew, R. and Martin, D. (eds), *Methods in Human Geography: A Guide for Students Doing a Research Project.* Longman, Harlow, pp. 46–56.

Foote, K.E. and Hubner, D.J. (1995) Error, accuracy, and precision. *The Geographer's Craft Project.* Department of Geography, University of Texas at Austin. <http://www.utexas.edu/depts/grg/gcraft/notes/error/error_f.html>

Fotheringham, A.S. and Wong, D.W.S. (1991) The modifiable areal unit problem in multivariate statistical analysis. *Environment and Planning A*, **23**: 1025–1044.

Frankfort-Nachmias, C. and Nachmias, D. (1996) *Research Methods in the Social Sciences.* Edward Arnold, London.

Frazier, J.W. (1981) Pragmatism: geography and the real world. In Harvey, M.E. and Holly, B.P. (eds), *Themes in Geographic Thought.* Croom Helm, London, pp. 61–72.

Freedman, D., Pisani, R. and Purves, R. (1978) *Statistics.* W.W. Norton, London.

Gale, F. (1992) A view of the world through the eyes of a cultural geographer. In Rogers, A., Viles, H. and Goudie, A. (eds), *The Student's Companion to Geography.* Blackwell, Oxford, pp. 21–24.

Gardner, M.J. and Altman, D.G. (1989) *Statistics with Confidence.* British Medical Journal, London.

Geary, R.C. (1968) The contiguity ratio and statistical mapping. In Berry, B.J.L. and Marble, D.K. (eds), *Spatial Analysis: A Reader in Statistical Geography.* Prentice Hall, Englewood Cliffs, NJ, pp. 461–478.

Golledge, R.G. (1981) Misconceptions, misinterpretations, and misrepresentations of behavioral approaches in human geography. *Environment and Planning A*, **13**: 1315–1344.

Golledge, R.G. and Rushton, G. (1984) A review of analytical behavioral research in geography. In Herbert, D.T. and Johnston, R.J. (eds), *Geography and the Urban Environment*, **6**: pp. 1–43.

Golledge, R.G. and Stimson, R. (1997) *Spatial Behaviour: A Geographic Perspective.* Guilford Press, New York.

Goodchild, M.F. (1986) Spatial Autocorrelation. *Concepts and Techniques in Modern Geography*, **47**, Geo Books, Norwich.

Goodchild, M.F. (1991) *Introduction to Spatial Analysis*, Workshop presented at URISA 1991 Annual Meeting, San Francisco.

Goodchild, M.F. (1993) Data models and data quality: problems and prospects. In Goodchild, M.F., Parks, B.O. and Steyaert, L.T. (eds), *Environmental Modelling with GIS*, Oxford University Press, Oxford, pp. 94–103.

Gould, P. (1985) *The Geographer at Work*. Routledge, London.

Gould, P. (1994a) Guest essay: sharing a tradition – geographers from the enlightenment. *The Canadian Geographer*, **38**: 194–200.

Gould, P. (1994b) Reply. *The Canadian Geographer*, **38**: 209–214.

Government of PEI (1997) Province of Prince Edward Island, 23rd Statistical Review 1996, Fiscal Management Division, Dept of the Provincial Treasury, Charlottetown, PEI, Canada.

Graham, E. (1997) Philosophies underlying human geography research. In Flowerdew, R. and Martin, D. (eds), *Methods in Human Geography: A Guide for Students Doing a Research Project*. Longman, Harlow, pp. 6–30.

Gregory, D. (1986a) Positivism. In Johnston, R.J., Gregory, D. and Smith, D.M. (eds), *The Dictionary of Human Geography*. Blackwell, Oxford.

Gregory, D. (1986b) Critical rationalism. In Johnston, R.J., Gregory, D. and Smith, D.M. (eds), *The Dictionary of Human Geography*. Blackwell, Oxford.

Gregory, D. (1986c) Pragmatism. In Johnston, R.J., Gregory, D. and Smith, D.M. (eds), *The Dictionary of Human Geography*. Blackwell, Oxford.

Gregory, I., Southall, H.R. and Dorling, D. (in press) A century of poverty in England and Wales 1898–1998: A geographical analysis. In Bradshaw, J. and Sainsbury, R. (eds), *Getting the Measure of Poverty: The Early Legacy of Seebohan Rowntree*, Polity Press, Bath.

Griffith, D.A. and Amrhein, C.G. (1997a) *Statistical Analysis for Geographers*. Prentice Hall, Englewood Cliffs, NJ.

Griffith, D.A. and Amrhein, C.G. (1997b) *Multivariate Statistical Analysis for Geographers*. Prentice Hall, Englewood Cliffs, NJ.

Guelke, L. (1974) An idealist alternative in human geography. *Annals of the Association of American Geographers*, **64**: 193–202.

Guelke, L. (1981) Idealism. In Harvey, M.E. and Holly, B.P. (eds), *Themes in Geographic Thought*. Croom Helm, London, pp. 133–147.

Habermas, J. (1978) *Knowledge and Human Interests*. Heinemann, London.

Haggett, P. (1981) Geography. In Johnston, R.J. (ed.), *The Dictionary of Human Geography*. Blackwell, Oxford, p. 133.

Haggett, P. (1990) *The Geographer's Art*. Blackwell, Oxford.

Hammersley, D. (1995) *The Politics of Social Research*. Sage, London.

Hanson, S. (1992) Geography and feminism: Worlds in collision? *Annals of the Association of American Geographers*, **82**: 569–586.

Hartshorne, R. (1959) *Perspective on the Nature of Geography*. Rand McNally, Chicago.

Hartwig, F. and Dearing, B.E. (1979) *Exploratory Data Analysis*. Sage University Papers Series on Quantitative Applications in the Social Sciences 07-016. Sage, Newbury Park, CA.

Harvey, D. (1969) *Explanation in Geography*. Edward Arnold, London.

Harvey, D. (1973) *Social Justice and the City*. Edward Arnold, London.

Harvey, M.E. and Holly, B.P. (eds) (1981) *Themes in Geographic Thought*. Croom Helm, London.

Hearnshaw, H.M. and Unwin, D.J. (eds) (1994) *Visualisation in Geographical Information Systems*. Wiley, Chichester.

Heath, D. (1995) *An Introduction to Experimental Design and Statistics for Biology*. UCL Press, London.

Heuvelink, G.B.M. and Burrough, P.A. (1993) Error propagation in cartographic modelling using Boolean logic and continuous classification. *International Journal of Geographical Information Systems*, **7**: 231–246.

Hill, M.R. (1981) Positivism: a 'hidden' philosophy in geography. In Harvey, M.E. and Holly, B.P. (eds), *Themes in Geographic Thought*. Croom Helm, London, pp. 38–60.

Hinchliffe, S.J., Crang, M.A., Reimer, S.M. and Hudson, A.C. (1997) Software for qualitative research. 2. Some thoughts on 'aiding' analysis. *Environment and Planning A*, **29**: 1109–1124.

Hodgkiss, A.G. (1981) *Understanding Maps: A Systematic History of their Use and Development*. Dawson, Folkestone.

Holt-Jensen, A. (1988) *Geography: History and Concepts*, 2nd edition. Paul Chapman, London.

Homan, R. (1991) *The Ethics of Social Research*. Longman, Harlow.

Howard, R.B., Chase, S.D. and Rothman, M. (1973) An analysis of four measures of cognitive maps. In Preiser, W.F.E. (ed.), *Environmental Design Research*, **1**. Dowden, Hutchinson and Ross, Stroudsberg, PA, pp. 254–264.

ICPSR (1998) The ICPSR Data Archive. <http://www.icpsr.umich.edu/archive1.html>

Jackson, P. and Smith, S.J. (1984) *Exploring Social Geography*. Allen and Unwin, London.

Jacob, H. (1984) *Using Published Data*. Sage University Papers Series on Quantitative Applications in the Social Sciences 07-042, Sage, Newbury Park, CA.

Jacobson, R.D., Kitchin, R.M., Garling, T., Golledge, R.G. and Blades, M. (in press) Learning a complex urban route without sight. *Spatial Cognition and Computation*.

Johnston, R.J. (1985) Introduction: exploring the future of geography. In Johnston, R.J. (ed.), *The Future of Geography*. Methuen, London.

Johnston, R.J. (1986a) *Philosophy and Human Geography: An Introduction to Contemporary Approaches*, 2nd edition. Edward Arnold, London.

Johnston, R.J. (1986b) Ecological validity. In Johnston, R.J., Gregory, D. and Smith, D.M. (eds), *The Dictionary of Human Geography*. Blackwell, Oxford.

Johnston, R.J. (1991) *Geography and Geographers: Anglo-American Human Geography since 1945*, 4th edition. Edward Arnold, London.

Johnston, R.J., Gregory, D. and Smith, D.M. (1986) *The Dictionary of Human Geography*. Blackwell, Oxford.

Jones, C.B. (1997) *Geographical Information Systems and Computer Cartography*. Longman, Harlow.

Kanji, G.K. (1994) *100 Statistical Tests*. Sage, London.

Katz, C. (1994) Playing the field: questions of fieldwork in geography. *The Professional Geographer*, **46**: 67–72.

Kidder, L.H. (1981) *Research Methods in Social Relations*, 4th edition. Harcourt Brace, London.

Kiecolt, K.J. and Nathan, L.E. (1985) *Secondary Analysis of Survey Data*. Sage University Papers Series on Quantitative Applications in the Social Sciences 07-053, Sage, Newbury Park, CA.

Kish, L. (1965) *Survey Sampling*. John Wiley & Sons, London.

Kitchin, R.M. (1996) Methodological convergence in cognitive mapping research: investigating configurational knowledge. *Journal of Environmental Psychology*, **16**(3): 163–185.

Kitchin, R.M. (1997) Exploring spatial thought. *Environment and Behaviour*, **29**(1): 123–156.

Kitchin, R.M. (in press) Morals and ethics in geographical studies of disability. In Proctor, J. and Smith, D.M. (eds), *Geography and Ethics: Journeys in a Moral Terrain*. Routledge, London.

Kitchin, R.M., Blades, M. and Golledge, R.G. (1997) Understanding spatial concepts at the geographic scale without the use of vision. *Progress in Human Geography*, **21**: 225–242.

Kitchin, R.M., Jacobson, R.D., Golledge, R.G. and Blades, M. (1998) Belfast without sight: Exploring geographies of blindness. *Irish Geography*, **31**: 34–46

Lam, N. S.-N. (1983) Spatial interpolation methods: a review. *American Cartographer*, **10**: 129–149.

Lee, R.M. (1993) *Doing Research on Sensitive Topics*. Sage, London.

Lewis, O. (1966) The culture of poverty. *Scientific American*, **215**: 19–25.

Ley, D. (1977) Social geography in a taken-for-granted world. *Transactions of the Institute of British Geographers*, N.S., **2**: 498–512.

Little, J.A. and Rubin, D.B. (1987) *Statistical Analysis with Missing Data*. John Wiley, New York.

Livingstone, D. (1992) A brief history of geography. In Rogers, A., Viles, H. and Goudie, A. (eds), *The Student's Companion to Geography*. Blackwell, Oxford, pp. 27–35.

Lofland, J. and Lofland, L.H. (1995) *Analysing Social Settings: A Guide to Qualitative Observation and Analysis*. Wadsworth, Belmont, CA.

Longley, P., Goodchild, M.F., Maguire, D. and Rhind, D. (1999) *Geographical Information Systems: Principles, Techniques, Management, Applications*. John Wiley, New York.

Longley, P.A., Brooks, S.M., McDonnell, R. and MacMillan, B. (1998) *Geocomputation: A primer*. John Wiley and Sons, Chichester.

Loy, W. (ed.) (1987) US National Report to ICA, 1987. *The American Cartographer*, **14**(3).

Lyon, D. (1994) *Postmodernity*. Open University Press, Buckingham.

MacEachren, A.M. (1995) *How Maps Work: Representation, Visualization, and Design*. Guilford Press, London.

Mackinder, H.J. (1887) On the scope and methods of geography. *Proceedings of the Royal Geographical Society*, **9**: 141–160.

Makower, J. (ed.) (1986) *The Map Catalogue*. Vintage Books, New York.

Malec, M.A. (1993) *Essential Statistics for Social Research*. Westview Press, Boulder, CO.

Mark, D. (1999) Spatial representation: a cognitive view. In Longley, P., Goodchild, M.F., Maguire, D. and Rhind, D. (eds), *Geographical Information*

Systems: Principles, Techniques, Management, Applications. John Wiley, New York, pp. 81–89.

Marshall, C. and Rossman, G.B. (1995) *Designing Qualitative Research*, 2nd edition. Sage, London.

Martin, D. (1989) Mapping population from zone centroid locations. *Transactions of the Institute of British Geographers*, N.S., **14**: 90–97.

Martin, D. (1993) *The UK Census of Population 1991*. Concepts and Techniques in Modern Geography 56, Geo Books, Norwich.

Martin, D. (1996) *Geographic Information Systems: Socioeconomic Applications*. Routledge, London.

Martin, D. (1999) Spatial representation: the social scientist's perspective. In Longley, P., Goodchild, M.F., Maguire, D. and Rhind, D. (eds), *Geographical Information Systems: Principles, Techniques, Management, Applications*. John Wiley, New York, pp. 71–80.

Martin, D., Harris, J., Sadler, J. and Tate, N.J. (1998) Putting the census on the web: lessons from two case studies. *Area*, **30**(4): 311–320.

May, T. (1993) *Social Research: Issues, Methods and Process*. Open University Press, Buckingham.

McCloskey, M., Blythe, S. and Robertson, C. (1997) *QUERCUS: Statistics for Bioscientists, a Student Guidebook*. Edward Arnold, London.

McEwan, C. (1996) Paradise or pandemonium? West African landscapes in the travel accounts of Victorian women. *Journal of Historical Geography*, **22**: 68–83.

McLeay, C. (1995) Musical words, musical worlds: geographic imagery in the music of U2. *New Zealand Geographer*, **51**(2): 1–6.

MELResearch (1994) *Trends in Household Waste Arisings*. Final report to the DoE, London, August 1994.

Mesev, V. (1998) The use of census data in urban image classification. *Photogrammetric Engineering and Remote Sensing*, **64**: 431–438.

MIDAS (1998) *Supported Datasets*. <http://midas.ac.uk/datasets.html>

Mikkelsen, B. (1995) *Methods for Development Work and Research: A Guide for Practioners*. Sage, London.

Miles, M.B. and Huberman, A.M. (1992) *Qualitative Data Analysis: An Expanded Sourcebook*. Sage, London.

Miller, J.I. and Taylor, B.J. (1987) *The Thesis Writers Handbook*. Alcove Publishing Company, West Linn, OR.

Monkhouse, F.J. and Wilkinson, H.R. (1971) *Maps and Diagrams*. Methuen, London.

Monmonier, M.S. (1996) *How to Lie with Maps*, 2nd edition. University of Chicago Press, Chicago.

Moran, P.A.P. (1948) The interpretation of statistical maps. *Journal of the Royal Statistical Society Series B*, **10**: 234–251.

Mort, D. (1990) *Sources of Unofficial UK Statistics*. Gower, Aldershot.

Mort, D. (1992) *UK Statistics: A Guide for Business Users*. Ashgate, Aldershot.

Moss, P. (1995) Reflections on the 'gap' as part of the politics of research design. *Antipode*, **27**(1): 82–90.

Neave, H.R. (1976) The teaching of hypothesis testing. *BIAS*, **3**: 55–63.

Neave, H.R. (1978) *Statistical Tables*. George Allen & Unwin, London.

O'Connor, M. (1991) *Writing Successfully in Science*. Chapman and Hall, London.

Odland, J. (1988) *Spatial Autocorrelation*. Sage Publications, London.

Office for National Statistics (1996) *Guide to Official Statistics, 1996 Edition*. HMSO, London.

Oliver, M. (1992) Changing the social relations of research production. *Disability, Handicap and Society*, **7**: 101–114.

Openshaw, S. (1984) *The Modifiable Areal Unit Problem*. Concepts and Techniques in Modern Geography 38, Geo Books, Norwich.

Openshaw, S. (1995) *Census Users' Handbook*. GeoInformation International, Cambridge.

Openshaw, S. (1996) Developing GIS-relevant zone-based spatial analysis. In Longley, P. and Batty, M. (eds), *Spatial Analysis: Modelling in a GIS Environment*. GeoInformation International, Cambridge, pp. 57–73.

Openshaw, S. and Alvanides, S. (in press) Designing zoning systems for the representation of socio-economic data. In Frank, A. *et al.* (eds), *Life and Motion of Socio-economic Units*. Taylor and Francis, London.

Oppenheim, A.N. (1992) *Questionnaire Design, Interviewing and Attitude Measurement*. Pinter, London.

Parr, H. (in press) The politics of methodology in 'post-medical geography': mental health research and the interview. *Health and Place*.

Parsons, T. and Knight, P.G. (1995) *How to Do Your Dissertation in Geography and Related Disciplines*. Chapman and Hall, London.

Patton, M. (1990) *Qualitative Evaluation and Research Methods*, 2nd edition. Sage, London.

Peake, L. (1994) Proper words in proper places . . . or of young turks and old turkeys. *The Canadian Geographer*, **38**: 204–206.

Peet, R.J. and Lyons, J.V. (1981) Marxism: Dialectical materialism, social formations and geographic relations. In Harvey, M.E. and Holly, B.P. (eds), *Themes in Geographic Thought*. Croom Helm, London, pp. 187–206.

Peet, R. and Thrift, N. (1989) Political economy and human geography. In Peet, R. and Thrift, N. (eds), *New Models in Geography: The Political-Economy Perspective*, vol. 1, pp. 1–27.

Pett, M.A. (1997) *Nonparametric Statistics for Health Care Research*. Sage, London.

Philo, C. (1987) 'Fit localities for an asylum': the historical geography of the 'mad-business' in England viewed through the pages of *The Asylum*. *Journal of Historical Geography*, **13**, 398–415.

Pickles, J. (1985) *Phenomenology, Science and Geography: Spatiality and the Human Sciences*. Cambridge University Press, Cambridge.

Pile, S. (1993) Human agency and human-geography revisited – a critique of new models of the self. *Transactions of the Institute of British Geographers*, **18**: 122–139.

Poster, M. (1995) *The Second Media Age*. Polity, Oxford.

Press, W.H., Flannery, B.P., Teukolsky, S.A. and Vetterling, W.T. (1989) *Numerical Recipes – The Art of Scientific Computing*. Cambridge University Press, Cambridge.

Ragurman, K. (1994) Philosophical debates in human geography and their impact on graduate students. *Professional Geographer*, **46**: 242–249.

Reading, R., Haynes, R., Lovett, A., Langford, I., Sunnenberg, G., Gale, S., Zalavra, K., Thompson, R. and Rose, S. (1997) *Variations in Accident Rates in Pre-school Children in the Norwich Area*. Research Report No. 4, School of Health Policy and Practice, University of East Anglia, Norwich.

Rees, P. (1995) Putting the census on the researcher's desk. In Openshaw, S. (ed.), *Census Users' Handbook*. GeoInformation International, Cambridge, pp. 27–81.

Relph, E.C. (1976) *Place and Placelessness*. Pion, London.

Relph, E.C. (1981) Phenomenology. In Harvey, M.E. and Holly, B.P. (eds), *Themes in Geographic Thought*. Croom Helm, London, pp. 99–114.

Reynolds, H.T. (1984) *Analysis of Nominal Data*. Sage, Newbury Park, CA.

Rhind, D.W. (ed.) (1983) *A Census User's Handbook*. Methuen, London.

Rhind, D.W. (1990) Counting the people: the role of GIS. In Maguire, D.J., Goodchild, M.F. and Rhind, D.W. (eds), *Geographical Information Systems: Principles and Applications*. pp. 127–137.

Richardson, G.D. (1981) The appropriateness of using various Minkowskian metrics for repres-

enting cognitive configurations. *Environment and Planning A*, **13**: 475–485.

Robinson, A.H., Morrison, J.L., Muehrcke, P.C., Kimerling, A.J. and Guptil, S.C. (1995) *Elements of Cartography*. John Wiley, New York.

Robson, C. (1993) *Real World Research: A Resource for Social Scientists and Practitioner-Researchers*. Blackwell, Oxford.

Robson, C. (1994) *Experiment, Design and Statistics in Psychology*. Penguin Books, London.

Rorty, R. (1980) *Philosophy and the Mirror of Nature*. Blackwell, Oxford.

Rosenau, P.M. (1992) *Postmodernism and the Social Sciences: Insights, Inroads and Intrusions*. Princeton University Press, Chichester.

Rowntree, D. (1981) *Statistics without Tears: A Primer for Non-mathematicians*. Penguin Books, Harmondsworth.

Royle, S. (1998) St Helena as a Boer prisoner of war camp, 1900–1902: information from the Alice Stopford Green papers. *Journal of Historical Geography*, **24**: 53–68.

Rubenstrank, K. (1996) *Standardizing Metadata*. Interview with DBMS, <http://www.dnmsmag.com/int9602.html/>

Russ, R.C. and Schenkman, R.I. (1980) Editorial statement: Theory and method and their basis in psychological investigation. *Journal of Mind and Behavior*, **1**: 1–7.

Ryan, B.F. and Joiner, B.L. (1994) *MINITAB Handbook*. Duxbury Press, Belmont, CA.

Sample, P.L. (1996) Beginnings: participatory action research and adults with developmental disabilities. *Disability and Society*, **11**(3): 317–332.

Samuels, M.S. (1978) Existentialism and human geography. In Ley, D. and Samuels, M.S. (eds), *Humanistic Geography*. Croom Helm, London, pp. 283–296.

Samuels, M.S. (1981) An existential geography. In Harvey, M.E. and Holly, B.P. (eds), *Themes in Geo-graphic Thought*. Croom Helm, London, pp. 115–133.

Sayer, A. (1985) Realism and geography. In Johnston, R.J. (ed.), *The Future of Geography*. Methuen, London.

Sayer, A. (1992) *Method in Social Science: A Realist Approach*. Routledge, London.

Schaefer, F.K. (1953) Exceptionalism in geography: a methodological examination. *Annals of the Association of American Geographers*, **43**: 226–249.

Scott, J. (1990) *A Matter of Record: Documentary Sources in Social Research*. Polity Press, Cambridge.

Seamon, D. (1979) *A Geography of the Lifeworld*. Croom Helm, London.

Shaw, G. and Wheeler, D. (1994) *Statistical Techniques in Geographical Analysis*. David Fulton Publishers, London.

Shea, K.S. and McMaster, R.B. (1989) Cartographic generalization in a digital environment: when and how to generalize. *Proceedings AUTOCARTO 9*, ACSM/ASPRS, Baltimore, MD, pp. 56–67.

Sieber, J. and Stanley, B. (1988) Ethical and professional dimensions of socially sensitive research. *American Psychologist*, **43**: 49–55.

Silk, J. (1979) *Statistical Concepts in Geography*. Allen and Unwin, London.

Silverman, D. (1993) *Interpreting Qualitative Data: Methods for Analysing Talk, Text and Interaction*. Sage, London.

Smith, N.L. (1987) Towards the justification of claims in evaluation research. *Evaluation and Programme Planning*, **10**: 209–314.

Smith, S.J. (1981) Humanistic method in contemporary social geography. *Area*, **13**: 293–298.

Smith, S.J. (1984) Practising humanistic geography. *Annals of the Association of American Geographers*, **74**: 353–374.

Sokal, R. and Rohlf, F.J. (1995) *Biometry: The Principles and Practice of Statistics in Biological Research*. W.H. Freeman, New York.

Southworth, M. and Southworth, S. (1982) *Maps: A Visual Survey and Design Guide*. Little and Brown, Boston, MA.

Stevens, S.S. (1946) On the theory of scales of measurement. *Science*, **103**: 677–680.

Stoddart, D. (1991) Do we need a feminist historiography of geography – and if we do, what should it be? *Transactions of the Institute of British Geographers*, **16**: 484–487.

Strachan, A.J., Unwin, D.J. and Hickin, W. (1993) *Getting Started with Workstation ARC/INFO*. CCP/USDU, University of Sheffield, Sheffield.

Strauss, A. and Corbin, J. (1990) *Basics of Qualitative Research: Grounded Theory, Procedures and Techniques*. Sage, Newbury Park, CA.

Sui, D.Z. and Wheeler, J.O. (1993) The location of office space in the metropolitan service economy of the United States, 1985–1990. *Professional Geographer*, **45**: 33–43.

Taylor, R. (1988) Social scientific research on the 'troubles' in Northern Ireland. *Economic and Social Review*, **19**: 123–145.

Tesch, R. (1990) *Qualitative Research: Analysis Types and Software Tools*. Falmer, London.

Thayer, H.S. (1973) *Meaning and Action: A Study of American Pragmatism*. Bobbs-Merrill, Indianapolis, IN.

The Data Archive (1998a) Major studies held by the Data Archive <http://biron.essex.ac.uk/cgi-bin/biron?msform>

The Data Archive (1998b) Resources for postgraduate students <http://dawww.essex.ac.uk/introduction/postgrad.html>

Thrower, N. (1996) *Maps and Civilization: Cartography, Culture and Society*. University of Chicago Press, Chicago.

Tobler, W.R. (1970) A computer movie simulating urban growth in the Detroit region. *Economic Geography*, **46**: 234–240.

Tobler, W.R. (1979) Smooth pycnophylactic interpolation for geographical regions. *Journal of the American Statistical Association*, **74**: 519–530.

Tobler, W.R. (1988) Resolution, resampling and all that. In Mounsey, H. and Tomlinson, R. (eds), *Building Databases for Global Science*. Taylor and Francis, London, pp. 129–137.

Townsend, P., Phillimore, P. and Beattie, A. (1988) *Health and Deprivation: Inequality and the North*. Croom Helm, London.

Tuan, Y.-F. (1971) Geography, phenomenology, and the study of human nature. *Canadian Geographer*, **15**: 181–192.

Tuan, Y.-F. (1974) *Topophilia*. Prentice-Hall, Englewood Cliffs, NJ.

Tufte, E.R. (1983) *The Visual Display of Quantitative Information*. Graphics Press, Cheshire, CT.

Turk, A.G. (1990) Towards an understanding of human–computer interaction aspects of Geographic Information Systems. *Cartography*, **19**: 31–60.

Turkle, S. (1996) *Life on the Screen: Identity in the Age of the Internet*. Weidenfeld and Nicolson, London.

Turner, M. (1984) Livestock in the agrarian economy of counties Down and Antrim from 1903 to the Famine. *Irish Economic and Social History*, **11**: 19–43.

Unwin, D. (1982) *Introductory Spatial Analysis*. Methuen, London.

Unwin, T. (1992) *The Place of Geography*. Longman, Harlow.

US Census Bureau (1996) *Census Catalog and Guide: 1996*. US Department of Commerce, Bureau of the Census, Washington, DC.

Walford, N. (1995) *Geographical Data Analysis*. John Wiley, Chichester.

Walmsley, D.J. and Lewis, G. (1993) *People and Environment*. Longman, London.

Waylen, P.R. and Snook, A.M. (1990) Patterns of regional success in the Football League, 1921–1987. *Area*, **22**: 353–367.

Weisberg, H.F. (1992) *Central Tendency and Variability*. Sage University Papers Series on Quantitative

Applications in the Social Sciences 07-083, Sage, Newbury Park, CA.

Weitzman, E.A. and Miles, M.B. (1995) *Computer Programs for Qualitative Data Analysis*. Sage, London.

Whyte, W.F. (1991) Introduction. In Whyte, G.F. (ed.), *Participatory Action Research*. Sage, London, pp. 7–15.

Williams, N.J., Sewel, J. and Twine, F. (1986) Council house allocation and tenant incomes. *Area*, **18**: 131–140.

Wittgenstein, L. (1921) *Tractatus Logico-Philosophicus*. Translated by Pears, D.F. and McGuinness, B.F. (1974), Routledge, London.

Wolcott, H. (1990) *Writing Up Qualitative Research*. Sage, London.

Wolcott, H. (1995) *The Art of Fieldwork*. AltaMira Press, Walnut Creek, CA.

Women in Geography Study Group (1997) *Feminist Geographies: Explorations in Diversity and Difference*. Longman, Harlow.

Wood, D. (1993) *The Power of Maps*. Routledge, London.

Worboys, M.F. (1995) *GIS, a Computing Perspective*. Taylor and Francis, London.

Wright, D.B. (1997) *Understanding Statistics: An Introduction for the Social Sciences*. Sage, London.

Wrigley, N. (1985) *Categorical Data Analysis for Geographers and Environmental Scientists*. Longman, London.

Wrigley, N. (1995) Revisiting the modifiable areal unit problem and the ecological fallacy. In Cliff, A.D., Gould, P.R., Hoare, A.G. and Thrift, N.J. (eds), *Diffusing Geography: Essays for Peter Haggett*. Blackwell, Oxford, pp. 49–71.

Wrigley, N., Holt, T., Steel, D. and Tranmer, M. (1996) Analysing, modelling and resolving the ecological fallacy. In Longley, P. and Batty, M. (eds), *Spatial Analysis: Modelling in a GIS Environment*. GeoInformation International, Cambridge, pp. 23–40.

Yeates, M. (1968) *An Introduction to Quantitative Analysis in Economic Geography*. McGraw-Hill, New York.

Yin, R.K. (1989) *Case Study Research: Design and Method*. Sage, Newbury Park, CA.

Yule, G.U. and Kendall, M.G. (1950) *An Introduction to the Theory of Statistics*. Griffin, London.

Index

access to data 39
action research 14, 23, 213, 225
alternative hypothesis *see* research hypothesis
analysis of variance (ANOVA) 113, 122–5, 126–7
analytical statements 8
anti-naturalist 19
anti-realism 24
approaches *see* philosophy
area of polygon 197, 199
ArcView GIS 187, 315
ARC/INFO 187, 315
ASCII file 71–2
autobiography 226
autonomy 8
average area per point 194

BASIC 76
behaviouralism 9–10, 13, 19, 20, 26
behaviourism 9
biographical approaches 225
biography 226
box-and-whisker plot 93–5, 96–7

C/C++ 76, 77–8
capitalism 14
case studies 213, 225
causal mechanisms 15
census 61, 65–7, 183
    geography, UK 66–7
    geography, US 67–9
central limit theorem 102
centre of polygon 198
Cook's distance 132
chi-square 113, 134, 139–41
choosing
    a topic 28–32, 33
    narrowing the focus 30–2
    generating data method 39–41
    data analysis method 41–2, 112–14
closed questions 48

conceptual model 33–4
cognitive processing 9
Cohen's kappa 181
computer programming 76–7
confidence level 58
constructivist 10
content analysis 51, 225
continuous variables 47
coping with problems 291
copyright 185
correlation *see* Pearsons Product Moment; Spearmans Rank
    Order
covert research 34–5, 221–2
costs 39, 59–60
critical values 110, 112
cultural ideologies 24

data
    accuracy 180
    analysis 186
    archives 62–5
    attribute 165, 186
    categorisation 235, 238, 239–45
    checking 75, 81
    cleaning 73–5
    coding 70–1, 222–3, 239
    entry 70, 71–3
    errors 73–5, 178, 179–82, 185
    format 72–3
    missing 75–6
    pre-processing 70–6, 81–2
    precision 180
    presentation of results 186
    quality 61, 178–82
    related 109
    safeguarding 75
    spatial 165–6, 178, 179–82
    standards 182, 185
    storage 75
    systems 186–7

transformation 105–7
unrelated 109
database management system (DBMS) 168
deconstruction 16, 17
deductive approach 14, 19, 23, 34
defining geography 1, 3–4, 6
defining human geography 4
density 194
description 2, 223
dialectics 14
diaries 226
digitising 185
discrete variables 47
distribution *see entries for specific distributions*

ecological fallacy 34–5, 177–8
emancipatory research 18
empirical questions 7
empirical regularity 15
empiricism 7, 8, 9, 19, 20
epistemology 6, 18
error/confusion matrix 181
essences 11
ethics *see* research
ethnography 12, 13–14, 15, 212, 222, 224–5
Excel 76
exclusivity 8
explanation 2, 23
explanatory concepts 33
exploration 2
exploratory data analysis (EDA) 70
existentialism 12, 21
extensive properties 45
extensive research 15–16
external relations 247

F-statistic 116, 123
falsification 8, 23
feminism 17–19, 22, 23, 24, 25, 215
field notes 234
film 227
focus of study 30–2
focus groups 41
formal logic 8
FORTRAN 76
fractal analysis 201

GASP 187, 316
gatekeeper 39
Geary's *c* 202–6
GIS 156, 159, 164–87
    buffering 170–2
    connectivity functions 170
    context operations 172
    database construction 184–5
    data model 166

data structure 166–8, 169
defining 164–5
network analysis 172
overlay 170
project planning 184–7
reclass 170
uses of 165–8
visibility analysis 172, 173
geographical imagination 227
geography
    defined 3–4
    philosophical approaches *see* philosophy
    types of study 5
grand theory 16, 24
ground truth 181
grounded theory 229
group research 42

histogram 83, 86–7, 88–9
historical materialism 15, 21
human agency 14, 15, 26
humanistic approaches 14, 25
hypotheses 34, 109–10
hypothesis tests *see* significance tests *and under individual tests*

idealism 12–13, 21
ideology 6, 23
idiographic science 3
IDRISI 187, 315
implicit links 248
imputation 76
inductive approach 19, 34, 220
initial data analysis (IDA) 70–107, 108, 114–15
independent research 291–2
intensive research 15
internal relations 247
interpreting tests 152–5
interval measurement 45–7
interviews 10, 12, 212, 230–4
    conducting 215–16
    critical appraisal 219
    medium 216
    recording 218
    recruitment 216–17
    style 216
    types 213–15

Kolmogorov-Smirnov test 113, 141–2, 143
kriging 174
Kruskal-Wallis test 113, 147, 149, 150, 151
kurtosis 206

language 17, 37
lateral thinking 248
laws 8, 11

letters 226
levels of measurement 45–7
line length 195–7, 198
literary sources 226
local base statistics 66–7
location quotients 121

Mann-Whitney U test 113, 142, 144–6
Marxism 14–15, 23, 26
MapInfo 187, 315
maps 156–64
    distortion 159–64
    drawing 282–4
    generalisation 163–4
    history of 157
    projection 160–3
    Ptolemy 157
    scale 160
    types 158–9
mathematical operators 45
mean 83, 84–6
mean centre 187–9
median 83, 84–6
measurement 45–7
mental construction 11
metadata 61
metaphysics 7, 11
methodology 6
metric data 47
MINITAB 71, 73, 76–82, 109, 114, 315
mode 83, 84–6
model construction 108
modifiable areal unit problem 177–8, 179–80
modernity 16, 17, 24
moral code 35
Moran's I 202–6
music 227

naturalist 19
nearest neighbour index 194–6
negative case analysis 220
negative philosophy 7
nominal measurement 45–6
nomothetic science 3
non-parametric tests 47, 109, 113, 134, 139–55
non-response error 52–3
normal distribution 58, 101–2, 103, 104, 109
normality 114
normative questions 7
NUD-IST 257–69, 315
    annotating data 260, 264
    attach 265, 266
    categorisation 260–1, 263–4
    connection 266–9
    data preparation 258
    importing data 259–60
    merge 265, 266

new project 259
    searching 265
    tree 261–3, 266
null hypothesis 109–10, 111

objectivity 8, 23, 25
observation 14, 212, 219–24
    coding schemes 222, 223
    conducting 222–3, 224
    descriptive 223
    holistic accounts 222
    participation 221–2
    straight 220–1
one-tailed test 109, 111–12
ontology 6
open questions 48
optical mark reader 71
oral defence 287
ordinal measurement 45–6

paintings 227
parametric tests 47, 109, 113, 115–38
participant observation 12, 13, 14, 15, 36–7
participatory action research 25
Pearson's product moment correlation coefficient 113, 125, 128–9
perimeter of polygon 197
Perkal epsilon bands 180–1
personal involvement 29–30
phenomenology 9, 10–12, 13, 20, 40
philosophy 1, 6–26 see also entries for individual philosophical approaches
    choosing a 19–26
    critical 6, 14–19
    empirical-analytical 6, 7–9
    historical-hermeneutic 6, 9–14
    reasons for 4–5
photographs 227
pilot study 42–3, 218–19
planning a project 28–44
point data 187–95
Poisson distribution 101–2
political correctness 37
political economy 15, 26
population 53–4
population census see census
positivism 7–9, 14, 19, 20, 25, 40
    critical rationalism 8
    logical positivism 8
postmodernism 16–17, 22, 24
poststructuralism 17, 22, 37
power relations 219
practice 32–4
pragmatism 13–14, 21, 23
primary data 39, 40
    qualitative 212–24
    quantitative 47–60

probability 95, 97–104
   density function (pdf) 102–3
   distributions 100–4
   factorials 98–100
   laws of 98–9
   tree 100
probing and prompting 217
programming 76–8

qualitative
   approaches 211–12
   data 40, 41
Qualitative data analysis 229–56
   corroboration 251–3
   classification 230, 234–5, 238–9, 242, 244
   computer-based 257–69
   connection 230, 235–6, 246–51
   data management 243
   description 230, 231, 233–4
   interpretation 251
   master sheet 243
   matrix 248
   quantitative analysis 253–6
   sorted category 243, 245
   splitting and splicing 244–6, 247
   transcription and annotation 236–8, 239, 240–1
quantification 8, 9
quantitative data 40, 41
quartiles 95–6
questionnaire 7, 9, 41, 43
   coding 49–51, 70–1
   design 48–53
questions 31, 32, 33, 49, 50, 51–2, 217

range 92–3, 94
raster data 166–8, 169
ratio measurement 45–7
realism 15–16, 21, 24
reflexivity 19
regression analysis 129–38
related *t*-test *see* *t*-test
reliability 34–5
remote sensing data 185
research
   definition 1
   design 34–9
   ethics 35–7
   geographic 3–4
   hypothesis 110–11
   management 42–3
   motivation for 2
   practical considerations 38–9, 59–60, 291
   problems, coping with 291
   planning 28–44
   reasons for 2
   strategy 1
   topic choice 21–33

sampling 53–60
   distribution 56–8
   fraction 58
   frame 55, 58–9
   judgemental 54–5
   population 48, 54
   quota 54–5
   representativeness 54
   size 59–60
   simple random 55–6
   stratified random 55–6
   systematic 55–6
   snowball 54, 217
SAS 76
SASPAC 67, 315–16
secondary data 39, 40
   qualitative 225–7
   quantitative 60–9
   sources 61–5, 225–7
   spatial 182–4
sense of place 11
sensitive topics 37–8, 291
set theory 95, 97–8
sexism in research 18, 37
shape measure 198–9, 200, 201
significance
   levels 109, 110–11
   tests 109, 112–14, 152, 154–5
   see also parametric; non-parametric tests; *and individual tests*
sinuosity 195–7, 198
situated knowledge 23, 24
small area statistics 67
sources of information 30–1
spatial behaviour 9, 10
spatial data
   accuracy 178–82
   analysis 168, 170–3 *see also* GIS
   autocorrelation 199–209
   digitising of 165, 185
   integration 173–5
   interpolation 173–5
   representation 175–7
   scanning of 185
   sources of 182–4
   transformation 173–5
spatial statistics 187–209
Spearman's rank order correlation coefficient 149–52, 153, 154
spreadsheet 73
SPSS 76
standard deviation 92–3, 94
standard deviation ellipse 189–94
standard distance 189, 190
standard error 58, 132
statistical inference 108
statistical tests 45, 108–55

stem-and-leaf plots 87, 90–2
structuration 14
structure-agency debate 26
student's *t* distribution 58
subjectivity 11, 12, 17, 25
submitting project 287–8
Surpop 67, 316
surveys 47–8
systematic statements 8

t-test 111, 113
   related *t*-test 120–2
   unrelated *t*-test 115–20
taboo topics 14, 38
talk, presenting a 286–8
test statistics 109, 110–11 *see also under individual tests*
text editor 71–3
Theisson polygons 174
theory 1, 32–4
timetable 39, 43
Tissot's indicatrix 162
transactionalist 10, 12
trust 219
Tukey test 124
two-tailed test 111–12, 109

understanding 2, 10, 11, 16, 23
UNIX 71, 73, 76
unrelated *t*-test *see t*-test

validity 34–5
value-free research 13, 23, 29
variables 49
variance 92–3, 94, 114
variogram 201
vector data 166–8, 169

Venn-Euler diagrams 98
verification 7–8, 23

Wilcoxon signed ranks test 113, 146–7, 148
World-Wide-Web 62–4
writing-up 270–86
   abstract 278–9
   appendices 286
   audience 271–3
   conclusion 280
   contents list 281
   developing your argument 276
   diagrams 282–4
   discussion 280
   documentation, styles of 285
   editing 276–7
   getting started 273–4
   introduction 278–9
   literature review 279
   knowing when to stop 271
   maps 282–4
   methodology 279–80
   presentation 281–5
   qualitative 285
   quotes 285
   referencing 285–6
   results 280
   statistical formulae 281–2
   structure 274–6
   style 275
   tables 282–3
   title 278
   word limit 277–8
   writing process 273

Yates' Continuity correction 139

Z-score 102, 104, 106